Biocatalysis: Enzymatic Basics and Applications

Biocatalysis: Enzymatic Basics and Applications

Edited by Naomi Wilson

Syrawood
PUBLISHING HOUSE

New York

Published by Syrawood Publishing House,
750 Third Avenue, 9ᵗʰ Floor,
New York, NY 10017, USA
www.syrawoodpublishinghouse.com

Biocatalysis: Enzymatic Basics and Applications
Edited by Naomi Wilson

International Standard Book Number: 978-1-64740-389-8 (Hardback)

Cataloging-in-publication Data

Biocatalysis : enzymatic basics and applications / edited by Naomi Wilson.
    p. cm.
Includes bibliographical references and index.
ISBN 978-1-64740-389-8
1. Biocatalysis. 2. Enzymes. 3. Catalysis. 4. Biomedical engineering. I. Wilson, Naomi.
TP248.65.E59 B56 2023
660.634--dc23

TABLE OF CONTENTS

PREFACE

Biocatalysis refers to the process of increasing the rate of a chemical reaction by adding a substance, known as a biocatalyst. Enzymes and hormones are types of biocatalysts. Enzymes are substances that function as catalysts in living organisms, controlling the pace at which the chemical reactions take place without changing the enzyme itself. They play a critical role in the catalysis of various reactions, such as the making of alcohols through fermentation process, and cheese through the breakdown of milk proteins. In large-scale chemical synthesis, the usage of enzymes is increasing and an in-depth knowledge of reaction mechanisms of enzymes will be helpful in drug target and delivery systems. This book is a valuable compilation of topics centered on the enzymatic basics and applications of biocatalysis. It is meant for students who are looking for an elaborate reference text on enzymes and biocatalysis. A number of latest researches have been included to keep the readers up-to-date with the global concepts in this area of study.

The researches compiled throughout the book are authentic and of high quality, combining several disciplines and from very diverse regions from around the world. Drawing on the contributions of many researchers from diverse countries, the book's objective is to provide the readers with the latest achievements in the area of research. This book will surely be a source of knowledge to all interested and researching the field.

In the end, I would like to express my deep sense of gratitude to all the authors for meeting the set deadlines in completing and submitting their research chapters. I would also like to thank the publisher for the support offered to us throughout the course of the book. Finally, I extend my sincere thanks to my family for being a constant source of inspiration and encouragement.

Editor

Cofactor F_{420}-Dependent Enzymes: An Under-Explored Resource for Asymmetric Redox Biocatalysis

Mihir V. Shah [1,†]**, James Antoney** [1,2,†]◉**, Suk Woo Kang** [1,2]**, Andrew C. Warden** [1]◉**,
Carol J. Hartley** [1]◉**, Hadi Nazem-Bokaee** [1]**, Colin J. Jackson** [1,2] **and Colin Scott** [1,*]◉

1   CSIRO Synthetic Biology Future Science Platform, Canberra 2601, Australia; Mihir.Shah@csiro.au (M.V.S.);
    James.Antoney@anu.edu.au (J.A.); Suk.Kang@anu.edu.au (S.W.K.); andrew.warden@csiro.au (A.C.W.);
    Carol.Hartley@csiro.au (C.J.H.); Hadi.Nazem-Bokaee@csiro.au (H.N.-B.); colin.jackson@anu.edu.au (C.J.J.)
2   Research School of Chemistry, Australian National University, Canberra 2601, Australia
*   Correspondence: colin.scott@csiro.au
†   M.V.S. and J.A. contributed equally to this review.

Abstract: The asymmetric reduction of enoates, imines and ketones are among the most important reactions in biocatalysis. These reactions are routinely conducted using enzymes that use nicotinamide cofactors as reductants. The deazaflavin cofactor F_{420} also has electrochemical properties that make it suitable as an alternative to nicotinamide cofactors for use in asymmetric reduction reactions. However, cofactor F_{420}-dependent enzymes remain under-explored as a resource for biocatalysis. This review considers the cofactor F_{420}-dependent enzyme families with the greatest potential for the discovery of new biocatalysts: the flavin/deazaflavin-dependent oxidoreductases (FDORs) and the luciferase-like hydride transferases (LLHTs). The characterized F_{420}-dependent reductions that have the potential for adaptation for biocatalysis are discussed, and the enzymes best suited for use in the reduction of oxidized cofactor F_{420} to allow cofactor recycling in situ are considered. Further discussed are the recent advances in the production of cofactor F_{420} and its functional analog F_O-5′-phosphate, which remains an impediment to the adoption of this family of enzymes for industrial biocatalytic processes. Finally, the prospects for the use of this cofactor and dependent enzymes as a resource for industrial biocatalysis are discussed.

Keywords: cofactor F_{420}; deazaflavin; oxidoreductase; hydride transfer; hydrogenation; asymmetric synthesis; cofactor biosynthesis

1. Introduction

Enzymes that catalyze the asymmetric reduction of activated double bonds are among the most important in biocatalysis, allowing access to chiral amines from imines (C=N), *sec*-alcohols from ketones C=O), and enantiopure products derived from enoates (C=C). To date, the reduction of imines, ketones and enoates has been achieved largely using enzymes that draw their reducing potential from the nicotinamide cofactors NADH and NADPH; e.g., imine reductases, ketoreductases and Old Yellow Enzymes [1–4]. However, there has been recent interest in an alternative reductive cofactor, cofactor F_{420} (8-hydroxy-5-deazaflavin) [5,6].

Cofactor F_{420} is a deazaflavin that is structurally similar to flavins (Figure 1), with a notable difference at position 5 of the isoalloxazine ring, which is a nitrogen in flavins and a carbon in deazaflavins. Additionally, while C-7 and C-8 are methylated in riboflavin, they are not in cofactor F_{420}: C-7 is hydroxylated and C-8 is unsubstituted. These structural differences cause significant differences in the electrochemical properties of cofactor F_{420} and flavins: a -360–340 mV the redox mid-point potential of cofactor F_{420} is not only lower than that of the flavins (-205 mV to -220 mV), but it is also lower than that of the nicotinamides (-320 mV) [7]. Additionally, as a consequence of the substitution of N-5 for a carbon, cofactor F_{420} cannot form a semiquinone (Figure 1), which means that unlike other flavins, cofactor F_{420} can only perform two-electron reductions.

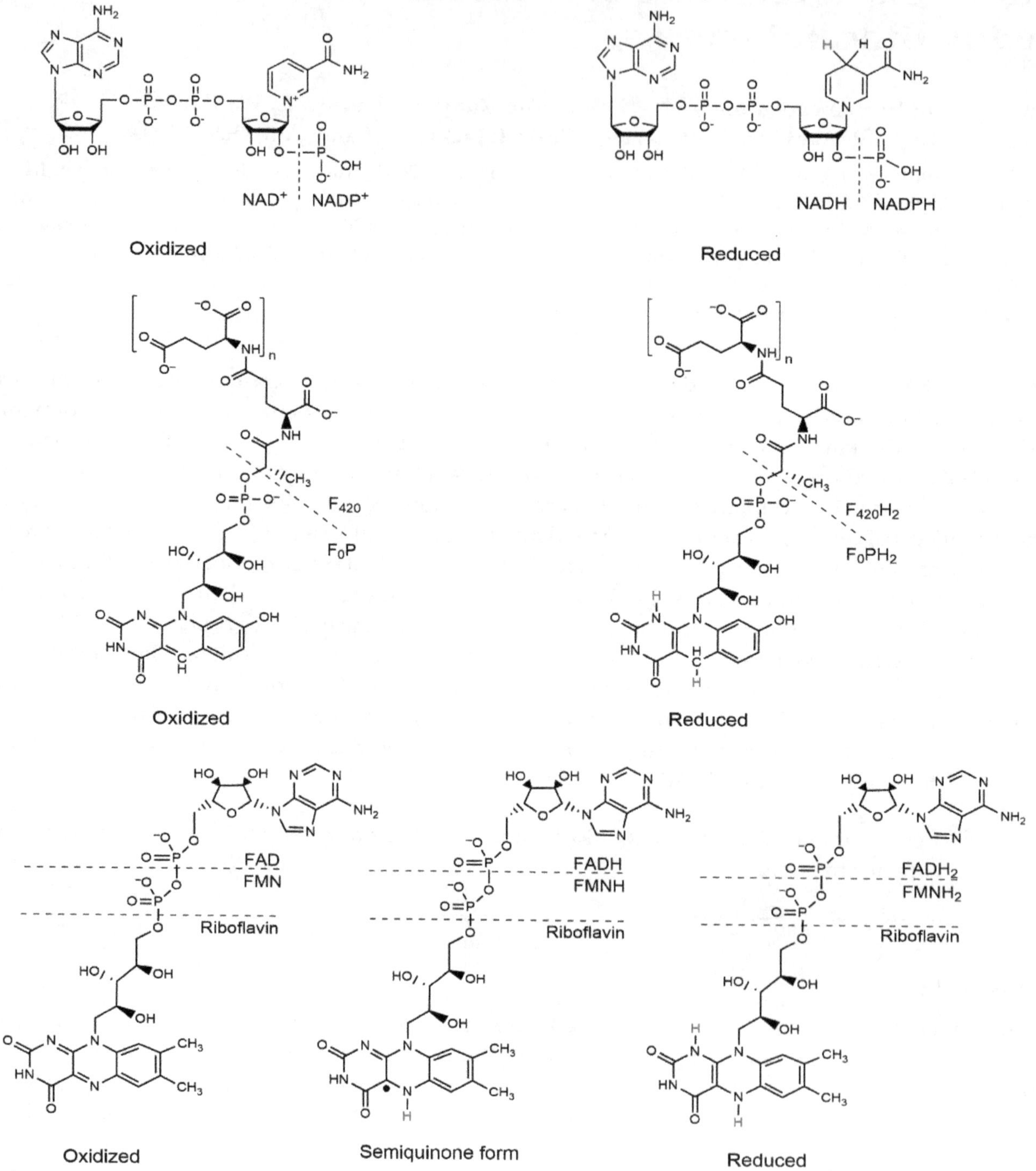

Figure 1. The structures of NAD(P) (top), cofactor F_{420} and its synthetic analog F_OP (center) and common flavins (riboflavin, FMN and FAD; bottom). The oxidized and reduced forms are shown, as is the flavin semiquinone. Dashed lines indicate the differences in the structures of F_OP and cofactor F_{420}, and riboflavin, FMN and FAD.

Cofactor F_{420} was originally described in methanogenic archaea, where it plays a pivotal role in methanogenesis [8,9]. Cofactor F_{420} has since been described in a range of soil bacteria supporting a range of metabolic activities, including catabolism of recalcitrant molecules (such as picric acid) and the production of secondary metabolites, such as antibiotics [7]. A comprehensive review of the biochemistry and physiological roles of cofactor F_{420} was recently published by Greening and coworkers [7]. This review considers the potential of F_{420}-dependent enzymes in industrial biocatalysis, focusing on the enzyme families relevant to biocatalytic applications and the reactions that they catalysis. Cofactor recycling strategies and cofactor production are also discussed, with a focus on the prospects for achieving low-cost production at scale in the latter case.

2. Families of F_{420}-Dependent Enzymes Relevant to Biocatalysis

With respect to their prospective biocatalytic applications, the two most important families of F_{420}-dependent enzymes are the Flavin/Deazaflavin Oxidoreductase (FDOR) and Luciferase-Like Hydride Transferase (LLHT) families, albeit F_{420}-dependent enzyme from other families have also been shown to have catalytic activities of interest (e.g., TomJ, the imine reducing flavin-dependent monooxygenase or OxyR, the tetracycline oxidoreductase) [10,11]. The FDOR and LLHT families are large and contain highly diverse flavin/deazaflavin-dependent enzymes. In both families, there are enzymes with preferences for flavins, such as flavin mononucleotide (FMN) and flavin adenine dinucleotide (FAD), as well as those that use cofactor F_{420} [12,13]. Moreover, there are F_{420}-dependent FDORs that have been shown to be able to promiscuously bind FMN and use it in oxidation reactions [14]. In this section, the FDOR and LLHT families and the classes of reaction that they catalyze are discussed.

2.1. The FDOR Superfamily

The FDOR superfamily (PFAM Clan CL0336) can be broadly divided into two groups: the FDOR-As (which includes a sub-group called the FDOR-AAs) and the FDOR-Bs. The FDOR-As are restricted to *Actinobacteria* and *Chloroflexi* and to date no FDOR-As have been described that use cofactors other than F_{420} [7,12]. The FDOR-Bs are found in a broader range of bacterial genera than the FDOR-A enzymes, and in addition to F_{420}-dependent enzymes, this group also includes heme oxygenases, flavin-sequestering proteins, pyridoxine 5′ oxidases and a number of proteins of unknown function [12,15–17]. Both groups of FDOR are highly diverse, with many homologs often found within a single bacterial genome (e.g., *Mycobacterium smegmatis* has 28 FDORs) [18]. In addition, the majority of the enzymes of this family are yet to be characterized with respect to either their biochemical or physiological function, and therefore the FDORs represent a currently under-explored source of enzymes for biocatalysis.

The FDOR enzymes share a characteristic split β-barrel fold that forms part of the cofactor-binding pocket. The majority of the protein sequences of enzymes currently identified as belonging to this family are small single-domain proteins. The topologies of the two FDOR subgroups are broadly similar (Figure 2), with the split-barrel core composed of 7–8 strands and with 4–5 helices interspersed. All FDOR-Bs studied so far have been demonstrated to be dimeric, with stands β2, β3, β5 and β6 making up the core of the dimer interface (Figure 2). In structures of full-length FDOR-As solved to date, the N-terminal helix (if present) lies on the opposite face of the beta sheet to that in FDOR-Bs. Thus, the N-terminus occupies part of the dimer interface region and prevents interaction between the sheets of adjacent monomers. In contrast to the FDOR-Bs, the oligomerization state of the FDOR-As is more varied. While a number of FDOR-As have been determined to be monomeric [18], the deazaflavin-dependent nitroreductase (DDN) from *M. tuberculosis* forms soluble aggregates through the amphipathic N-terminal helix [19]. DDN and the FDOR-AA subgroup have been shown to be membrane-associated [20–22], and FDOR-AAs have been associated with fatty acid metabolism [12]. No structures of FDOR-AAs have been solved to date.

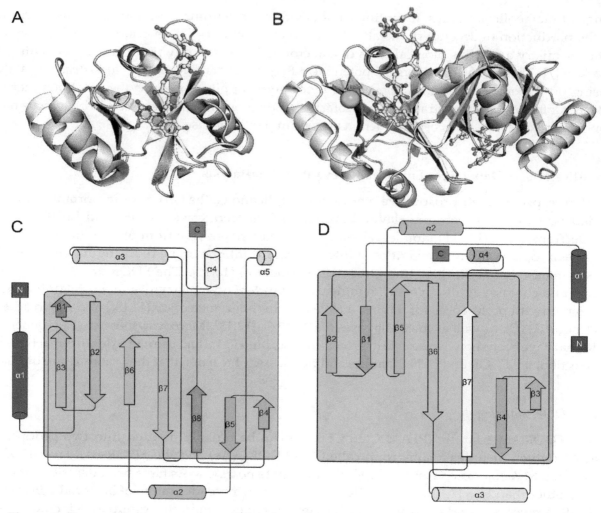

Figure 2. Representative structures of F_{420}-dependent FDOR-A (PDB: 3R5Z, panels **A** and **C**) and FDOR-B (PDB: 5JAB, panels **B** and **D**). Both are predominantly composed of a single β-sheet forming a split barrel. The N-terminal helices are spatially displaced between the two families, falling on opposite faces of the β-sheet.

2.2. The LLHT Family:

The LLHT family form part of the Luciferase-Like Monooxygenase family (PFAM PF00296). They adopt an $(\alpha/\beta)_8$ TIM-barrel fold with three insertion regions, IS1–4 (Figure 3). IS1 contains a short loop and forms part of the substrate cleft. IS2 contains two antiparallel β-strands, and IS3 contains a helical bundle at the C-terminus of the β-barrel and contains the remainder of the substrate-binding pocket (Figure 3). All structures solved to date from the LLHT family contain a non-prolyl *cis* peptide in β3 [23–26]. Recent phylogenetic reconstructions have shown that the F_{420}-dependent LLHTs form two clades: the F_{420}-dependent reductases and the F_{420}-depented dehydrogenases [27]. The F_{420}-reductases contain methylenetetrahydromethanopterin reductases (MERs), which catalyze the reversable, ring-opening cleavage of a carbon-nitrogen bond during the biosynthesis of folate in some archaea [28–30]. The F_{420}-dependent dehydrogenases can be further divided into three subgroups. The first contains F_{420}-dependent secondary alcohol dehydrogenases (ADFs) and the hydroxymycolic acid reductase from *M. tuberculosis* [31]. The second contains the F_{420}-dependent glucose-6-phosphate dehydrogenases (FGDs) from *Mycobacteria* and *Rhodococcus*, while the third appear to be more general sugar-phosphate dehydrogenases [27]. In contrast to the heterodimeric structure of bacterial luciferase, the F_{420}-dependent dehydrogenases form homodimers with the dimer interface burying a relatively large portion of the surface area of the monomers ($\approx$2000 Å^2, roughly 15% of the total surface area) [24–26]. A number of enzymes involved in the F_{420}-dependent degradation of nitroaromatic explosives, such as picrate and 2,4-dinitroanisole, appear to belong to the LLHT family as well [32,33].

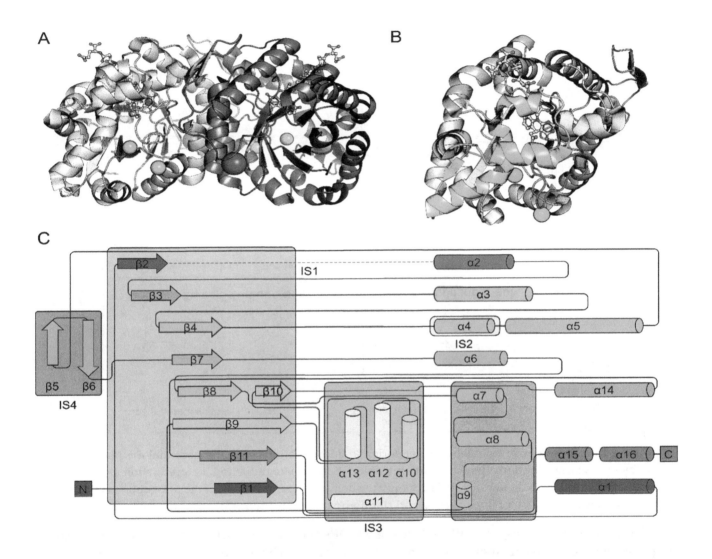

Figure 3. Structure of representative luciferase-like hydride transferase (LLHT) (PDB: 1RHC). (**A**) A 3D representation of the biologically relevant dimer (panel **A**). Monomer of an LLHT with insertion sequences IS1–4 highlighted, along with the helical bundle composed of α7–9 (panel **B**). Topology diagram showing (α/β)$_8$ fold with insertion sequences highlighted: IS1, red; IS2, orange; IS3, light green, IS4, pink. The helical bundle of α7–9 is highlighted in purple (panel **C**).

2.3. Cofactor F_{420}-Dependent Reactions with Relevance to Biocatalysis

From the perspective of biocatalysis, cofactor F_{420}-dependent enzymes catalyze a number of key reductions including the reduction of enoates, imines, ketones and nitro-groups (Table 1; Figure 4).

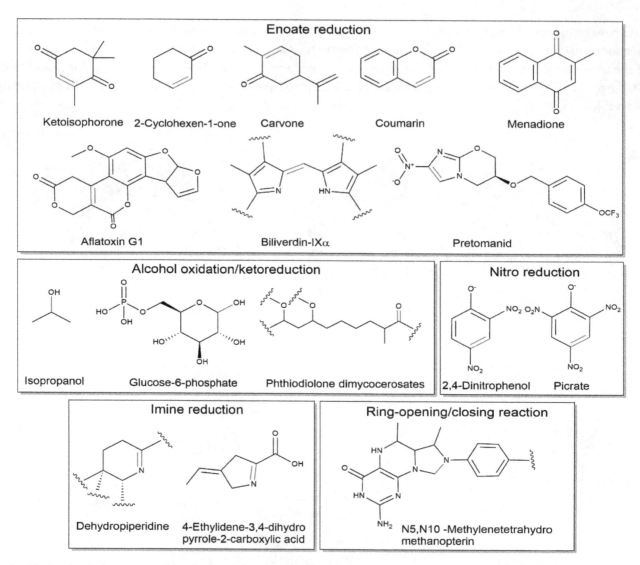

Figure 4. Representative cofactor F_{420}-dependent oxidoreductions with the potential for adaptation to biocatalytic applications. Included are: nitroreduction, enoate reduction, ketoreduction and imine reduction (from top to bottom). For clarity, only the dehydropiperidine ring of the thiopeptide is shown and partial structures for biliverdin-Ixα and phthiodiolone dimycocerosates are shown.

For enoate reductions, a small number of FDORs have been studied. However, the substrate range for most of these enzymes is yet to be fully elucidated. The ability of the mycobacterial FDORs to reduce activated C=C double bonds was first identified when DDN was shown to be responsible for activating the bicyclic nitroimidazole PA-824 in *M. tuberculosis*. These enzymes were then shown to also reduce enoates in aflatoxins, coumarins, furanocoumarins and quinones [6,12,14,16,34–38]. Recent studies have shown that these enzymes are promiscuous and can use cyclohexen-1-one, malachite green and a wide range of other activated ene compounds as substrates [35]. However, there have been a few FDOR studies to date that have examined their kinetic properties and stereospecificity. In one of these studies, FDORs from *Mycobacterium hassiacum* (FDR-Mha) and *Rhodococcus jostii* RHA1 (FDR-Rh1 and FDR-Rh2) were shown to reduce a range of structurally diverse enoates with conversions ranging from 12 to >99% and *e.e.* values of up to >99% [6]. Interestingly, it has been proposed that both the hydride and proton transfer from $F_{420}H_2$ in these reactions was directed to the same face of the activated double bond (Figure 5), which results in the opposite enantioselectivity compared to that of the FMN-dependent Old Yellow Enzyme family of enoate reductases [6]. This suggests that the F_{420}-dependent FDORs may provide a stereocomplementary enoate reductase toolbox. However, other studies suggest that protonation of the substrate is mediated by solvent or an enzyme side-chain (as it is in Old Yellow Enzyme) [37]. Further structure/function studies are needed to fully understand the mechanistic diversity of this family of enzymes.

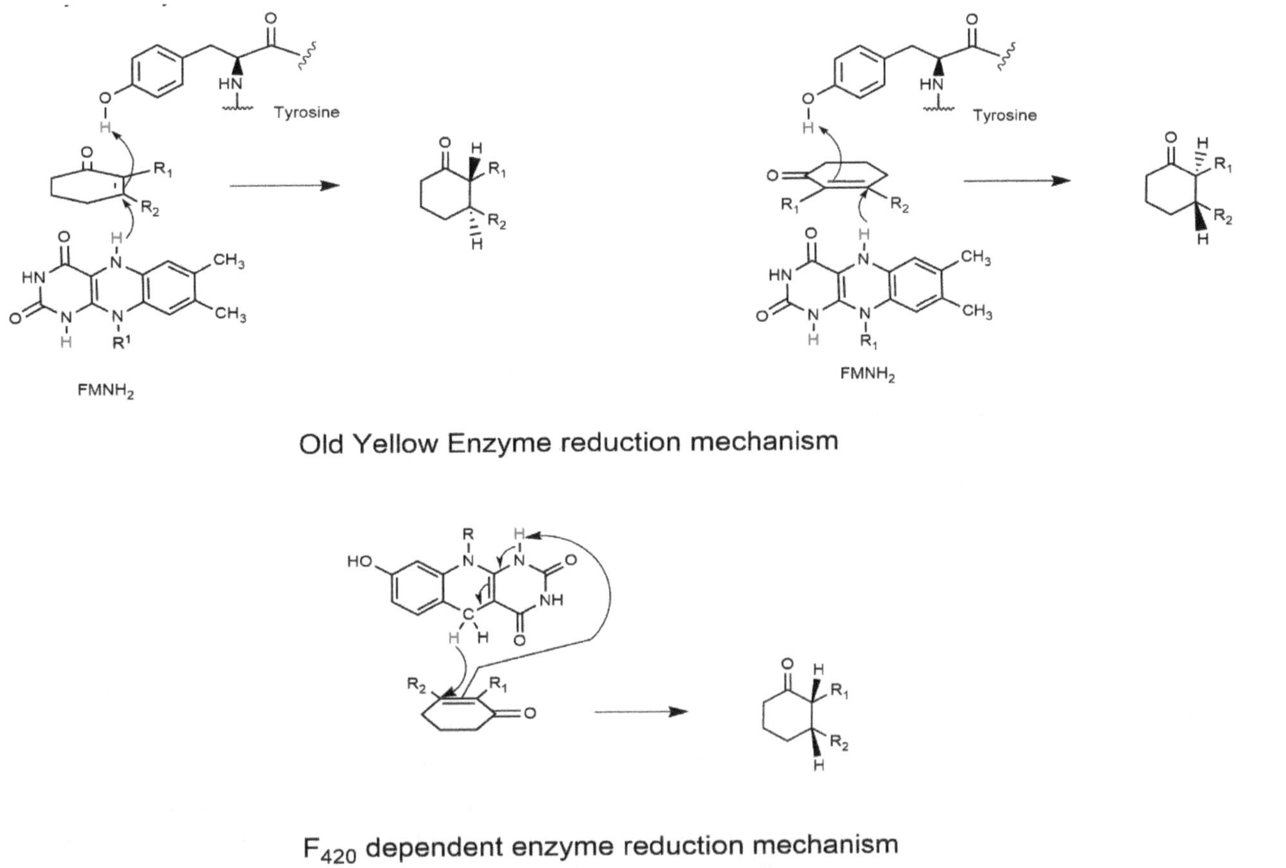

Old Yellow Enzyme reduction mechanism

F_{420} dependent enzyme reduction mechanism

Figure 5. Enoate reduction by a flavin-dependent enzyme (Old Yellow Enzyme) and the proposed mechanism for cofactor F_{420}-dependent reduction. Notably the mechanism of reduction yields *trans*-hydrogenation products for Old Yellow Enzyme and *cis*-hydrogenation products for the F_{420}-dependent enzymes.

The LLHT family contains several enzymes with alcohol oxidase or ketoreductase activity (Table 1; Figure 4). The F_{420}-dependent glucose-6-phosphate dehydrogenases of several species have been investigated [25,26,39]. Although an extensive survey of their substrate ranges has yet to be conducted, it has been demonstrated that glucose is a substrate for the *Rhodococcus jostii* RHA1 enzymes [26]. An F_{420}-dependent alcohol dehydrogenase (ADH) from *Methanogenium liminatans* has been shown to catalyze the oxidation of the short chain aliphatic alcohols 2-propanol, 2-butanol and 2-pentanol (85, 49 and 23.1 s^{-1} k_{cat}, 2.2, 1.2 and 7.2 mM K_M respectively) [40], but it was unable to oxidize primary alcohols, polyols or secondary alcohols with more than five carbons. It is unclear whether these alcohol oxidations are reversible, but in the oxidative direction, these reactions provide enzymes that can be used to recycle reduced cofactor F_{420} (see Section 4). Alcohol oxidation can also be used to produce ketones as intermediates in biocatalytic cascades that can then be used in subsequent reactions, such those catalyzed by transaminases or amine dehydrogenases in chiral amine synthesis [1,41–43] or by ketoreductases or alcohol dehydrogenases in chiral *sec*-alcohol synthesis (i.e., deracemization or stereoinversion of *sec*-alcohols). This approach can be achieved in a one pot cascade if different cofactors are used for the oxidation and reduction (Figure 6) [44].

At least one F_{420}-dependent ketoreductase has been described. The mycobacterial F_{420}-dependent phthiodiolone ketoreductase catalyzes a key reduction in the production of phthiocerol dimycocerosate, a diacylated polyketide found in the mycobacterial cell wall [45]. Although the physiological role of this enzyme has been elucidated, biochemical studies of the catalytic properties and substrate range are required to assess this enzymes' potential for use as a biocatalyst.

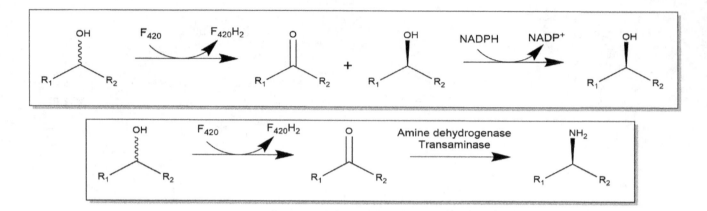

Figure 6. Proposed scheme for one-pot, enzyme cascades for deracemization/steroinversion of *sec*-alcohols (top) and chiral amine synthesis (bottom) using cofactor F_{420}-dependent alcohol oxidation.

F_{420}-dependent enzymes have also been shown to reduce imines (Table 1; Figure 4). An FDOR fromr *Streptomyces tateyamensis* (TpnL) is responsible for the reduction of dehydropiperidine in the piperidine-containing series *a* group of thiopeptide antibiotics produced in this bacterium (Figure 4). TpnL was identified as the F_{420}-dependent dehydropiperidine reductase responsible for the reduction of dehydropiperidine ring in thiostrepton A to produce the piperidine ring in the core macrocycle of thiostrepton A [45]. TpnL activity was affected by substrate inhibition at concentrations higher than 2 μM of thiostrepton A, preventing the measurement of the K_M, but its k_{cat}/K_M was measured at 2.80×10^4 M^{-1} S^{-1} [45]. The substrates for phthiodiolone ketoreductase and TpnL are large secondary metabolites and, as yet, it is unclear if it will accept smaller substrates or substrates with larger/smaller heterocycles (e.g., dehydropyrroles).

Another F_{420}-dependent imine reductase (TomJ) has been described from *Streptomyces achromogenes* that reduces the imine in 4-ethylidene-3,4-dehydropyrrole-2-carboxylic acid during the production of the secondary metabolite tomaymycin, which has been shown to have potentially interesting pharmaceutical properties [11]. Additionally, the reduction of a prochiral dihydropyrrole to a pyrrole is a reaction with a number of biocatalytic applications [5].

Nitroreductases have the potential application in the reduction of a prochiral nitro group to form a chiral amine [46]. The LLHT family F_{420}-dependent nitroreductase Npd from *Rhodococcus* catalyzes the two-electron reduction of two nitro groups in picric acid during catabolism of the explosive TNT (Table 1; Figure 4) [47]. While this stops short of reducing the nitro group to an amine, this catalytic activity may contribute to a reductive cascade that achieves this conversion.

The final class of reaction for consideration in this review is the unusual, reversable ring-opening/ closing reaction catalyzed by the MERs (Figure 4; Table 1). This reaction is required for folate biosynthesis in some archaea [23,28–30]. However, ring-closing reactions of this type could be used for producing N-containing heterocycles, which are intermediates in the synthesis of numerous pharmaceuticals [48,49]. The promiscuity of the MERs has not yet been investigated, and so the potential to re-engineer these enzymes is not fully understood.

Table 1. Characterized F_{420}-dependent enzymes with activities that could be adapted for biocatalytic applications.

Reaction	Family	Reference
Enoate reduction		
Aflatoxins	FDOR	[14,18,34]
Coumarins	FDOR	[14,34,35]
Quinones	FDOR	[36]
Biliverdin reduction	FDOR	[12,16]
Nitroimidazoles	FDOR	[36]
Cyclohexenones	FDOR	[6,34,38]
Citral/Neral/Geranial	FDOR	[6]
Carvone	FDOR	[6]
Ketoisophorone	FDOR	[6]
Alcohol oxidation/ketoreduction		
Glucose-6-phosphate	LLHT	[26,50]
Phthiodiolone dimycocerosate	LLHT	[51]
Isopropanol	LLHT	[40]
Imine reductions		
Dehydropiperidine (in thiopeptins)	FDOR	[45]
4-ethylidene-3,4-dihydropyrrole-2-carboxylic acid	Flavin-dependent monooxygenase	[11]
Nitroreductions		
Picrate	LLHT	[47,50]
2,4-DNP	LLHT	[48,50]
Ring opening/closing		
C-N bond cleavage/formation in methylenetetrahydromethanopterin	LLHT	[23,28–30]

3. Cofactor Recycling for Cofactor F_{420}

Cofactor recycling is essential for the practical application of the F_{420}-dependent enzymatic processes in biocatalysis. There are various strategies for cofactor regeneration for NADH and NADPH, including enzymatic, chemical, electrochemical and photochemical methods [52]. In this section, the potential enzymes for the regeneration of cofactor F_{420} are discussed. As most of the industrially relevant F_{420}-dependent reactions are asymmetric reductions, F_{420}-dependent oxidases are required for cofactor regeneration. Figure 7 shows the characterized enzymes that catalyze F_{420}-dependent oxidations that could be applied in cofactor F_{420} reduction.

Emulating methods developed for nicotinamide cofactors, both formate dehydrogenase (FDH) and glucose 6-phosphate dehydrogenase (G6PD) enzymes are attractive enzymatic routes for cofactor reduction both in vitro [53–56] and in vivo [57,58]. Fortunately, F_{420}-dependent G6PDs and FDHs have been identified and characterized. The F_{420}-dependent G6PD from *Mycobacteria* (FGD) is one potential cofactor F_{420}-recycling enzyme. FGD is the only enzyme in these bacteria known to reduce oxidized cofactor F_{420}. The intracellular concentration of G6P in *Mycobacteria* is up to 100-fold higher than it is in *E. coli*, which provides a ready source of reducing power for F_{420}-dependent reduction reactions [59]. FGD from *Rhodococcus jostii* and *Mycobacterium smegmatis* have been studied and expressed in *E. coli*, both the enzymes were stable in in vitro assays [26,39,60]. Both FGDs have been expressed in engineered *E. coli* producing cofactor F_{420} together with FDORs [38,59] FGDs have been shown to efficiently regenerate reduced cofactor F_{420} both in vivo and in vitro. However, the cost of the glucose-6-phosphate and the need to separate reaction products from the accumulated FGD byproduct (6-phosphoglucono-D-lactone) may prove to be impediments for the adoption of FGD as a recycling system for cofactor F_{420} in the in vitro biotransformations.

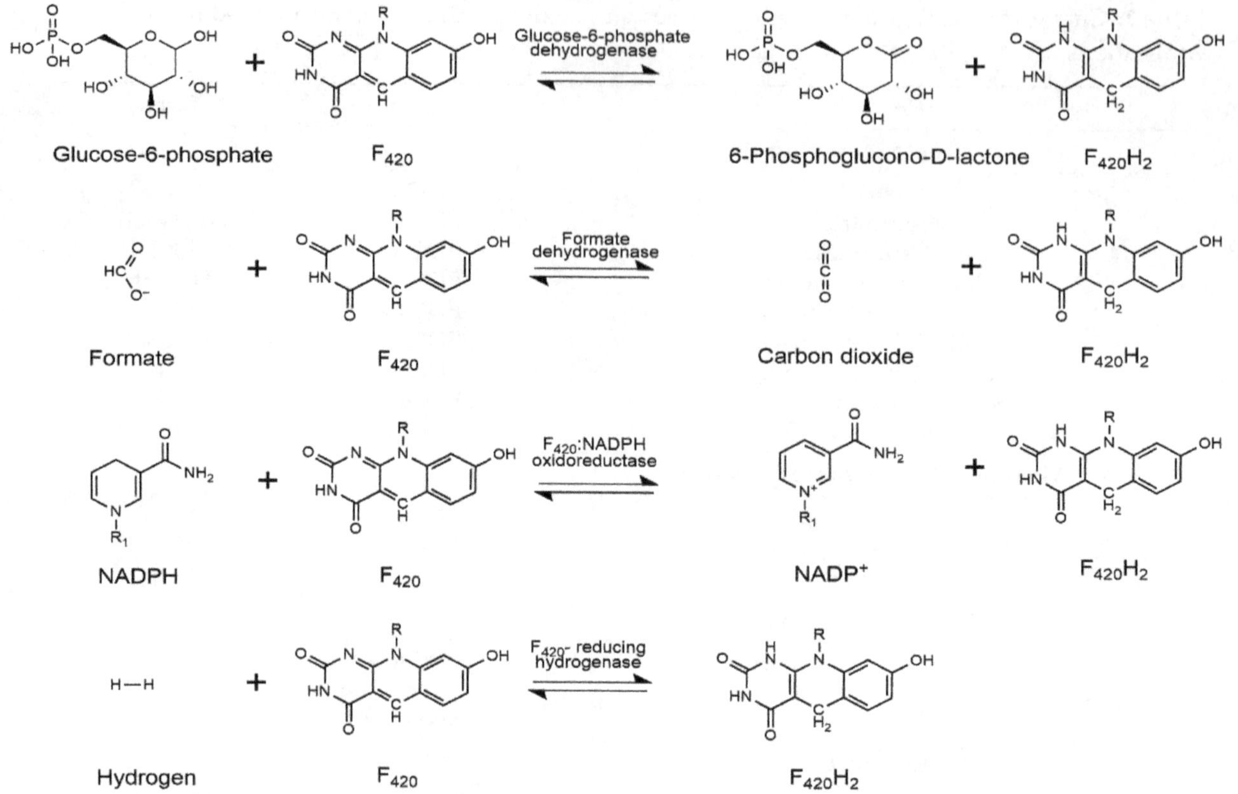

Figure 7. Cofactor F_{420}-dependent oxidation reactions that could be exploited to produce reduced cofactor F_{420}.

Formate is an excellent reductant for cofactor recycling, with FDH-dependent cofactor reduction yielding carbon dioxide, a volatile byproduct that can be easily removed from the reaction mixture, thereby simplifying the downstream processing of the product of interest. Additionally, formate is a low-cost reagent, leading to favorable process economics. Most methanogens have the capability to use formate as sole electron donor using F_{420}-dependent formate dehydrogenase [61]. The soluble F_{420}-dependent FDH from *Methanobacterium formicium* has been expressed in *E. coli* [62], purified and studied in vitro with the reduction of 41.2 μmol of F_{420} min^{-1} mg^{-1} of FDH, with non-covalently bound FAD required for optimal activity [8]. *Methanobacterium ruminantium* FDH reduces cofactor F_{420} at a much slower rate than *M. formicium*: 0.11 μmol of F_{420} min^{-1} mg^{-1} of FDH [8]. As yet, the use of F_{420}-dependnt FDHs for in vitro cofactor recycling has been sparsely studied. However, as these enzymes are soluble and can be heterologously expressed, they represent a promising system for use in cofactor F_{420}-dependent biocatalytic processes.

Another potential recycling system for cofactor F_{420} is the F_{420}:NADPH oxidoreductase (Fno), which couples the reduction of cofactor F_{420} with oxidation of NADPH. Methanogenic archaea use this enzyme to transfer reducing equivalents from hydrogenases to produce NADPH via F_{420}, while in bacteria it functions in the opposite direction, that is, to provide the cell with reduced F_{420} via NADPH [63]. Fno is also required for the production of reduced F_{420} for tetracycline production in *Streptomyces* [63]. The Fno enzymes from the thermophilic bacteria *Thermobifida fusca* and the thermophilic archaeon *Archeoglobus fulgidus* have been expressed in *E. coli* [64,65]. These enzymes are thermostable, with their highest activity observed at 65 °C. As the redox midpoint potentials of NADP and cofactor F_{420} are very similar, it is perhaps unsurprising that pH has a significant influence on the equilibrium of the reaction, with the reduction of NADP$^+$ favored at high pH (8–10) and the reduction of F_{420} favored at low pH (4–6) [64,65]. The Fno *Streptomyces griseus* has also been purified and characterized, and also displayed pH-dependent reaction directionality [66]. Fno may be an excellent enzyme for the in vivo reduction of cofactor F_{420}, where NADPH would be provided from central metabolism. However, for its use as a cofactor F_{420} recycling enzyme in vitro, Fno would need to be coupled with an NADPH regenerating enzyme, such as an NADPH-dependent formate dehydrogenase [67]. This added complexity and cost may limit the use of Fno-dependent cofactor F_{420} recycling in vivo.

Hydrogenotrophic archaea, including methanogens and sulfate-reducing archaea, possess an essential, cofactor F_{420}-dependent hydrogenase (FhrAGB) [68–71]. These nickel/iron enzymes could potentially be used in vivo to allow the direct H_2-dependent reduction of cofactor F_{420}. However, as these heterododecameric enzymes have complex cofactor requirements (four [4Fe 4S] clusters, and NiFe center and FAD), are oxygen-sensitive and tend to aggregate [71], it is unclear if they can be made suitable for in vitro use.

4. Cofactor Production

The lack of a scalable production system for cofactor F_{420} has been noted as a major impediment to the adoption of F_{420}-dependent enzymes by industry [5]. Cofactor F_{420} is available as a research reagent (http://www.gecco-biotech.com/), but its production at scale is not yet economic. In fact, most research laboratories with an interest in cofactor F_{420}-dependent enzymes synthesize and purify the cofactor themselves using slow-growing F_{420} producing microorganisms, most commonly methanogens and actinobacteria (Table 2). The economic production of cofactor F_{420} at large scale is not feasible using natural producers as they are ill-suited to industrial fermentation and generally lack the genetic tools required to improve cofactor F_{420} yield.

Table 2. Published production systems for cofactor F_{420}.

Source	F_{420} Yield (µmol/g Cell Weight)	Growth Conditions	Ref
Methanobacterium thermoautotrophicum	0.42 [a,c]	Grown at 60 °C using complex media in fermenter, under pressurized hydrogen	[9]
Methanobacterium formicium	0.27 [a,c]	Grown at 37 °C using complex media in fermenters	[9]
Methanospirillum hungatii	0.41 [a,c]	Grown at 37 °C using complex media in fermenters	[9]
Methanobacterium strain M.o.H	0.53 [a,c]	Grown at 40 °C using complex media in fermenters	[9]
Methanobacterium thermoautotrophicum	1.7 [e]	Grown using complex media in fermenters, under pressurized hydrogen gas	[73]
Streptomyces flocculus	0.62 [e]	Grown using complex media in fermenters	[73]
Streptomyces coelicolor	0.04 [e]	Grown using complex media in fermenters	[73]
Streptomyces griseus	0.008 [a,c]	Growth conditions not mentioned in the publication	[74]
Rhodococcus rhodochrous	0.11 [e]	Grown using complex media in fermenters	[73]
Mycobacterium smegmatis	0.30 [e]	Grown using complex media in fermenters	[73]
Mycobacterium smegmatis	3.0 [d]	Overexpression of F_{420} pathway genes, cultivation in complex media at 37 °C in shake flasks	[72]
Escherichia coli	0.38 [b]	Overexpressing F_{420} pathway genes, grown in minimal media at 30 °C in shake flasks.	[59]

[a] Mol weight of F_{420} with 1 glutamate tail is 773.6 Da, which was used to convert values published as µg of F_{420}, noting that micro-organisms produce mixture of F_{420} with different number of glutamates (1–9) attached. [b] Concentration estimated through absorbance at 420 nm and using extinction coefficient of 41.4 mM^{-1} cm^{-1} [73]. [c] F_{420} concentration per g of wet cell weight. [d] Concentration of F_{420} not mentioned in the publication, but F_{420} yield was stated to be 10 times higher than wild-type M. smegmatis. [e] Concentration estimated through absorbance at 400 nm and using extinction coefficient of 25.7 mM^{-1} cm^{-1} [74].

Recently, there have been significant advances towards the scalable production of the cofactor for F_{420}-dependent enzymes. M. smegmatis has been engineered to overexpress the biosynthetic genes for cofactor F_{420} production, leading to a substantial improvement in yields (Table 2) [72]. However, M. smegmatis is not ideally suited as a fermentation organism as it is slow growing, forms clumps during cultivation and is not recognized as GRAS (generally regarded as safe). More recently, the biosynthetic pathway for cofactor F_{420} has been successfully transplanted to *E. coli* [59], allowing the heterologous production of the cofactor at levels similar to those of the natural F_{420} producers (Table 2) [59], accumulated to 0.38 μmol of F_{420} per gram of dry cells.

There is scope to further improve the production of F_{420} in *E. coli*. Cofactor F_{420} does not appear to be toxic to *E. coli* [59], which suggests that there is little interaction between F_{420} and the enzymes *E. coli* (although this is yet to be confirmed experimentally). The thermodynamics of cofactor F_{420} production are favorable (Appendix A), suggesting that there are no major thermodynamic impediments to improving yield. Interestingly, the first dedicated step of cofactor F_{420} production (catalyzed by CofC/FbiD) is not energetically favorable and may consequently be sensitive to intracellular metabolite concentrations. In addition to the engineering considerations that this may impose, it may also be responsible for the biochemical diversity of this step in different microorganisms. In different microbes, the CofC/FbiD-dependent step uses 2-phospholactate [75], 3-phosphoglycerate [76] or phosphoenolpyruvate [59] as a substrate, which may reflect the relative abundance of those metabolites in various bacteria and archaea and the thermodynamic constraints on this step.

Another recent advance is the production of a synthetic analog of cofactor F_{420}, called F_O-5'-phosphate (F_OP). F_OP was derived from F_O, the metabolic precursor of cofactor F_{420}, which is phosphorylated using an engineered riboflavin kinase [38]. F_OP has also been shown to function as an active cofactor for cofactor F_{420}-dependent enzymes activities, albeit there is a penalty in the rates of these reactions [38]. Drenth and coworkers prepared F_O by chemical synthesis, using a method developed by Hossain et al. [77]. However, it is likely that the engineered kinase for the phosphorylation of F_O could be introduced into an organism that over-produces F_O allowing for the production of F_OP by fermentation. This semisynthetic pathway would have the advantage that it needs only two biosynthetic steps, instead of the four steps needed for cofactor F_{420} production, and demands less metabolic input from the native host metabolism (e.g., no glutamate is required) [38]. The production of F_OP also opens the possibility of making deazaflavin analogs of FMN and FAD, which would be electrochemically more like F_{420} than flavins, but may still bind FMN and FAD- dependent enzymes and potentially allow access to new chemistry with already well-characterized enzymes.

5. Prospects

Reduced cofactor F_{420} is electrochemically well suited for biocatalytic applications, and the small number of F_{420}-dependent enzymes characterized to date show promise as potential biocatalysts (as discussed above). However, before these enzymes can be widely and effectively used as biocatalysts, further research is needed to better characterize them as the biochemistry of cofactor F_{420}-dependent enzymes remains under-explored. The LLHT and FDOR families are a rich source of highly diverse enzymes with considerable potential for biocatalysis, albeit much of the research to date has focused on the physiological roles of these enzymes, rather than their in vitro enzymology. Although some of these enzymes have been shown to have small molecule substrates, those involved with secondary metabolite biosynthesis tend to act on high molecular weight substrates and it is not yet clear whether they will accept lower molecular weight molecules.

To be cost competitive, cofactor F_{420} needs to have effective recycling systems. The enzymes for cofactor recycling have already been identified, although there have been a few studies investigating their performance in this role. Moreover, alternative cofactor recycling strategies, such electrochemical or photochemical recycling, have not yet been investigated for cofactor F_{420}. The production of cofactor F_{420} at scale and at low cost remains a roadblock for the use of these enzymes by industry. However, considerable progress has been made on this front in the last few years and it is likely that low cost cofactor F_{420}, or F_{420} surrogates, will soon be available. Additionally, the availability of

F_{420}-producing bacteria with tools for facile genetic manipulation, along with a growing number of empirically determined protein structures, opens up the prospect of improving this class of enzymes using in vitro evolution and rational design. It is notable that there is still some uncertainty concerning the mechanistic detail of F_{420}-dependent reactions, which need to be addressed through a detailed structure/function analysis to enable a rational design of these enzymes.

Author Contributions: Writing—original draft preparation, M.V.S., J.A., S.W.K., H.N.-B., C.J.J. and C.S.; writing— review and editing, M.V.S., J.A., S.W.K., H.N.-B., A.C.W., C.J.H., C.J.J. and C.S.

Abbreviations

5AD: 5′-Deoxyadenosine; 5ARPD: 5-amino-6-(D-ribitylamino)uracil; 5ARPD4HB: 5-amino-5-(4-hydroxybenzyl)-6-(D-ribitylimino)-5,6-dihydrouracil; d_{F420}-0: Dehydro coenzyme F_{420}-0 (oxidized); EPPG: Enolpyruvyl-diphospho-5′-guanosine; F_o: 7,8-Didemethyl-8-hydroxy-5-deazariboflavin; F_{420}-0: Coenzyme F_{420}-0 (oxidized); F_{420}-1: Coenzyme F_{420}-1 (oxidized); F_{420}-2: Coenzyme F_{420}-2 (oxidized); F_{420}-3: Coenzyme F_{420}-3 (oxidized); FMN: Flavin mononucleotide (oxidized); $FMNH_2$: Flavin mononucleotide (reduced); GDP: Guanosine diphosphate; GMP: Guanosine monophosphate; Glu: L-Glutamate; GTP: Guanosine triphosphate; H^+: Proton; ImiAce: 2-iminoacetate or Dehydroglycine; Met: L-Methionine; NH_4: Ammonium; PEP: Phosphoenolpyruvate; Pi: Phosphate; PPi: Diphosphate; SAMe: S-Adenosyl-L-methionine; Tyr: L-Tyrosine.

Appendix A Thermodynamics of F420 Biosynthesis

The thermodynamic properties of each of the steps in cofactor F_{420} biosynthesis were estimated to evaluate the feasibility of increasing the production of the cofactor in an engineered microorganism. The pathway assembled by Bashiri et al. [59] in *E. coli* was used (i.e., PEP was used as substrate for CofC). The standard transformed Gibbs free energy ($\Delta_r G^t$) of each step were calculated under the physiological conditions (25 °C, pH 7, and ionic concentration of 0.25 M) as described elsewhere [78,79]. The overall Gibbs free energy (ΔG^t) was then calculated by summing up all individual $\Delta_r G^t$ (Table A1). The Gibbs free energy of metabolite formation ($\Delta_f G$) for each metabolite in the pathway was obtained (Supplementary Information) from comprehensive lists of metabolites whose $\Delta_f G$ were estimated using a group contribution method [80,81]. The $\Delta_f G$ for each metabolite was then converted into its transformed type ($\Delta_f G^t$) method of Alberty [78]. The data were collected from relevant biochemical databases and the literature for any metabolite with missing $\Delta_f G$ [82–84]. Owing to possessing different protonation states, the inconsistencies in $\Delta_f G$ of certain metabolites such as the glutamates in F_{420}-n among databases and the literature are inevitable. Thus, $\Delta_r G^t$ for reactions containing metabolites with varying $\Delta_f G$ were calculated considering the differences in their $\Delta_f G$ leading to the generation of a total of four sets of $\Delta_r G^t$. Finally, the mean and standard deviations were calculated for these sets to yield the variation in each reaction as well as in the overall pathway (Table A1).

The data shown in Table A1 confirms that the overall cofactor F_{420} biosynthesis pathway is thermodynamically feasible under the given conditions. However, certain steps in this pathway impose a thermodynamic barrier with respect to the physiological conditions examined. For example, CofC seems to be one of the major thermodynamically unfavorable steps in the whole pathway possibly due to the energy-dependent synthesis of EPPG, one of the precursors for making F_{420}. CofG/H combined appears to be the most thermodynamically favorable step in the whole pathway driving the biosynthesis of F_0, the other key precursor for F_{420} biosynthesis. Interestingly, the formation of F_{420}-2 molecule seems to be the most favorable step among other F_{420} molecules downstream of the pathway. It should be noted that the thermodynamic calculations were only performed up to three steps of F_{420} molecule production (i.e., F_{420}-3) largely because of the high levels of inconsistencies of the data available for $\Delta_f G$ of higher F_{420} molecules.

Table A1. Standard transformed Gibbs free energy of reaction ($\Delta_r G^t$), for the F_{420} biosynthesis pathway, calculated based on Gibbs free energy of metabolite formation ($\Delta_f G^t$) calculated at 25 °C, pH of 7, and ionic concentration of 0.25 M.

Enzyme	Reaction [a]	$\Delta_r G^t$ (kJ) [b]
CofC/FbiD	PEP + GTP → EPPG + PP_i [d]	+71.27(±67)
CofG/FbiC	5ARPD + Tyr + SAMe → 5ARPD4HB + ImAcet + Met + 5AD	−1192.39(±0) [c]
CofH/FbiC	5ARPD4HB + SAMe → F_O + NH_4^+ + Met + 5AD	+71.90(±36) [c]
CofD/FbiA	F_O + EPPG → d_{F420}-0 + GMP	−31.3(±128)
CofX/FbiB	d_{F420}-0 + $FMNH_2$ → F_{420}-0 + FMN	−74.59(±87)
CofE/FbiB	F_{420}-0 + GTP + Glu → F_{420}-1 + GDP + P_i	−7.50(±24)
CofE/FbiB	F_{420}-1 + GTP + Glu → F_{420}-2 + GDP + P_i	−39.44(±35)
CofE/FbiB	F_{420}-2 + GTP + Glu → F_{420}-3 + GDP + P_i	−21.99(±38)
Overall	PEP + 5ARPD + Tyr + (2) SAMe + $FMNH_2$ + (3) Glu + (4) GTP → F_{420}-3 + (2) Met + (2) 5AD + ImAcet + NH_4^+ + FMN + (3) GDP + (3) P_i + GMP + PP_i	−1224.05(±82)

[a] For simplicity, protons were omitted in these equations and subsequent calculations as the $\Delta_f G^t$ of a proton under the set conditions is ~0.08 kJ. However, all $\Delta_r G^t$ calculations are based on a balanced equation. [b] The mean values of four sets and their standard deviations in parenthesis shown for each reaction. [c] $\Delta_f G$ of 5ARPD4HB has only been reported in MetaCyc inferred by computational analysis. Including it in the calculations of $\Delta_r G^t$ for CofG and CofH results in −225.88(±0) and −894.62(±36), respectively. [d] Hydrolysis of PP_i ($H_3P_2O_7^{3-}$ + H_2O → 2 HPO_4^{2-} + H^+) yields a $\Delta_r G^t$ of ~17 kJ/mole, resulting in less than 2% change in the overall $\Delta_r G^t$.

References

1. Patil, M.D.; Grogan, G.; Bommarius, A.; Yun, H. Oxidoreductase-catalyzed synthesis of chiral amines. *ACS Catal.* **2018**, *8*, 10985–11015. [CrossRef]

2. Toogood, H.S.; Scrutton, N.S. New developments in 'ene'-reductase catalysed biological hydrogenations. *Curr. Opin. Chem. Biol.* **2014**, *19*, 107–115. [CrossRef] [PubMed]

3. Cosgrove, S.C.; Brzezniak, A.; France, S.P.; Ramsden, J.I.; Mangas-Sanchez, J.; Montgomery, S.L.; Heath, R.S.; Turner, N.J. Imine reductases, reductive aminases, and amine oxidases for the synthesis of chiral amines: Discovery, characterization, and synthetic applications. In *Enzymes in Synthetic Biology*; Scrutton, N., Ed.; Elsevier Academic Press Inc.: San Diego, CA, USA, 2018; Volume 608, pp. 131–149.

4. Bai, D.Y.; He, J.Y.; Ouyang, B.; Huang, J.; Wang, P. Biocatalytic asymmetric synthesis of chiral aryl alcohols. *Prog. Chem.* **2017**, *29*, 491–501.

5. Taylor, M.; Scott, C.; Grogan, G. F-420-dependent enzymes-potential for applications in biotechnology. *Trends Biotechnol.* **2013**, *31*, 63–64. [CrossRef]

6. Mathew, S.; Trajkovic, M.; Kumar, H.; Nguyen, Q.-T.; Fraaije, M.W. Enantio- and regioselective *ene*-reductions using $F_{420}H_2$-dependent enzymes. *Chem. Commun.* **2018**, *54*, 11208–11211. [CrossRef]

7. Greening, C.; Ahmed, F.H.; Mohamed, A.E.; Lee, B.M.; Pandey, G.; Warden, A.C.; Scott, C.; Oakeshott, J.G.; Taylor, M.C.; Jackson, C.J. Physiology, biochemistry, and applications of F_{420}- and Fo-dependent redox reactions. *Microbiol. Mol. Biol. Rev.* **2016**, *80*, 451–493. [CrossRef]

8. Tzing, S.F.; Bryant, M.P.; Wolfe, R.S. Factor 420-dependent pyridine nucleotide-linked formate metabolism of *Methanobacterium ruminantium*. *J. Bacteriol.* **1975**, *121*, 192–196.

9. Eirich, L.D.; Vogels, G.D.; Wolfe, R.S. Distribution of coenzyme F_{420} and properties of its hydrolytic fragments. *J. Bacteriol.* **1979**, *140*, 20–27.

10. Wang, P.; Bashiri, G.; Gao, X.; Sawaya, M.R.; Tang, Y. Uncovering the enzymes that catalyze the final steps in oxytetracycline biosynthesis. *J. Am. Chem. Soc.* **2013**, *135*, 7138–7141. [CrossRef]

11. Li, W.; Chou, S.C.; Khullar, A.; Gerratana, B. Cloning and characterization of the biosynthetic gene cluster for tomaymycin, an SJG-136 monomeric analog. *Appl. Environ. Microbiol.* **2009**, *75*, 2958–2963. [CrossRef]

12. Ahmed, F.H.; Carr, P.D.; Lee, B.M.; Afriat-Jurnou, L.; Mohamed, A.E.; Hong, N.-S.; Flanagan, J.; Taylor, M.C.; Greening, C.; Jackson, C.J. Sequence–structure–function classification of a catalytically diverse oxidoreductase superfamily in *mycobacteria*. *J. Mol. Biol.* **2015**, *427*, 3554–3571. [CrossRef] [PubMed]

13. Selengut, J.D.; Haft, D.H. Unexpected abundance of coenzyme F_{420}-dependent enzymes in *Mycobacterium tuberculosis* and other actinobacteria. *J. Bacteriol.* **2010**, *192*, 5788–5798. [CrossRef] [PubMed]

14. Lapalikar, G.V.; Taylor, M.C.; Warden, A.C.; Onagi, H.; Hennessy, J.E.; Mulder, R.J.; Scott, C.; Brown, S.E.; Russell, R.J.; Easton, C.J.; et al. Cofactor promiscuity among F_{420}-dependent reductases enables them to catalyse both oxidation and reduction of the same substrate. *Catal. Sci. Technol.* **2012**, *2*, 1560–1567. [CrossRef]

15. Harold, L.K.; Antoney, J.; Ahmed, F.H.; Hards, K.; Carr, P.D.; Rapson, T.; Greening, C.; Jackson, C.J.; Cook, G.M. FAD-sequestering proteins protect Mycobacteria against hypoxic and oxidative stress. *J. Biol. Chem.* **2019**, *294*, 2903–2912. [CrossRef] [PubMed]

16. Ahmed, F.H.; Mohamed, A.E.; Carr, P.D.; Lee, B.M.; Condic-Jurkic, K.; O'Mara, M.L.; Jackson, C.J. Rv2074 is a novel $F_{420}H_2$-dependent biliverdin reductase in *Mycobacterium tuberculosis*. *Protein Sci.* **2016**, *25*, 1692–1709. [CrossRef] [PubMed]

17. Mashalidis, E.H.; Mukherjee, T.; Śledź, P.; Matak-Vinković, D.; Boshoff, H.I.; Abell, C.; Barry, C.E. Rv2607 from *Mycobacterium tuberculosis* is a pyridoxine 5'-phosphate oxidase with unusual substrate specificity. *PLoS ONE* **2011**, *6*, e27643. [CrossRef] [PubMed]

18. Taylor, M.C.; Jackson, C.J.; Tattersall, D.B.; French, N.; Peat, T.S.; Newman, J.; Briggs, L.J.; Lapalikar, G.V.; Campbell, P.M.; Scott, C.; et al. Identification and characterization of two families of $F_{420}H_2$-dependent reductases from *Mycobacteria* that catalyse aflatoxin degradation. *Mol. Microbiol.* **2010**, *78*, 561–575. [CrossRef]

19. Cellitti, S.E.; Shaffer, J.; Jones, D.H.; Mukherjee, T.; Gurumurthy, M.; Bursulaya, B.; Boshoff, H.I.; Choi, I.; Nayyar, A.; Lee, Y.S.; et al. Structure of DDN, the deazaflavin-dependent nitroreductase from *Mycobacterium tuberculosis* involved in bioreductive activation of PA-824. *Structure* **2012**, *20*, 101–112. [CrossRef]

20. De Souza, G.A.; Leversen, N.A.; Målen, H.; Wiker, H.G. Bacterial proteins with cleaved or uncleaved signal peptides of the general secretory pathway. *J. Proteom.* **2011**, *75*, 502–510. [CrossRef]

21. He, Z.; De Buck, J. Cell wall proteome analysis of *Mycobacterium smegmatis* strain mc2 155. *BMC Microbiol.* **2010**, *10*, 121. [CrossRef]

22. Sinha, S.; Kosalai, K.; Arora, S.; Namane, A.; Sharma, P.; Gaikwad, A.N.; Brodin, P.; Cole, S.T. Immunogenic membrane-associated proteins of *Mycobacterium tuberculosis* revealed by proteomics. *Microbiology* **2005**, *151*, 2411–2419. [CrossRef] [PubMed]

23. Aufhammer, S.W.; Warkentin, E.; Ermler, U.; Hagemeier, C.H.; Thauer, R.K.; Shima, S. Crystal structure of methylenetetrahydromethanopterin reductase (MER) in complex with coenzyme F_{420}: Architecture of the F_{420}/FMN binding site of enzymes within the nonprolyl *cis*-peptide containing bacterial luciferase family. *Protein Sci.* **2005**, *14*, 1840–1849. [CrossRef] [PubMed]

24. Aufhammer, S.W.; Warkentin, E.; Berk, H.; Shima, S.; Thauer, R.K.; Ermler, U. Coenzyme binding in F_{420}-dependent secondary alcohol dehydrogenase, a member of the bacterial luciferase family. *Structure* **2004**, *12*, 361–370. [CrossRef] [PubMed]

25. Bashiri, G.; Squire, C.J.; Moreland, N.J.; Baker, E.N. Crystal structures of F_{420}-dependent glucose-6-phosphate dehydrogenase FGD1 involved in the activation of the anti-tuberculosis drug candidate PA-824 reveal the basis of coenzyme and substrate binding. *J. Biol. Chem.* **2008**, *283*, 17531–17541. [CrossRef] [PubMed]

26. Nguyen, Q.T.; Trinco, G.; Binda, C.; Mattevi, A.; Fraaije, M.W. Discovery and characterization of an F_{420}-dependent glucose-6-phosphate dehydrogenase (Rh-FGD1) from *rhodococcus jostii* rha1. *Appl. Microbiol. Biotechnol.* **2017**, *101*, 2831–2842. [CrossRef] [PubMed]

27. Mascotti, M.L.; Kumar, H.; Nguyen, Q.-T.; Ayub, M.J.; Fraaije, M.W. Reconstructing the evolutionary history of F_{420}-dependent dehydrogenases. *Sci. Rep.* **2018**, *8*, 17571. [CrossRef] [PubMed]

28. Ceh, K.; Demmer, U.; Warkentin, E.; Moll, J.; Thauer, R.K.; Shima, S.; Ermler, U. Structural basis of the hydride transfer mechanism in F_{420}-dependent methylenetetrahydromethanopterin dehydrogenase. *Biochemistry* **2009**, *48*, 10098–10105. [CrossRef]

29. Shima, S.; Warkentin, E.; Grabarse, W.; Sordel, M.; Wicke, M.; Thauer, R.K.; Ermler, U. Structure of coenzyme F_{420} dependent methylenetetrahydromethanopterin reductase from two methanogenic archaea. *J. Mol. Biol.* **2000**, *300*, 935–950. [CrossRef]

30. Vaupel, M.; Thauer, R.K. Coenzyme F_{420} dependent N-5, N-10-methylenetetrahydromethanopterin reductase (MER) from *Methanobacterium thermautotrophicum* strain marburg: Cloning, sequencing, transcriptional analysis and functional expression in *Escherichia coli* of the *mer* gene. *Eur. J. Biochem.* **1995**, *231*, 773–778.

31. Purwantini, E.; Mukhopadhyay, B. Rv0132c of *Mycobacterium tuberculosis* encodes a coenzyme F_{420}-dependent hydroxymycolic acid dehydrogenase. *PLoS ONE* **2013**, *8*, e81985. [CrossRef]

32. Fida, T.T.; Palamuru, S.; Pandey, G.; Spain, J.C. Aerobic biodegradation of 2,4-dinitroanisole by *Nocardioides* sp. Strain js1661. *Appl. Environ. Microbiol.* **2014**, *80*, 7725–7731. [CrossRef] [PubMed]

33. Ebert, S.; Rieger, P.G.; Knackmuss, H.J. Function of coenzyme f_{420} in aerobic catabolism of 2,4,6-trinitrophenol and 2,4-dinitrophenol by *Nocardioides simplex* FJ2-1A. *J. Bacteriol.* **1999**, *181*, 2669–2674. [PubMed]

34. Lapalikar, G.V.; Taylor, M.C.; Warden, A.C.; Scott, C.; Russell, R.J.; Oakeshott, J.G. $F_{420}H_2$-dependent degradation of aflatoxin and other furanocoumarins is widespread throughout the *Actinomycetales*. *PLoS ONE* **2012**, *7*, e30114. [CrossRef] [PubMed]

35. Greening, C.; Jirapanjawat, T.; Afroze, S.; Ney, B.; Scott, C.; Pandey, G.; Lee, B.M.; Russell, R.J.; Jackson, C.J.; Oakeshott, J.G.; et al. Mycobacterial $F_{420}H_2$-dependent reductases promiscuously reduce diverse compounds through a common mechanism. *Front. Microbiol.* **2017**, *8*. [CrossRef]

36. Gurumurthy, M.; Rao, M.; Mukherjee, T.; Rao, S.P.S.; Boshoff, H.I.; Dick, T.; Barry, C.E.; Manjunatha, U.H. A novel F_{420}-dependent anti-oxidant mechanism protects *Mycobacterium tuberculosis* against oxidative stress and bactericidal agents. *Mol. Microbiol.* **2013**, *87*, 744–755. [CrossRef]

37. Mohamed, A.E.; Ahmed, F.H.; Arulmozhiraja, S.; Lin, C.Y.; Taylor, M.C.; Krausz, E.R.; Jackson, C.J.; Coote, M.L. Protonation state of $F_{420}H_2$ in the prodrug-activating deazaflavin dependent nitroreductase (DDN) from *Mycobacterium tuberculosis*. *Mol. BioSys.* **2016**, *12*, 1110–1113. [CrossRef]

38. Drenth, J.; Trajkovic, M.; Fraaije, M.W. Chemoenzymatic synthesis of an unnatural deazaflavin cofactor that can fuel F_{420}-dependent enzymes. *ACS Catal.* **2019**, *9*, 6435–6443. [CrossRef]

39. Purwantini, E.; Daniels, L. Molecular analysis of the gene encoding F_{420}-dependent glucose-6-phosphate dehydrogenase from *Mycobacterium smegmatis*. *J. Bacteriol.* **1998**, *180*, 2212–2219.

40. Bleicher, K.; Winter, J. Purification and properties of F_{420}-and $NADP^+$-dependent alcohol dehydrogenases of *Methanogenium liminatans* and *Methanobacterium palustre*, specific for secondary alcohols. *Europ. J. Biochem.* **1991**, *200*, 43–51. [CrossRef]

41. Knaus, T.; Cariati, L.; Masman, M.F.; Mutti, F.G. In vitro biocatalytic pathway design: Orthogonal network for the quantitative and stereospecific amination of alcohols. *Org. Biomol. Chem.* **2017**, *15*, 8313–8325.

42. Guo, F.; Berglund, P. Transaminase biocatalysis: Optimization and application. *Green Chem.* **2017**, *19*, 333–360. [CrossRef]

43. Adams, J.P.; Brown, M.J.B.; Diaz-Rodriguez, A.; Lloyd, R.C.; Roiban, G.D. Biocatalysis: A pharma perspective. *Adv. Synth. Catal.* **2019**, *361*, 2421–2432. [CrossRef]

44. Musa, M.M.; Hollmann, F.; Mutti, F.G. Synthesis of enantiomerically pure alcohols and amines *via* biocatalytic deracemisation methods. *Cat. Sci. Technol.* **2019**, *9*, 10–1039. [CrossRef]

45. Ichikawa, H.; Bashiri, G.; Kelly, W.L. Biosynthesis of the thiopeptins and identification of an $F_{420}H_2$-dependent dehydropiperidine reductase. *J. Am. Chem. Soc.* **2018**, *140*, 10749–10756. [CrossRef]

46. Miller, A.F.; Park, J.T.; Ferguson, K.L.; Pitsawong, W.; Bommarius, A.S. Informing efforts to develop nitroreductase for amine production. *Molecules* **2018**, *23*, 22. [CrossRef]

47. Heiss, G.; Hofmann, K.W.; Trachtmann, N.; Walters, D.M.; Rouvière, P.; Knackmuss, H.J. *Npd* gene functions of *Rhodococcus (opacus) erythropolis* Hl PM-1 in the initial steps of 2,4,6-trinitrophenol degradation. *Microbiology* **2002**, *148*, 799–806. [CrossRef]

48. Xu, J.; Green, A.P.; Turner, N.J. Chemo-enzymatic synthesis of pyrazines and pyrroles. *Angew. Chem.-Int. Ed.* **2018**, *57*, 16760–16763. [CrossRef]

49. Busacca, C.A.; Fandrick, D.R.; Song, J.J.; Senanayake, C.H. The growing impact of catalysis in the pharmaceutical industry. *Adv. Synth. Catal.* **2011**, *353*, 1825–1864. [CrossRef]

50. Heiss, G.; Trachtmann, N.; Abe, Y.; Takeo, M.; Knackmuss, H.J. Homologous *npdgi* genes in 2,4-dinitrophenol- and 4-nitrophenol-degrading *Rhodococcus* spp. *Appl. Environ. Microbiol.* **2003**, *69*, 2748–2754. [CrossRef]

51. Purwantini, E.; Daniels, L.; Mukhopadhyay, B. $F_{420}H_2$ is required for phthiocerol dimycocerosate synthesis in *Mycobacteria*. *J. Bacteriol.* **2016**, *198*, 2020–2028. [CrossRef]

52. Wichmann, R.; Vasic-Racki, D. Cofactor regeneration at the lab scale. In *Technology Transfer in Biotechnology*; Springer: Berlin, Germany, 2005; pp. 225–260.

53. Tishkov, V.I.; Popov, V.O. Catalytic mechanism and application of formate dehydrogenase. *Biochemistry* **2004**, *69*, 1252. [CrossRef] [PubMed]

54. Eguchi, T.; Kuge, Y.; Inoue, K.; Yoshikawa, N.; Mochida, K.; Uwajima, T. NADPH regeneration by glucose dehydrogenase from *Gluconobacter scleroides* for L-leucovorin synthesis. *Biosci. Biotechnol. Biochem.* **1992**, *56*, 701–703. [CrossRef] [PubMed]

55. Demir, A.S.; Talpur, F.N.; Sopaci, B.; Kohring, G.-W.; Celik, A. Selective oxidation and reduction reactions with cofactor regeneration mediated by galactitol-, lactate-, and formate dehydrogenases immobilized on magnetic nanoparticles. *J. Biotechnol.* **2011**, *152*, 176–183. [CrossRef] [PubMed]

56. Wong, C.-H.; Whitesides, G.M. Enzyme-catalyzed organic synthesis: NAD(P)H cofactor regeneration by using glucose-6-phosphate and the glucose-5-phosphate dehydrogenase from *Leuconostoc mesenteroides*. *J. Am. Chem. Soc.* **1981**, *103*, 4890–4899. [CrossRef]

57. Lee, W.-H.; Park, J.-B.; Park, K.; Kim, M.-D.; Seo, J.-H. Enhanced production of ε-caprolactone by overexpression of NADPH-regenerating glucose 6-phosphate dehydrogenase in recombinant *Escherichia coli* harboring cyclohexanone monooxygenase gene. *Appl. Microbiol. Biotechnol.* **2007**, *76*, 329–338. [CrossRef] [PubMed]

58. Berrios-Rivera, S.J.; Bennett, G.N.; San, K.Y. Metabolic engineering of *Escherichia coli*: Increase of NADH availability by overexpressing an NAD$^+$-dependent formate dehydrogenase. *Metab. Eng.* **2002**, *4*, 217–229. [CrossRef]

59. Bashiri, G.; Antoney, J.; Jirgis, E.N.M.; Shah, M.V.; Ney, B.; Copp, J.; Stuteley, S.M.; Sreebhavan, S.; Palmer, B.; Middleditch, M.; et al. A revised biosynthetic pathway for the cofactor F$_{420}$ in prokaryotes. *Nat. Commun.* **2019**, *10*, 1558. [CrossRef]

60. Purwantini, E.; Daniels, L. Purification of a novel coenzyme F$_{420}$-dependent glucose-6-phosphate dehydrogenase from *Mycobacterium smegmatis*. *J. Bacteriol.* **1996**, *178*, 2861–2866. [CrossRef]

61. Costa, K.C.; Wong, P.M.; Wang, T.; Lie, T.J.; Dodsworth, J.A.; Swanson, I.; Burn, J.A.; Hackett, M.; Leigh, J.A. Protein complexing in a methanogen suggests electron bifurcation and electron delivery from formate to heterodisulfide reductase. *Proc. Natl. Acad. Sci. USA* **2010**, *107*, 11050–11055. [CrossRef]

62. Shuber, A.P.; Orr, E.C.; Recny, M.A.; Schendel, P.F.; May, H.D.; Schauer, N.L.; Ferry, J.G. Cloning, expression, and nucleotide sequence of the formate dehydrogenase genes from *Methanobacterium formicicum*. *J. Biol. Chem.* **1986**, *261*, 12942–12947.

63. Novotná, J.; Neužil, J.; Hoš?álek, Z. Spectrophotometric identification of 8-hydroxy-5-deazaflavin: Nadph oxidoreductase activity in *Streptomycetes* producing tetracyclines. *FEMS Microbiol. Lett.* **1989**, *59*, 241–245. [CrossRef]

64. Kumar, H.; Nguyen, Q.T.; Binda, C.; Mattevi, A.; Fraaije, M.W. Isolation and characterization of a thermostable F$_{420}$:NADPH oxidoreductase from *Thermobifida fusca*. *J. Biol. Chem.* **2017**, *292*, 10123–10130. [CrossRef] [PubMed]

65. Kunow, J.; Schwörer, B.; Stetter, K.O.; Thauer, R.K. A F$_{420}$-dependent NADP reductase in the extremely thermophilic sulfate-reducing *Archaeoglobus fulgidus*. *Arch. Microbiol.* **1993**, *160*, 199–205.

66. Eker, A.P.M.; Hessels, J.K.C.; Meerwaldt, R. Characterization of an 8-hydroxy-5-deazaflavin: NADPH oxidoreductase from *Streptomyces griseus*. *Biochim. Biophys. Acta* **1989**, *990*, 80–86. [CrossRef]

67. Seelbach, K.; Riebel, B.; Hummel, W.; Kula, M.R.; Tishkov, V.I.; Egorov, A.M.; Wandrey, C.; Kragl, U. A novel, efficient regenerating method of NADPH using a new formate dehydrogenase. *Tetrahedron Lett.* **1996**, *37*, 1377–1380. [CrossRef]

68. Alex, L.A.; Reeve, J.N.; Ormejohnson, W.H.; Walsh, C.T. Cloning, sequence determination, and expression of the genes encoding the subunits of the nickel-containing 8-hydroxy-5-deazaflavin reducing hydrogenase from *Methanobacterium thermoautotrophicum* delta-H. *Biochemistry* **1990**, *29*, 7237–7244. [CrossRef]

69. Tersteegen, A.; Hedderich, R. *Methanobacterium thermoautotrophicum* encodes two multisubunit membrane-bound NiFe hydrogenases - transcription of the operons and sequence analysis of the deduced proteins. *Eur. J. Biochem.* **1999**, *264*, 930–943. [CrossRef]

70. Hocking, W.P.; Stokke, R.; Roalkvam, I.; Steen, I.H. Identification of key components in the energy metabolism of the hyperthermophilic sulfate-reducing archaeon *archaeoglobus fulgidus* by transcriptome analyses. *Front. Microbiol.* **2014**, *5*, 20. [CrossRef]

71. Vitt, S.; Ma, K.; Warkentin, E.; Moll, J.; Pierik, A.J.; Shima, S.; Ermler, U. The F-420-reducing NiFe-hydrogenase complex from *methanothermobacter marburgensis*, the first x-ray structure of a group 3 family member. *J. Mol. Biol.* **2014**, *426*, 2813–2826. [CrossRef]

72. Bashiri, G.; Rehan, A.M.; Greenwood, D.R.; Dickson, J.M.; Baker, E.N. Metabolic engineering of cofactor F$_{420}$ production in *Mycobacterium smegmatis*. *PLoS ONE* **2010**, *5*, e15803. [CrossRef]

73. Isabelle, D.; Simpson, D.R.; Daniels, L. Large-scale production of coenzyme F$_{420}$-5,6 by using *Mycobacterium smegmatis*. *Appl. Environ. Microbiol.* **2002**, *68*, 5750–5755. [CrossRef] [PubMed]

74. Eker, A.P.M.; Pol, A.; van der Meyden, P.; Vogels, G.D. Purification and properties of 8-hydroxy-5-deazaflavin derivatives from *Streptomyces griseus*. *FEMS Microbiol. Lett.* **1980**, *8*, 161–165. [CrossRef]

75. Grochowski, L.L.; Xu, H.M.; White, R.H. Identification and characterization of the 2-phospho-L-lactate guanylyltransferase involved in coenzyme F-420 biosynthesis. *Biochemistry* **2008**, *47*, 3033–3037. [CrossRef] [PubMed]

76. Braga, D.; Lasta, D.; Hasan, M.; Guo, H.; Leichnitz, D.; Uzum, Z.; Richter, I.; Schalk, F.; Beemelmanns, C.; Hertweck, C.; et al. Metabolic pathway rerouting in *Paraburkholderia rhizoxinica* evolved long-overlooked derivatives of coenzyme F_{420}. *ACS Chem. Biol.* **2019**, 2088–2094. [CrossRef]

77. Hossain, M.S.; Le, C.Q.; Joseph, E.; Nguyen, T.Q.; Johnson-Winters, K.; Foss, F.W. Convenient synthesis of deazaflavin cofactor F_0 and its activity in F_{420}-dependent NADP reductase. *Organ. Biomol. Chem.* **2015**, *13*, 5082–5085. [CrossRef]

78. Alberty, R.A. Calculation of standard transformed formation properties of biochemical reactants and standard apparent reduction potentials of half reactions. *Arch. Biochem. Biophys.* **1998**, *358*, 25–39. [CrossRef]

79. Alberty, R.A. Calculation of standard transformed gibbs energies and standard transformed enthalpies of biochemical reactants. *Arch. Biochem. Biophys.* **1998**, *353*, 116–130. [CrossRef]

80. Benedict, M.N.; Gonnerman, M.C.; Metcalf, W.W.; Price, N.D. Genome-scale metabolic reconstruction and hypothesis testing in the methanogenic archaeon *Methanosarcina acetivorans* C2A. *J. Bacteriol.* **2012**, *194*, 855–865. [CrossRef]

81. Jankowski, M.D.; Henry, C.S.; Broadbelt, L.J.; Hatzimanikatis, V. Group contribution method for thermodynamic analysis of complex metabolic networks. *Biophys. J.* **2008**, *95*, 1487–1499. [CrossRef]

82. Henry, C.S.; DeJongh, M.; Best, A.A.; Frybarger, P.M.; Linsay, B.; Stevens, R.L. High-throughput generation, optimization and analysis of genome-scale metabolic models. *Nat. Biotechnol.* **2010**, *28*, 977–982. [CrossRef]

83. Nazem-Bokaee, H.; Gopalakrishnan, S.; Ferry, J.G.; Wood, T.K.; Maranas, C.D. Assessing methanotrophy and carbon fixation for biofuel production by *Methanosarcina acetivorans*. *Microb. Cell Fact.* **2016**, *15*, 10. [CrossRef] [PubMed]

84. Caspi, R.; Billington, R.; Fulcher, C.A.; Keseler, I.M.; Kothari, A.; Krummenacker, M.; Latendresse, M.; Midford, P.E.; Ong, Q.; Ong, W.K.; et al. The metacyc database of metabolic pathways and enzymes. *Nucleic Acids Res.* **2018**, *46*, D633–D639. [CrossRef] [PubMed]

2

Co-immobilization of an Enzyme System on a Metal-Organic Framework to Produce a More Effective Biocatalyst

Raneem Ahmad[ID], **Jordan Shanahan**, **Sydnie Rizaldo**, **Daniel S. Kissel** *[ID] **and Kari L. Stone** *[ID]

Department of Chemistry, Lewis University, Romeoville, IL 60446, USA; raneemahmad@lewisu.edu (R.A.); jordanjshanahan@lewisu.edu (J.S.); sydnierizaldo@lewisu.edu (S.R.)
* Correspondence: kisselda@lewisu.edu (D.S.K.); kstone1@lewisu.edu (K.L.S.)

Abstract: In many respects, enzymes offer advantages over traditional chemical processes due to their decreased energy requirements for function and inherent greener processing. However, significant barriers exist for the utilization of enzymes in industrial processes due to their limited stabilities and inability to operate over larger temperature and pH ranges. Immobilization of enzymes onto solid supports has gained attention as an alternative to traditional chemical processes due to enhanced enzymatic performance and stability. This study demonstrates the co-immobilization of glucose oxidase (GO_x) and horseradish peroxidase (HRP) as an enzyme system on Metal-Organic Frameworks (MOFs), UiO-66 and UiO-66-NH$_2$, that produces a more effective biocatalyst as shown by the oxidation of pyrogallol. The two MOFs utilized as solid supports for immobilization were chosen to investigate how modifications of the MOF linker affect stability at the enzyme/MOF interface and subsequent activity of the enzyme system. The enzymes work in concert with activation of HRP through the addition of glucose as a substrate for GO_x. Enzyme immobilization and leaching studies showed HRP/GO_x@UiO-66-NH$_2$ immobilized 6% more than HRP/GO_x@UiO-66, and leached only 36% of the immobilized enzymes over three days in the solution. The enzyme/MOF composites also showed increased enzyme activity in comparison with the free enzyme system: the composite HRP/GO_x@UiO-66-NH$_2$ displayed 189 U/mg activity and HRP/GO_x@UiO-66 showed 143 U/mg while the free enzyme showed 100 U/mg enzyme activity. This increase in stability and activity is due to the amine group of the MOF linker in HRP/GO_x@UiO-66-NH$_2$ enhancing electrostatic interactions at the enzyme/MOF interface, thereby producing the most stable biocatalyst material in solution. The HRP/GO_x@UiO-66-NH$_2$ also showed long-term stability in the solid state for over a month at room temperature.

Keywords: enzyme co-immobilization; metal-organic framework; biocatalysis

1. Introduction

Due to current advances in biotechnology, the commercial value of enzymes as biocatalysts has increased dramatically in recent years [1–4]. The use of biocatalysts for industrial applications is attractive because of the benefits offered by enzyme proteins, including high selectivities, diverse functionalities, and the promotion of greener chemistries [4]. However, the fragile nature of enzymes and narrow operating temperatures and pH ranges have hampered their use commercially. One strategy of preparing enzymes that are more robust to withstand pH, temperature, and/or organic solvents is by immobilizing enzymes on solid supports. In this way, enzymes may be altered that may increase their activity, specificity, or selectivity of the target reaction [5]. Industrially, there is a great deal of interest in the immobilization of enzymes onto solid supports, where an enzyme is attached to a physical surface or material by covalent and/or noncovalent interactions, which has been shown to enhance enzyme stabilities and activities [1,4,6,7]. Solid supports also offer potential for enzyme recovery and recyclability, which increases their commercial value [1,4,6,8–12]. Many reports investigating enzyme immobilization have shown utility for applications in organic syntheses, such as warfarin catalysis and even esterification reactions where immobilization was crucial for long term use [7,13–17]. Several of these immobilization studies explore techniques and methodologies surrounding enzymes immobilized on relatively new materials, such as Metal-Organic Frameworks (MOFs) [1]. These materials have proven to be capable of many different types of immobilizations, both noncovalent and covalent by nature, including surface attachment, cross linkage, covalent linkage and bonding, and encapsulation [16,18–24].

Metal-organic frameworks are coordination networks composed of organic linkers and inorganic nodes containing potential void space with large surface areas [1,25–29]. MOFs stand out as effective solid supports for immobilization due to their highly tunable organic linkers and potential void space, which are ideal for the design of strong guest–host interactions. The target MOFs, UiO-66 and UiO-66-NH$_2$ used in this study as solid supports for enzyme co-immobilization of HRP and GO$_x$, create new biocatalytic materials referred to as HRP/GO$_x$@UiO-66 and HRP/GO$_x$@UiO-66-NH$_2$. The Zr-based UiO MOFs were chosen for immobilization because of their high thermochemical stability, making them amongst the most stable MOFs known, as well as their high degree of tunability [30]. Encapsulation or diffusion into potential void space is typically the focus of enzyme immobilization studies. This requires the use of MOFs with large organic linkers that can be difficult to synthesize or expensive to purchase, and typically have much lower stabilities in solution. The highly tunable surface chemistry of MOFs, however, remains an under-utilized avenue for enzyme immobilization despite the simplicity in pre- and post-synthetic modifications that can enhance interactions at the enzyme/MOF interface. This is especially true for co-enzyme systems where surface immobilization has less probability of hindering enzyme activity. The UiO-66-NH$_2$ MOF, in particular, features an amine group off the organic linker that could serve as a site for covalent bond formation or increased hydrogen-bonding at the enzyme/MOF interface. Increasing the interfacial stability of the enzyme/MOF biocatalyst is expected to increase immobilization loading, thereby improving stability and catalytic activity.

Co-immobilization of enzymes on solid supports have been studied for various applications with multiple systems in the past decade [5,6,9–12,30–33]. Immobilization of two or more enzymes has its advantages and disadvantages. For systems that involve cascade reactions, co-immobilization offers a kinetic advantage where the first product is in close proximity to the second enzyme that is activated by the first product. This removes the potential lag time of the diffusion of substrate at high concentrations for the second enzyme, which allows the second enzyme to be activated at high concentrations from the first product [31]. Enzyme co-immobilizations have shown to be useful for two main tasks in catalysis: (1) to act as a pathway for intermediate reaction products to proceed directly to a second enzyme, and (2) to maximize intermediate and avoid the loss of intermediate due to diffusion or instability [12]. Co-immobilizing prevents high concentrations of intermediates or reaction products to inhibit the enzymes by restricting local concentration as discussed in the review by Betancor et al. on co-immobilized coupled enzymes [12]. While individual immobilization offers some of these advantages, the strategy of co-immobilization can have even more positive effects, such as fast conversion of substrate to product, easy separation of enzyme and product, and higher efficiency [9]. Furthermore, co-immobilized or multienzyme systems have shown better performance than single enzyme systems due to restriction in the diffusion of unstable intermediates [9]. Garcia-Galen et al. reviewed how immobilization strategies affect and many times improve enzyme performance. They describe how co-immobilization also presents some challenges: (1) immobilization of all enzymes must use the same immobilization strategy, and (2) the robustness of the overall biocatalyst will be determined by the least stable enzyme [31,34]. When co-immobilizing enzymes, there must be considerations of whether separate immobilization of individual enzymes or co-immobilization on the same solid support utilizing the same reaction conditions [31]. By using the same immobilization protocol for a multienzyme system, the researcher is assuming that immobilization of each will be similar, which may result in the loss of immobilization advantages, if this is not the case. A proposed strategy to avoid this problem is to co-immobilize the most stable enzymes, which was shown in a study done by Rios et al. [35]. The study showed that when lipase with different stabilities were co-immobilized, the most stable lipase dominated co-immobilization [35]. A benefit of our system in this study is that the surface of the MOF is tunable as we report differences in stabilities due to changes in MOF linker functionalities. The UiO-66 MOFs, in particular, can readily be functionalized post-synthetically to alter surface properties, as demonstrated by Marshall et al., which could allow for the immobilization of different individual enzymes [36]. Other disadvantages of co-immobilization or multienzyme systems are the randomness of the enzyme concentrations, the inability to use large multidomain enzymes, and the supports can be a diffusion barrier [9,12]. In an attempt to improve co-immobilization and multienzyme systems, a rational design has been recently developed by quantitative tools using yeast cell surface to study how multienzyme assemblies form, molecular crowding, and how to maximize enzyme density [37,38]. For co-immobilization of enzymes to work properly, both enzymes need to be immobilized on the solid support. This study demonstrates that our MOF system is suitable for enzyme co-immobilization.

The results presented herein focus on co-immobilization of two enzymes onto the surface of zirconium-based MOFs using a facile adsorption technique to create novel biocatalysts held together by strong, noncovalent interactions at the enzyme/MOF interface. The enzyme system in this investigation consists of glucose oxidase (GO_x) from *Aspergillus niger* and horseradish peroxidase (HRP). Glucose oxidase is available commercially and is used for a wide range of applications in chemical industries related to biosensors, biofuel cells, food and beverage production, and textile manufacturing [7,39]. Horseradish peroxidase is a common enzyme used for organic syntheses that is often coupled with other enzymes for immunoassays, chemiluminescent assays, and water treatment assays [40,41]. These two enzymes are known to work cooperatively in systems where sugar is oxidized by GO_x to produce hydrogen peroxide, which activates the HRP enzyme that in turn oxidizes an organic substrate [41,42].

In relation to this paper, the co-immobilization of glucose oxidase (GO_x) and horseradish peroxidase (HRP) generates hydrogen peroxide in situ by the oxidation of glucose from GO_x to activate HRP. The two enzymes have been co-immobilized in previous studies for applications such as catalysis in organic syntheses and biosensing [9,38,43–52]. Coordination polymer formation was previously utilized by Jia and colleagues to co-immobilize HRP/GO_x by binding a nucleotide and a metal, guanine and copper, for a biocompatible composite in glucose biosensing technology [46]. The two enzymes were also investigated by Zhu et al. for glucose biosensing by co-immobilization on carbon nanotube electrodes, and by Chen et al. for bisphenol A detection by co-immobilization on a copper-based MOF [24,30]. Gustafsson and colleagues have reported co-immobilization by using dendronized polymer and mesoporous silica nanoparticle layers, which showed stability for at least two weeks [44]. Furthermore, studies that attempted to spatially control the co-immobilization of HRP/GO_x utilizing micelles and inorganic nanocrystal–protein complexes reported similar enhanced activities and stabilities [43,53]. The coupled enzymes have also been co-immobilized on DNA-directed assemblies by conjugate covalent DNA oligonucleotides to oxidize Ampex Red [54]. While immobilizations on scaffolds such as DNA have shown to allow for spatial control, this method is costly and only stable under standard conditions [38].

MOFs have proven to be a more tunable solid support for HRP/GO_x in studies that investigate factors influencing immobilization such as changes in surface area, magnetic properties, and, in this work, linker functionalization [8,47,48]. Chen et al. showed how magnetic nanoparticles can be used in co-immobilization of HRP/GO_x to tune layers of enzymes on the MOF HKUST-1 [48]. Synthesis studies reported by Lou et al. have immobilized soybean epoxide hydrolase (SEH) onto UiO-66-NH_2 to synthesize vicinal diols by a crosslinking technique [13]. Additionally, Wu et al. have reported a single immobilization by encapsulating a three enzyme/MOF composite to synthesize purpurogallin using the

MOF ZIF-8, HRP, cytochrome c (Cyt c), and *Candida antarctica* lipase B (CALB) [47]. This investigation, however, demonstrates that effective immobilization of the HRP/GO_x coenzyme system can be accomplished through adsorption to the MOF surface by simply incubating in a buffered solution for 24 hours at cold temperatures. While prior investigations have focused on producing novel biocatalysts using more complex techniques, this paper focuses on understanding the effect of linker functionalization on adsorption at the enzyme/MOF interface, and how it affects immobilization, stability, and activity in a bi-enzymatic system.

2. Results and Discussion

2.1. The Biocatalytic Enzyme System

A biocatalytic system was created to verify co-immobilization and test the catalytic activity of HRP/GO_x@UiO-66 and HRP/GO_x@UiO-66-NH_2 in the production of purpurogallin from pyrogallol. Purpurogallin has been shown to have inhibiting properties and anti-inflammatory effects potentially useful in the pharmaceutical industry [55–58]. Pyrogallol is known for its auto-oxidation properties, and is often used to synthesize purpurogallin using a catalyst [59]. Purpurogallin has been synthesized from pyrogallol using reconstituted forms of HRP containing synthetic hemes in the presence of H_2O_2 [60]. This biocatalytic system, however, utilizes the oxidation of glucose to catalyze the oxidation of pyrogallol to purpurogallin in the presence of HRP. Figure 1 shows the overall procedure used with the enzyme/MOF biocatalytic materials investigated. The reaction mixture initially consists of glucose, pyrogallol, and the enzyme/MOF composite (HRP/GO_x@UiO-66 or HRP/GO_x@UiO-66-NH_2). Glucose is oxidized by GO_x immobilized onto the enzyme/MOF composite to produce hydrogen peroxide. Hydrogen peroxide activates HRP in the enzyme/MOF composite, which subsequently oxidizes pyrogallol to purpurogallin. Pyrogallol is colorless in solution, but oxidation to purpurogallin creates a yellow-brown solution with a λ_{max} of 420 nm in the visible range [59]. The appearance of purpurogallin can therefore be tracked by monitoring increases in absorbance at 420 nm over time. Eventually, as it becomes insoluble, the purpurogallin will precipitate out of solution as an orange solid overnight.

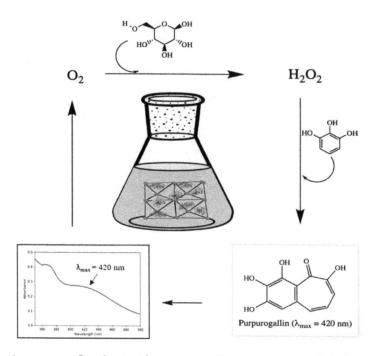

Figure 1. Biocatalytic system. Synthesis of purpurogallin from pyrogallol and glucose by catalysis using HRP/GO_x@UiO-66 or HRP/GO_x@UiO-66-NH_2. The appearance of purpurogallin is monitored spectroscopically in the visible region by tracking increases in absorbance at 420 nm.

In addition to using the biocatalytic system shown in Figure 1, protein quantification was performed using the Bradford assay to study the difference in MOF linker functionalization on enzyme immobilization and composite stability. The Bradford assay is widely used to quantify protein by binding to a dye reagent, causing a blue shift at 595 nm [61]. This assay was used to quantify the amount of enzyme immobilized onto the MOF surface from the solution during immobilization.

2.2. Zeta Potential Characterization

The MOFs used in this investigation possess uniquely dynamic surface charge characteristics with terminal carboxylic acid struts across their surfaces. These surface charges are important for enzyme loading, and can vary depending on the pH and composition of the buffered solution used during immobilization. Zeta potential measurements of solutions containing pure MOFs and enzyme/MOF composites, therefore, can indirectly probe surface charge and stability of MOF dispersions providing useful insights into differences observed in immobilization and leaching [62]. The higher zeta potentials observed in UiO-66-NH$_2$ (-26.67 mV) compared with UiO-66 (-18.00 mV) reflect a slightly more stable dispersion in the buffered solution utilized for immobilization. After enzyme loading, there is a sharp decrease in zeta potential observed for all enzyme/MOF composites investigated, as shown in Table 1. This is congruent with the pIs reported for HRP and GO$_x$, 6.35 and 4.64 respectfully, which indicate a negative charge in the pH = 7 buffered solution.

Table 1. Zeta potentials of all composite materials.

Composite	Zeta Potential (mV)
HRP/GO$_x$@UiO-66-NH$_2$	-3.13 (s = ± 8.73)
HRP@UiO-66-NH$_2$	-8.46 (s = ± 0.14)
GO$_x$@UiO-66-NH$_2$	-10.00 (s = ± 5.64)
UiO-66-NH$_2$	-26.67 (s = ± 6.68)
HRP/GO$_x$@UiO-66	-10.80 (s = ± 4.86)
HRP@UiO-66	-7.97 (s = ± 4.89)
GO$_x$@UiO-66	-3.38 (s = ± 4.50)
UiO-66	-18.00 (s = ± 5.96)

2.3. FT-IR Characterization

The FT-IR spectra of the pure MOFs UiO-66 and UiO-66-NH$_2$ are shown in Figure 2b. Both spectra have characteristic peaks at 1580, 1390, 723, and 670 cm^{-1} for UiO-66 and 1565, 1365, 760 and 656 cm^{-1} for UiO-66-NH$_2$. These peaks correspond to asymmetric O–C–O stretching, O–C–O symmetric stretching, C–C ring stretching and Zr-O stretching, respectively [63]. The spectra for UiO-66-NH$_2$ show additional peaks at 3317, 3444, 1416 and 1252 cm^{-1}, which correspond to the asymmetric and symmetric N–H stretching, N–H bending and the C–N stretch from the aromatic amine, respectively [64].

2.4. SEM of Enzyme/MOF Composites

SEM images of pure UiO-66 and UiO-66-NH$_2$ powder in Figure 3a,b show distorted octahedrons with aggregates ranging from 400 to 1000 nm in size. This is consistent with morphologies and sizes commonly reported for these MOFs [58,65,66]. Enzyme immobilization has no perceivable effects on MOF morphology in UiO-66 or UiO-66-NH$_2$ at 17,000× magnification. This indicates enzyme immobilization did not disrupt the MOF architectures, and that enzymes did not aggregate in appreciable quantities in the MOF suspension.

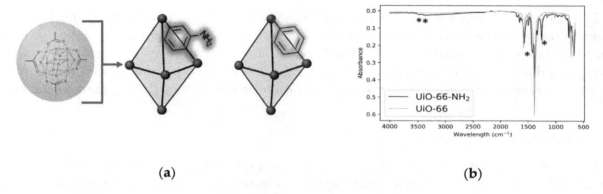

(a)   **(b)**

Figure 2. FT-IR spectra of UiO-66 and UiO-66-NH$_2$. (**a**) Representation of the Secondary Building Unit (SBU) of UiO MOFs with cartoons of the octahedrons showing differences in linker functionalization for UiO-66 vs. UiO-66-NH$_2$. (**b**) The FT-IR Spectra of UiO-66 and of UiO-66-NH$_2$. * The characterization peaks of synthesized MOF UiO-66-NH$_2$.

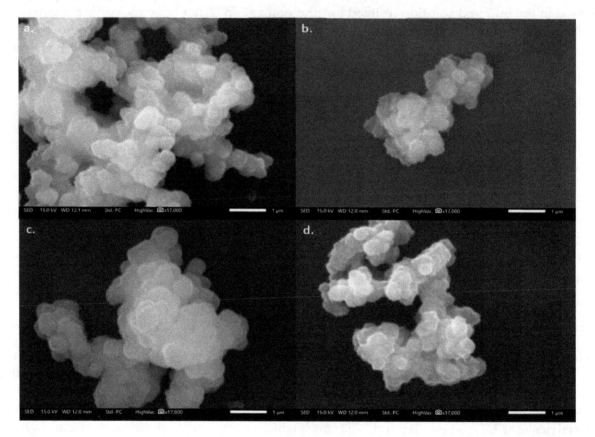

Figure 3. SEM images of MOF and enzyme/MOF composites. (**a**) UiO-66 aggregate cluster, (**b**) UiO-66-NH$_2$ aggregate cluster, (**c**) HRP/GO@UiO-66 aggregate cluster, (**d**) HRP/GO@UiO-66-NH$_2$ aggregate cluster. Both enzyme/MOF composites show clusters composed of irregular octahedrons consistent with MOF images, indicating retention of MOF morphology. All images were recorded at 17,000× magnification.

2.5. Purpurogallin Synthesis

Initially, enzymes were singly immobilized and tested using the formation of purpurogallin. This was done to assure that the enzyme HRP was immobilizing onto the MOF, since HRP is crucial after activation from peroxide to oxidize the organic substrate. This was also performed using GO$_x$ to assure that the oxidation is from the enzyme HRP rather than auto-oxidation of pyrogallol. The single immobilized composites, GO$_x$@UiO-66-NH$_2$ and GO$_x$@UiO-66, showed little to no absorbance, which is expected since HRP is not present to catalyze the oxidation. In the presence of hydrogen peroxide in the solution, single immobilized HRP@UiO-66-NH$_2$ composite did show greater absorbance values by 4% compared with the HRP@UiO-66 composite. The observation of the amino functionalized enzyme/MOF system performing better is also observed with the bi-enzymatic system. Both enzyme/MOF composites, HRP/GO$_x$@UiO-66-NH$_2$ and HRP/GO$_x$@UiO-66, showed enhanced catalysis of pyrogallol to purpurogallin compared with solutions containing free HRP and GO$_x$. The HRP/GO$_x$@UiO-66-NH$_2$ biocatalyst, which showed the greatest enzyme immobilization, produced higher absorbance values at 420 nm using the biocatalytic system, indicating greater production of purpurogallin over time. The free enzyme activity was studied by utilizing our enzyme/MOF immobilization data to replicate the enzyme concentration in the MOF to be only free enzymes in the solution for purpurogallin synthesis. Furthermore, this enzyme/MOF composite showed the greatest stability during leaching studies where each of the biocatalysts were placed in buffered solution over long periods of time. As shown in Figure 4a, the composite HRP/GO$_x$@UiO-66-NH$_2$ excels in catalyzing the synthesis of purpurogallin from pyrogallol and glucose. This is especially noteworthy considering it validates immobilization of two cooperative enzymes onto a solid support. In addition, the current data from our study indicate the parameter for expressed activity, as standardized by Lafuente et al., is increased percent enzyme activity [67]. In Table 2, the enzyme activity was calculated from the increasing absorbance of purpurogallin formation. Both composites, HRP/GO$_x$@UiO-66-NH$_2$ and HRP/GO$_x$@UiO-66, showed an increase in enzyme activity of 189 and 143 U/mg, respectively. The composite HRP/GO$_x$@UiO-66-NH$_2$ showed an 88.6% activity increase, while composite HRP/GO$_x$@UiO-66 showed only 42% activity increase compared to free enzyme HRP/GO$_x$ displaying 100 U/mg of enzyme activity. To confirm that this observed increase in activity is indeed enzymatic and not simple auto-oxidation of pyrogallol, controls with only UiO-66-NH$_2$ and UiO-66 were run using the biocatalytic system, which produced almost no absorbance at 420 nm as shown in Figure 4a,b. A ^{1}H NMR taken of the purpurogallin product synthesized from HRP/GO$_x$@UiO-66-NH$_2$ using the biocatalytic system shown in Figure 5 verifies the presence of purpurogallin formation in the solution. This also shows the clean conversion of substrate to product using the enzyme/MOF composite as a potential biocatalyst for synthesis of organic substances.

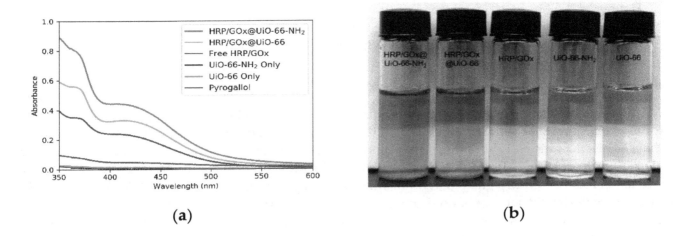

(a)   (b)

Figure 4. Purpurogallin synthesis. (**a**) UV/vis absorbance values of solutions after synthesis using the biocatalytic system for all conditions investigated. The appearance of purpurogallin is tracked by monitoring absorbance at 420 nm over time, indicating HRP/GO$_x$@UiO-66-NH$_2$ is the most catalytically robust system. (**b**) Images of synthesized purpurogallin in the solution using the biocatalytic system for all conditions investigated. The appearance of a yellow-brown color indicates the presence of purpurogallin in the solution.

Table 2. Percent immobilization and leaching of all composite materials.

Composite	% Immobilization [a]	% Leached from Composite [b]	Enzyme Activity (U/mg) [d]
HRP/GO$_x$@UiO-66-NH$_2$	9.91 (s = ±0.033)	35.6 (s = ±0.025)	189
HRP/GO$_x$@UiO-66	3.91 (s = ±0.007)	100 (s = ±0.045)	143
HRP/GO$_x$	0	6 [c]	100
UiO-66-NH$_2$	0	0	0
UiO-66	0	0	0

[a] Percent immobilization was calculated as mass percent of enzyme immobilized from solution relative to the mass of MOF present in the solution. [b] Percent leached was calculated as mass percent of enzyme leached from the composite relative to the amount of enzyme immobilized onto the MOF. [c] Free enzyme leaching was measured as the amount of enzyme that "disappears" presumably from structural denaturation. [d] Calculated from absorbance data in activity units (U per mg) of contained enzymes.

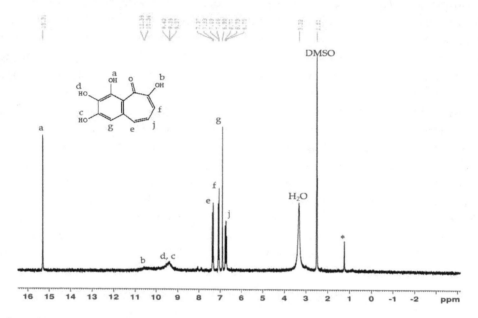

Figure 5. ^{1}H NMR spectrum of the product from HRP/GO$_x$@UiO-66-NH$_2$ catalysis. The presence of 2,3,4,6-tetrahydroxy-5H-benzocyclohepten-5-one (also known as purpurogallin) confirms product formation in the solution from pyrogallol and glucose using the HRP/GO$_x$@UiO-66-NH$_2$ biocatalyst. * indicates hydrocarbon impurity.

2.6. Enzyme/MOF Immobilization and Leaching

Although both UiO-66 and UiO-66-NH$_2$ showed co-immobilization of both enzymes, UiO-66-NH$_2$ exhibited 6% greater enzyme immobilization than UiO-66, as shown in Table 2. This is likely due to enhanced electrostatic and H-bonding interactions imparted by the amine group attached to the organic linker in UiO-66-NH$_2$. Although maximum immobilization occurred with 24 hours of incubation, more data are required to signify immobilization percent yield. Nonetheless, the HRP/GO$_x$@UiO-66 composite material did show enhanced catalysis using the biocatalytic system, however, this composite lacks long-term stability in solution. Both enzymes have completely leached from UiO-66 by the end of the three-day leaching study, as show in Table 2. Further investigation of the HRP/GO$_x$@UiO-66 stability showed that enzyme immobilization did not last over 24 hours in buffered solution. The discrepancy in solution stability between HRP/GO$_x$@UiO-66 and HRP/GO$_x$@UiO-66-NH$_2$ is likely due to weaker interactions at the enzyme/MOF interface as a result of differences in the surface of UiO-66 compared with UiO-66-NH$_2$. In Figure 6, the immobilization (a) and leaching (b) data demonstrate the enhanced stability from immobilization onto UiO-66-NH$_2$ in the HRP/GO$_x$@UiO-66-NH$_2$ composite. In addition to a higher percent immobilization (9.9%), the HRP/GO$_x$@UiO-66-NH$_2$ only loses 36% of enzyme loading after three days in buffered solution. Conversley, the HRP/GO$_x$@UiO-66 composite shows lower enzyme loading (3.9% immobilization) and loses 100% of the adsorbed enzymes after only 24 hours in buffered solution. As shown in Table 2, the results for percent immobilization and composite stability correlate well with enzyme activity for the formation of purpurogallin, showing immobilization efficiency and indicating greater stability and enzyme loading increases enzyme activity. This study did not investigate individual enzyme loading. However, Lou et al. performed single immobilized soybean epoxide hydrolase (SEH) onto $UiO-66-NH_2$ to synthesize vicinal diols [13]. In the Lou et al. report, a shift in the N-H bending vibration of $UiO-66-NH_2$ was observed using XRD and FT-IR after immobilization with SEH via crosslinking. Lou et al. also reported high immobilization loading and solution stability for their $SEH@UiO-66-NH_2$ biocatalyst [13]; however, we are able to achieve comparable immobilization loading and composite stability using only surface adsorption to the MOF support. Based on our results for both UiO-66 and $UiO-66-NH_2$, as well as the results reported by Lou et al., it is apparent that MOF linker functionality plays a crucial role in maximizing loading and catalytic activity by enzyme immobilization on MOF supports.

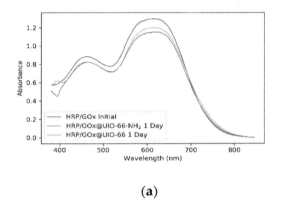

(a)

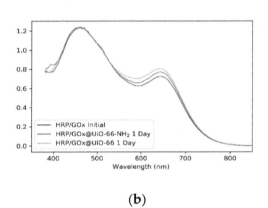

(b)

Figure 6. Immobilization and leaching in all composite materials. (**a**) UV/vis spectra of pure enzyme and enzyme/MOF solutions after 24 h showing a more dramatic decrease in absorbance at 595 nm for $HRP/GO_x@UiO-66-NH_2$ due to higher enzyme immobilization; (**b**) Leaching of enzymes to buffered solution from enzyme/MOF composites showing $HRP/GO_x@UiO-66$ leaching more enzyme to solution in comparison with $HRP/GO_x@UiO-66-NH_2$.

3. Materials and Methods

3.1. Materials

MOPSO buffer was prepared by diluting Bioworld 0.2 M MOPSO buffer with a pH of 6.5 to 50 mM buffered solution. Solid glucose oxidase from *Aspergillus niger* 145,200 U/g, solid β-D-Glucose, and solid pyrogallol were all purchased from Sigma Aldrich. Solid horseradish peroxidase 190 U/mL was purchased from Tokyo Chemical Industry (TCI). The reaction mixture of purpurogallin synthesis consisted of the following glucose to pyrogallol ratio in millimoles: 2:0.95 and 10 mg of enzyme/MOF composite, and 0.05 mg/mL of free enzyme solution for controls.

3.2. Methods

SEM images were collected on a JEOL JCM-7000 SEM. MOF suspensions were deposited on atomically flat silica wafers and were precoated using 10 nm Au nanoparticles to reduce charging. Using the immobilized 166:50 MOF-enzyme ratio of each composite, a working solution of 1 mg/mL of the composite material in methanol was prepared. After sonicating the working solution, 100 μL of it was further diluted in 8 mL of methanol. To prepare the disk sample, 11 μL of the diluted solution of each composite was dropped on the disk individually. Fourier-transform infrared attenuated total resonance spectroscopy (FT-IR-ATR) was used to characterize UiO-66 and $UiO-66-NH_2$ using a CARY 630 FT-IR Spectrometer. Zeta potentials of the MOFs—UiO-66 and $UiO-66-NH_2$—and the enzyme/MOF composites—HRP/GO$_x$@UiO-66 and HRP/GO$_x$@UiO-66-NH$_2$—were measured using a Malvern Zetasizer Nano ZS. All solutions were prepped by dissolving 2 mg of MOF and 20 μL of total enzymes into 15 mL of 0.2 M MOPSO buffer with a pH of 6.5.

3.3. Synthesis of UiO-66 and UiO-66-NH$_2$

UiO-66 synthesis was completed as a one pot synthesis in a Teflon-lined reaction vessel based on a solvothermal synthesis reported by Shearer et al. [68]. Zirconium (IV) chloride (1.98 mmol) and terephthalic acid (1.98 mmol) precursors were dissolved in 34 mL DMF in the reaction vessel. An addition of 340 μL of concentrated hydrochloric acid moderator was then added to improve ligand solvent exchange. The solution was placed in a 120 °C oven and heated for 24 h. The resulting precipitate was washed 3 times with DMF, and once with methanol, before being collected by vacuum filtration. This process was repeated for the synthesis of UiO-66-NH$_2$ substituting 2-aminoterephthalic acid for terephthalic acid. Functional groups of the MOFs composites were identified using Fourier transform infrared attenuated total resonance (FT-IR-ATR) spectroscopy on a CARY 630 FT-IR spectrometer.

3.4. Enzyme Immobilization and Characterization

Enzymes were immobilized with MOFs in a MOPSO buffer system in an Erlenmeyer flask by shaking while incubating for 24 hours at cold temperatures maintained using gel ice packs. The following MOF to enzyme mass ratios (in mg) were utilized to study the immobilization capacity of both MOFs: 77:34, 166:50, and 30:45. All enzyme solutions consisted of 23% GO$_x$, 23% HRP, (both from 5 mg/mL stock enzyme solution) in 50 mM MOPSO buffer. In 10 mL of 50 mM MOPSO, pH 6.5 buffered solution, 5 mg of GO$_x$/HRP@MOF (at the mass ratios listed above) were shaken at room temperature for 3 days. Before and after incubation, the Bradford assay was performed on the centrifuged enzyme solution for spectrophotometric quantification using a Persee T8-DS Double Beam UV-Vis spectrometer to determine percent immobilization and percent leaching. The Bradford assay was carried out using Bio-rad Dye Protein Assay Dye Reagent Concentrate, and Bio-rad Lyophilized Bovine γ-Globin was used to form a standard concentration calibration curve.

4. Conclusions

The results of this investigation indicate that HRP/GO$_x$@UiO-66-NH$_2$ is the better biocatalyst for pyrogallol oxidation by activity and stability measurements. The enhanced catalytic activity observed in both enzyme/MOF systems studied show that enzyme immobilization onto metal-organic frameworks can increase reactivity for cooperative enzyme systems. The effect of linker functionalization on the enzyme/MOF interface also shows that the amine-functionalized MOF linker in HRP/GOx@ UiO-66-NH$_2$ enhances loading, reactivity, and stability when compared with HRP/GOx@UiO-66. HRP/GOx@UiO-66-NH$_2$ and HRP/GO$_x$@UiO-66 showed increased enzyme activity in comparison to free enzymes. The increase in enzyme loading and composite stability is likely due to the amine group increasing electrostatic attractive forces on the surface of the MOF for the enzymes used in this investigation. While the HRP/GO$_x$@UiO-66 also showed enhanced catalysis compared to free enzymes in the solution, it is not stable enough to maintain enzyme loading and catalytic enhancement over time.

Despite the low degree of spatial control, enzyme/MOF composite systems provide an adaptable, low cost material that can function effectively as a green biocatalyst for synthesis reactions, such as the one utilized in this investigation. The enzyme/MOF composite materials prepared in this study are both valid biocatalysts potentially useful for synthesizing organic compounds in different industries, such as pharmaceuticals. Schwartz and colleagues have shown a tradeoff between stability and activity when immobilizing enzymes by multipoint covalent attachment to a polymer [69]. While the enzymes in this study are not in multipoint covalent attachment, however, similar results were observed. After a month of storage, the measured absorbance at the λ_{max} of the purpurogallin for HRP/GO$_x$@UiO-66-NH$_2$ composite decreased by 37%, while HRP/GO$_x$@UiO-66 decreased by 44%. This study demonstrates that an analysis of structure and function relationships can be utilized to design better biocatalysts. Under investigation in our laboratory is determining the relationship between enzyme dynamics and its catalytic activity with the intent on preparing even more effective biocatalysts.

Author Contributions: Conceptualization, K.L.S. and D.S.K.; methodology, K.L.S. and D.S.K.; formal analysis, K.L.S. and D.S.K.; investigation, R.A., J.S. and S.R.; writing—original draft preparation, R.A.; writing—review and editing, R.A., K.L.S. and D.S.K.; supervision, K.L.S. and D.S.K.; project administration, K.L.S. and D.S.K.; All authors have read and agreed to the published version of the manuscript.

Acknowledgments: The authors would like to acknowledge the McCrone Group for the use of their SEM.

References

1. Zhou, H.-C. Enzyme–MOF (metal–organic framework) composites. *Royal Society of Chemistry* **2017**.
2. Rehm, B.H.A. Enzyme Engineering for in Situ Immobilization. *Molecules* **2016**, *21*. [CrossRef] [PubMed]
3. Shih, Y.H.; Lo, S.H.; Yang, N.S.; Singco, B.; Cheng, Y.J.; Wu, C.Y.; Chang, I.H.; Huang, H.Y.; Lin, C.H. Trypsin-Immobilized Metal–Organic Framework as a Biocatalyst in Proteomics Analysis. *ChemPlusChem* **2012**, *77*, 982–986. [CrossRef]
4. Ahmed, E.; Ismail, C.Z.D. Industrial Applications of Enzymes: Recent Advances, Techniques, and Outlooks. *Catalysts* **2018**, *8*, 238.
5. Rodrigues, R.C.; Ortiz, C.; Berenguer-Murcia, Á.; Torres, R.; Fernández-Lafuente, R. Modifying enzyme activity and selectivity by immobilization. *Chem. Soc. Rev.* **2013**, *42*, 6290–6307. [CrossRef]
6. Ki-Hyun Kim, A.D. Recent advances in enzyme immobilization techniques: Metal-organic frameworks as novel substrates. *Elev. Coord. Chem. Rev.* **2016**, *322*, 30–40.
7. Sheldon, R.A. Role of Biocatalysis in Sustainable Chemistry. *ACS Chem. Rev.* **2017**, *118*, 801–838. [CrossRef]
8. Gkaniatsou, E.; Sicard, C.; Ricoux, R.; Mahy, J.P.; Steunou, N.; Serre, C. Metal–organic frameworks: A novel host platform for enzymatic catalysis and detection. *Mater. Horizons* **2017**, *5*, 55–63. [CrossRef]
9. Ren, S.; Li, C.; Jiao, X.; Jia, S.; Jiang, Y.; Bilal, M.; Cui, J. Recent progress in multienzymes co-immobilization and multienzyme system applications. *Chem. Eng. J.* **2019**, *373*, 1254–1278. [CrossRef]
10. Homaei, A.A.; Sariri, R.; Vianello, F.; Stevanato, R. Enzyme immobilization: An update. *J. Chem. Biol.* **2013**, *6*, 185–205. [CrossRef]
11. Jesionowski, T.; Zdarta, J.; Krajewska, B. Enzyme immobilization by adsorption: A review. *Adsorption* **2014**, *20*, 801–821. [CrossRef]
12. Betancor, L.; Luckarift, H.R. Co-immobilized coupled enzyme systems in biotechnology. *Biotechnol. Genet. Eng. Rev.* **2013**, *27*, 95–114. [CrossRef] [PubMed]
13. Cao, S.L.; Yue, D.M.; Li, X.H.; Smith, T.J.; Li, N.; Zong, M.H.; Wu, H.; Ma, Y.Z.; Lou, W.Y. Novel Nano-/Micro-Biocatalyst: Soybean Epoxide Hydrolase Immobilized on UiO-66-NH2 MOF for Efficient Biosynthesis of Enantiopure (R)-1, 2-Octanediol in Deep Eutectic Solvents. *ACS Sustain. Chem. Eng.* **2016**, *4*, 3586–3595. [CrossRef]
14. Shieh, F.K.; Wang, S.C.; Yen, C.I.; Wu, C.C.; Dutta, S.; Chou, L.Y.; Morabito, J.V.; Hu, P.; Hsu, M.H.; Wu, K.C.; et al. Imparting Functionality to Biocatalysts via Embedding Enzymes into Nanoporous Materials by a de Novo Approach: Size-Selective Sheltering of Catalase in Metal–Organic Framework Microcrystals. *J. Am. Chem. Soc.* **2015**, *137*, 4276–4279. [CrossRef] [PubMed]
15. Liu, W.L.; Yang, N.S.; Chen, Y.T.; Lirio, S.; Wu, C.Y.; Lin, C.H.; Huang, H.Y. Lipase-Supported Metal–Organic Framework Bioreactor Catalyzes Warfarin Synthesis. *Chem. Eur. J.* **2015**, *21*, 115–119. [CrossRef] [PubMed]
16. Fan, S.; Liang, B.; Xiao, X.; Bai, L.; Tang, X.; Lojou, E.; Cosnier, S.; Liu, A. Controllable Display of Sequential Enzymes on Yeast Surface with Enhanced Biocatalytic Activity toward Efficient Enzymatic Biofuel Cells. *J. Am. Chem. Soc.* **2020**, *4*, 3222–3230. [CrossRef] [PubMed]
17. Liu, W.L.; Wu, C.Y.; Chen, C.Y.; Singco, B.; Lin, C.H.; Huang, H.Y. Fast Multipoint Immobilized MOF Bioreactor. *Chem. Eur. J. Biocatal.* **2014**, *20*, 8923–8928. [CrossRef]
18. Jung, S.; Kim, Y.; Kim, S.J.; Kwon, T.H.; Huh, S.; Park, S. Bio-functionalization of metal–organic frameworks by covalent protein conjugation. *ChemComm* **2011**, *47*, 2904–2906. [CrossRef]

19. Zhou, H.-C. Coupling two enzymes into a tandem nanoreactor utilizing a hierarchically structured MOF. *RCS Chem. Sci.* **2016**, *7*. [CrossRef]

20. Patra, S.; Crespo, T.H.; Permyakova, A.; Sicard, C.; Serre, C.; Chaussé, A.; Steunou, N.; Legrand, L. Design of metal organic framework–enzyme based bioelectrodes as a novel and highly sensitive biosensing platform. *J. Mater. Chem. B* **2015**, *3*. [CrossRef]

21. Majewski, M.B.; Howarth, A.J.; Li, P.; Wasielewski, M.R.; Hupp, J.T.; Farha, O.K. Enzyme encapsulation in metal–organic frameworks for applications in catalysis. *RCS CrystEngComm* **2017**. [CrossRef]

22. Distefano, M.D. Site-Specific, Covalent Attachment of Proteins to a Solid Surface. *Bioconjugate Chem.* **2006**, *17*, 967–974. [CrossRef]

23. Morten Meldal, S.S. Recent advances in covalent, site-specific protein immobilization. *F1000Research* **2016**, *5*, 1–11.

24. Wang, X.; Lu, X.; Wu, L.; Chen, J. 3D metal-organic framework as highly efficient biosensing platform for ultrasensitive and rapid detection of bisphenol A. *Biosens. Bioelectron.* **2015**, *65*, 295–301. [CrossRef] [PubMed]

25. Ashlee, J.; Howarth, Y.L. Chemical, thermal and mechanical stabilities of metal–organic frameworks. *Nat. Rev. Mater.* **2016**. [CrossRef]

26. Liang, K. Biomimetic mineralization of metal-organic frameworks as protective coatings for biomacromolecules. *Nat. Commun.* **2015**, *6*. [CrossRef] [PubMed]

27. Patterson, J.P. Direct Observation of Amorphous Precursor Phases in the Nucleation of Protein–Metal–Organic Frameworks. *J. Am. Chem. Soc.* **2020**, 1433–1442. [CrossRef]

28. Kenneth, J.; Balkus, J. Hybrid materials for immobilization of MP-11 catalyst. *Top. Catal.* **2006**, *36*. [CrossRef]

29. Feng, D.; Liu, T.F.; Su, J.; Bosch, M.; Wei, Z.; Wan, W.; Yuan, D.; Chen, Y.P.; Wang, X.; Wang, K.; et al. Stable metal-organic frameworks containing single-molecule traps for enzyme encapsulation. *Nat. Commun.* **2015**, *6*. [CrossRef]

30. Zhu, L.; Yang, R.; Zhai, J.; Tian, C. Bienzymatic glucose biosensor based on co-immobilization of peroxidase and glucose oxidase on a carbon nanotubes electrode. *Biosens. Bioelectron.* **2007**, *23*, 528–535. [CrossRef]

31. Garcia-Galan, C.; Berenguer-Murcia, Á.; Fernandez-Lafuente, R.; Rodrigues, R.C. Potential of Different Enzyme Immobilization Strategies to Improve Enzyme Performance. *Adv. Synth. Catal.* **2011**, *353*, 2885–2904. [CrossRef]

32. Pitzalis, F.; Monduzzi, M.; Salis, A. A bienzymatic biocatalyst constituted by glucose oxidase and Horseradish peroxidase immobilized on ordered mesoporous silica. *Microporous Mesoporous Mater.* **2017**, *241*, 145–154. [CrossRef]

33. Barbosa, O.; Torres, R.; Ortiz, C.; Berenguer-Murcia, A.; Rodrigues, R.C.; Fernandez-Lafuente, R. Rodrigues, Roberto Fernandez-Lafuentee. Heterofunctional Supports in Enzyme Immobilization: From Traditional Immobilization Protocols to Opportunities in Tunning Enzyme Properties. *Biomacromolecules* **2013**, *8*, 2433–2462. [CrossRef] [PubMed]

34. Don, A.; Cowana, R.F.-L. Enhancing the functional properties of thermophilic enzymes by chemical modification and immobilization. *Enzyme Microb. Technol.* **2011**, *49*, 326–346.

35. Rios, N.S.; Arana-Peña, S.; Mendez-Sanchez, C.; Ortiz, C.; Gonçalves, L.R.; Fernandez-Lafuente, R. Gonçalves, and Roberto Fernandez-Lafuente Reuse of Lipase from Pseudomonas fluorescens via Its Step-by-Step Coimmobilization on Glyoxyl-Octyl Agarose Beads with Least Stable Lipases. *Catalysts* **2019**, *9*, 487. [CrossRef]

36. Marshall, R.J.; Forgan, R.S. Postsynthetic Modification of Zirconium Metal-Organic Frameworks. *Eur. J. Inorg. Chem.* **2016**, 4310–4331. [CrossRef]

37. Smith, M.R.; Gao, H.; Prabhu, P.; Bugada, L.F.; Roth, C.; Mutukuri, D.; Yee, C.M.; Lee, L.; Ziff, R.M.; Lee, J.K.; et al. Elucidating structure–performance relationships in whole-cell cooperative enzyme catalysis. *Nat. Catal.* **2019**, *2*, 809–819. [CrossRef]

38. Bugada, L.F.; Smith, M.R.; Wen, F. Engineering Spatially Organized Multienzyme Assemblies for Complex Chemical Transformation. *ACS Catal.* **2018**, *8*, 7898–7906. [CrossRef]

39. Ananthanarayan, L. Glucose oxidase—An overview. *Biotechnol. Adv.* **2009**, *27*, 489–501. [CrossRef]

40. Veitch, N.C. Horseradish peroxidase: A modern view of a classic enzyme. *Phytochemistry* **2004**, *65*, 249–259. [CrossRef]

41. Gao, F.; Guo, Y.; Fan, X.; Hu, M.; Li, S.; Zhai, Q.; Jiang, Y.; Wang, X. Enhancing the catalytic performance of chloroperoxidase by coimmobilization with glucose oxidase on magnetic graphene oxide. *Biochem. Eng. J.* **2018**, *143*, 101–109. [CrossRef]

42. Liu, Y. Hemin@metal–organic framework with peroxidaselike activity and its application to glucose detection. *RCS Catal. Sci. Technol.* **2013**, *3*, 2761–2768. [CrossRef]

43. Jia, F.; Zhang, Y.; Narasimhan, B.; Mallapragada, S.K. Block Copolymer-Quantum Dot Micelles for Multienzyme Colocalization. *ACS Langmuir* **2012**, *28*, 17389–17395. [CrossRef]

44. Gustafsson, H.; Küchler, A.; Holmberg, K.; Walde, P. Co-immobilization of enzymes with the help of a dendronized polymer and mesoporous silica nanoparticles. *RCS Mater. Chem. B* **2012**, 1–3. [CrossRef] [PubMed]

45. Yang, K.L. Combined cross-linked enzyme aggregates of horseradish peroxidase and glucose oxidase for catalyzing cascade chemical reactions. *Enzyme Microbial Technol.* **2017**, *100*, 52–59. [CrossRef]

46. Memon, A.H.; Ding, R.; Yuan, Q.; Liang, H.; Wei, Y. Coordination of GMP ligand with Cu to enhance the multiple enzymes stability and substrate specificity by co-immobilization process. *Biochem. Eng. J.* **2018**, *136*, 102–108. [CrossRef]

47. Wu, X.; Yang, C.; Ge, J. Green synthesis of enzyme/ metal-organic framework composites with high stability in protein denaturing solvents. *Bioresour. Bioprocess.* **2017**, *4*. [CrossRef]

48. Chen, S.; Wen, L.; Svec, F.; Tan, T.; Lv, Y. Magnetic metal–organic frameworks as scaffolds for spatial co-location and positional assembly of multi-enzyme systems enabling enhanced cascade biocatalysis. *RCS Adv.* **2017**, *7*, 21205–21213. [CrossRef]

49. Vriezema, D.M.; Garcia, P.M.; Sancho Oltra, N.; Hatzakis, N.S.; Kuiper, S.M.; Nolte, R.J.; Rowan, A.E.; van Hest, J.C. Positional Assembly of Enzymes in Polymersome Nanoreactors for Cascade Reactions. *Angew. Chem. Commun. Nanoreactors* **2007**, *46*, 7378–7382. [CrossRef]

50. Kazenwadel, F. Synthetic enzyme supercomplexes: Coimmobilization of enzyme cascades. *RSC Anal. Methods* **2015**, *7*, 4030–4037. [CrossRef]

51. van de Velde, F.; Lourenço, N.D.; Bakker, M.; van Rantwijk, F.; Sheldon, R.A. Sheldon. Improved Operational Stability of Peroxidases by Coimmobilization with Glucose Oxidase. *Biotechnol. Bioeng.* **2000**, *69*, 286–291. [CrossRef]

52. Jia, F.; Narasimhan, B.; Mallapragada, S.K. Biomimetic Multienzyme Complexes Based on Nanoscale Platforms. *AIChE* **2012**, *59*. [CrossRef]

53. Li, Z.; Zhang, Y.; Su, Y.; Ouyang, P.; Ge, J.; Liu, Z. Spatial co-localization of multi-enzymes by inorganic nanocrystal–protein complexes. *RCS ChemComm* **2014**, *50*, 12465–12468. [CrossRef] [PubMed]

54. Müller, J.; Niemeyer, C.M. DNA-directed assembly of artificial multienzyme complexes. *Biochem. Biophys. Res. Commun.* **2008**, *377*, 62–67. [CrossRef]

55. Park, H.Y. Purpurogallin exerts anti-inflammatory effects in lipopolysaccharide-stimulated BV2 microglial cells through the inactivation of the NF-κB and MAPK signaling pathways. *Int. J. Mol. Med.* **2013**, *32*, 1171–1178. [CrossRef]

56. Bilal, M.; Mehmood, S.; Rasheed, T.; Iqbal, H. Bio-Catalysis and Biomedical Perspectives of Magnetic Nanoparticles as Versatile Carriers. *Magnetochemistry* **2019**, *5*. [CrossRef]

57. Bosio, V.E.; Islan, G.A.; Martínez, Y.N.; Durán, N.; Castro, G.R. Nanodevices for the immobilization of therapeutic enzymes. *Crit. Rev. Biotechnol.* **2016**, *36*, 447–464. [CrossRef]

58. Wang, W.; Wang, L.; Li, Y.; Liu, S.; Xie, Z.; Jing, X. Nanoscale Polymer Metal–Organic Framework Hybrids for Effective Photothermal Therapy of Colon Cancers. *Adv. Mater.* **2016**, *28*, 9320–9325. [CrossRef]

59. Ramasarma, T. New insights of superoxide dismutase inhibition of pyrogallol autoxidation. *Mol. Cell. Biochem.* **2014**, *400*, 277–285. [CrossRef]

60. Fernandez, M.; Frydman, R.B.; Hurst, J.; Buldain, G. Structure/activity relationships in porphobilinogen oxygenase and horseradish peroxidase an analysis using synthetic hemins. *Eur. J. Biochem.* **1993**, *218*, 251–259. [CrossRef]

61. Bradford, M.M. A rapid and sensitive method for the quantitation of microgram quantities of protein utilizing the principle of protein-dye binding. *Elsevier Anal. Biochem.* **1976**, *72*, 248–254. [CrossRef]

62. Salopek, B.; Krasic, D.; Filipovic, S. Measurement and application of zeta-potential. *Rudarsko-geolosko-Naftni Zbornik* **1992**, *4*, 147–151.

63. Guo, X. Facile synthesis of morphology- and size-controlled zirconium metal-organic framework UiO-66: The role of hydrofluoric acid in crystallization. *CrystEngComm* **2015**, *17*, 6434–6440. [CrossRef]

64. Wu, L. Electronic effects of ligand substitution on metal–organic framework photocatalysts: The case study of UiO-66. *PCCP* **2015**, *17*, 117–121. [CrossRef]

65. Wang, S. Amino-functionalized Zr-MOF nanoparticles for adsorption of CO2 and CH4. *Int. J. Smart Nano Mater.* **2013**, *4*, 72–82. [CrossRef]

66. Zhou, J.J.; Wang, R.; Liu, X.L.; Peng, F.M.; Li, C.H.; Teng, F.; Yuan, Y.P. In situ growth of CdS nanoparticles on UiO-66 metal-organic framework octahedrons for enhanced photocatalytic hydrogen production under visible light irradiation. *Appl. Surf. Sci.* **2015**, *346*, 278–283. [CrossRef]

67. Boudrant, J.; Woodley, J.M.; Fernandez-Lafuente, R. Parameters necessary to define an immobilized enzyme preparation. *Process Biochem.* **2019**, *90*, 66–80. [CrossRef]

68. Lillerud, K.P. Tuned to Perfection: Ironing out the Defects in Metal-Organic Framework UiO-66. *Chem. Mater.* **2016**, *14*, 4068–4071. [CrossRef]

69. Weltz, J.S.; Kienle, D.F.; Schwartz, D.K.; Kaar, J.L. Reduced Enzyme Dynamics upon Multipoint Covalent Immobilization Leads to Stability-Activity Trade-off. *J. Am. Chem. Soc.* **2020**, *142*, 3463–3471. [CrossRef]

3

Recent Advances in Enzymatic and Chemoenzymatic Cascade Processes

Noelia Losada-Garcia [1], Zaida Cabrera [2], Paulina Urrutia [2], Carla Garcia-Sanz [1], Alicia Andreu [1] and Jose M. Palomo [1,*] ID

[1]   Department of Biocatalysis, Institute of Catalysis (CSIC), Marie Curie 2, Cantoblanco, Campus UAM, 28049 Madrid, Spain; n.losada@csic.es (N.L.-G.); carla.garciasanz@estudiante.uam.es (C.G.-S.); alicia.andreu@estudiante.uam.es (A.A.)
[2]   School of Biochemical Engineering, Pontificia Universidad Católica de Valparaíso, Avda. Brasil, 2085 Valparaíso, Chile; zaida.cabrera@pucv.cl (Z.C.); paulina.urrutia@pucv.cl (P.U.)
*   Correspondence: josempalomo@icp.csic.es

Abstract: Cascade reactions have been described as efficient and universal tools, and are of substantial interest in synthetic organic chemistry. This review article provides an overview of the novel and recent achievements in enzyme cascade processes catalyzed by multi-enzymatic or chemoenzymatic systems. The examples here selected collect the advances related to the application of the sequential use of enzymes in natural or genetically modified combination; second, the important combination of enzymes and metal complex systems, and finally we described the application of biocatalytic biohybrid systems on in situ catalytic solid-phase as a novel strategy. Examples of efficient and interesting enzymatic catalytic cascade processes in organic chemistry, in the production of important industrial products, such as the designing of novel biosensors or bio-chemocatalytic systems for medicinal chemistry application, are discussed

Keywords: enzymes; cascade catalysis; organometallic; biocatalysis; chemoenzymatic processes

---

1. Introduction

Cascade reactions, typically defined as a consecutive series of chemical reactions proceeding in a concurrent fashion, have attracted the research community's attention in the last few years.

One of the main areas where this strategy plays a pivotal role is in nature, with the biosynthesis of natural products [1,2].

Systems where different enzymes, homogeneous organometallic complexes, are combined successively in a one-pot or tandem processes, in different manners, have been described as successful catalysts in different chemical applications [3,4].

These processes present advantages when compared to the typical single reaction, such as atom economy, step-saving, and therefore high yield and efficiency of the chemical process [5,6]. Additionally, the use of enzymes as catalysts, conforming the cascade system, leads to a more sustainable and environmentally benign process. However, these systems have some disadvantages, such as the fact that the enzymes have different pH and optimum temperatures, so it is necessary to develop strategies that exceed these requirements.

The application of molecular biology techniques for the creation of improved enzymes [7], solid-support technology in order to obtain heterogeneous more efficient cascades [8,9], or the combination of enzymes with novel nanozymes systems, with enzyme-like activities [10,11], have made more efficient cascade systems.

Hence, we focus this review article on the most recent advances achieved in the development and application of multi-enzymatic systems, based on three different approaches: enzymatic cascade systems, chemoenzymatic cascades, or a third strategy based on multi-enzymatic cascade on solid phase (Figure 1).

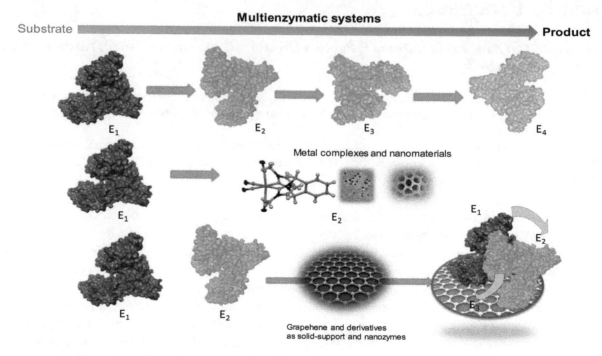

Figure 1. Different concepts of multi-enzymatic cascade processes. E: Enzyme.

2. Enzymatic Cascade Processes

The application of multiple enzymes as a single system in a one-pot process has important advantages in the development of cascade processes [12].

Multi-enzymatic reactions have the benefit of being biodegradable, highly selective, and compatible with each other within certain operating conditions. The application of multiple enzymes as a single system in a one-pot process has the important advantages of improving overall synthetic efficiency, and avoiding intermediate purification steps, with the consequent reduction of waste generation [13]. Further benefits include handling unstable intermediates, and the favorable shift of reaction equilibrium [14].

Despite the advantages associated with the use of multi-enzymatic reactions, their application in industry require strategies for reducing the costs of biocatalyst generation, improving biocatalyst stability, use of low-cost substrates, and enabling economically sustainable continuous cascade operation [15]. Developments in protein engineering provide excellent opportunities for designing and constructing novel industrial biocatalysts, with improved functional properties to match process demands. Among these properties are higher catalytic activity, stability, substrate specificity, improved expression levels, and additionally, the creation of non-natural activities [16]. On the other hand, the improvement of enzyme stability, through their co-immobilization is also gaining scientific popularity, as reflected in several recent reviews, especially because the immobilization of cascade enzymes in a co-localized form offers the advantage that the product of one enzyme reaction is promptly transported to the next biocatalyst, improving overall performance [17,18].

Since a major part of multi-enzymatic cascade reactions incorporate the requirement of stoichiometric amounts of expensive cofactors, which are easily degraded under certain conditions (presence of oxygen, extremes of pH, or high temperature), the achievement of efficient regeneration and reuse of these compounds is a key factor in the development of an economically viable process [19]. One traditional approach to regenerating the cofactor is the use of a second enzyme;

however, the necessity for sacrificial co-substrates, as well as an additional enzyme, is unfavorable. Consequently, self-sufficient one-pot reactions that show high atom economy and high molecular selectivity are pursued [20].

Cascade enzyme reactions have been utilized with different purposes, especially in the development of a more-efficient synthesis of chemical compounds, such as bioactive oligosaccharides, glycosylated molecules, or amoniacid derivatives [21–26].

2.1. Glyco-Enzymes

Chemical synthesis of rare sugars (monosaccharides and their derivatives that rarely exist in nature) usually results in low yield, complicated steps, or/and toxic reagents, and enzymatic processes have represented an alternative way. The production of ketoses by an aldolase-catalyzed reaction of dihydroxyacetone phosphate (DHAP) and L-/D-glyceraldehyde, and the further dephosphatation with acid phosphatase, has been described [27]. DHAP may be produced from glycerol 3-phosphate through oxidation by glycerol phosphate oxidase, and the by-product hydrogen peroxide must be degraded into water and oxygen by a catalase [28,29]. Considering that glycerol 3-phosphate and L/D-glyceraldehyde are still either expensive and/or relatively unstable, their production from glycerol has been investigated [29,30].

Recently, Li et al. [30] introduced the one-pot synthesis of a series of rare ketohexoses, using glycerol as the sole carbon source, by a four-enzyme cascade reaction (Figure 2). Glycerol was initially phosphorylated by the acid phosphatase from *Shigella flexneri* (PhoN-Sf), and further oxidized by a recombinant glycerol phosphate oxidase (GPO). Oxidation of glycerol to D- and L-glyceraldehyde was evaluated using alditol oxidase (AldO). For the C–C bonding formation, different aldolases from different sources (L-rhamnulose-1-phosphate aldolase (RhaD), L-fuculose-1-phosphate aldolase (FucA), D-fructose-1,6-bisphosphate aldolase (FruA)L-rhamnulose-1-phosphate) were tested. Each aldolase was evaluated using AldO or a dehydrogenase, obtaining six systems. The same PhoN-Sf removed the phosphate group in the aldol adduct and phosphate was recycled in the system. The reaction conditions and multienzyme ratios were extensively studied and optimized. According to the enzymes utilized, D-sorbose/D-allulose, L-fructose, L-tagatose/L-fructose, and L-sorbose were obtained. This study provides a useful method for rare ketose synthesis on a multi-milligram to gram scale, starting from relatively inexpensive materials [30].

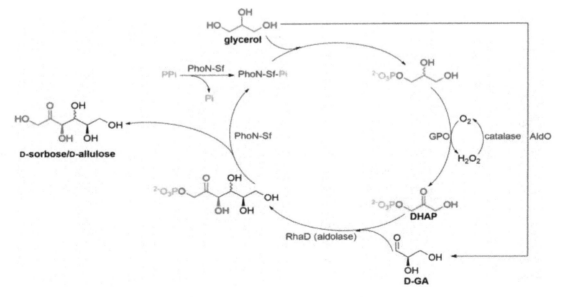

Figure 2. Multi-enzyme synthesis of D-sorbose/D-allulose from glycerol. PPi: pyrophosphate, Pi: inorganic phosphate. Reprinted with permission from Ref. [30].

D-allulose was also synthesized from inulin using a one-pot two-enzyme cascade system formed by an exo-inulinase and D-allulose 3-epimerase [31]. A fructose syrup containing D-glucose, D-fructose, and D-allulose in a 1:3:1 proportion at 67% yield was obtained.

The asymmetric synthesis of ketoses and related acyloin compounds in one step by transketolases (TK) is another interesting process [32]. These enzymes transfer an α-hydroxy carbonyl (ketol) group from a donor to an aldehyde acceptor, where the irreversible release of carbon dioxide from hydroxypyruvate (HPA) as donor kinetically drives the conversion. A challenge for the application of this reaction is the instability of HPA and the cost of expensive aldehyde precursors. In this sense, Lorilliere et al. [33] recently optimized the synthesis of natural (3S,4S)-ketoses (L-ribulose, D-tagatose, and L-psicose) from (2S)-hydroxyaldehyde using TK from *Geobacillus stearothermophilus*; the donor HPA as well as the acceptor substrates were generated in situ, in a one-pot procedure. HPA was obtained from L-serine and pyruvate by a transaminase-catalyzed reaction. Additionally, three different (2S)-α-hydroxylated aldehydes, L-glyceraldehyde, D-threose, and L-erythrose, were generated from the simple achiral compounds, glycolaldehyde and formaldehyde, by D-fructose-6-phosphate aldolase catalysis.

For the synthesis of more complex sugar structures, cellodextrins are linear β-1,4-gluco-oligosaccharides (DP ≤ 6) with attractive properties [34], which can be produced by depolymerization of cellulose [35] or bottom-up synthesis [36]. Recently, Nidetzky and coworkers [36] developed a three-enzyme glycoside phosphorylase cascade for the synthesis of soluble cellodextrins, with a degree of polymerization between three and six, using sucrose and glucose as substrates (Figure 3). The reactions involved iterative β-1,4-glucosylation of glucose from α-glucose 1-phosphate (αGlc1-P) donor that is formed in situ from sucrose and phosphate. Sucrose phosphorylase (ScP), cellobiose phosphorylase (CbP), and cellodextrin phosphorylase (CdP) were used.

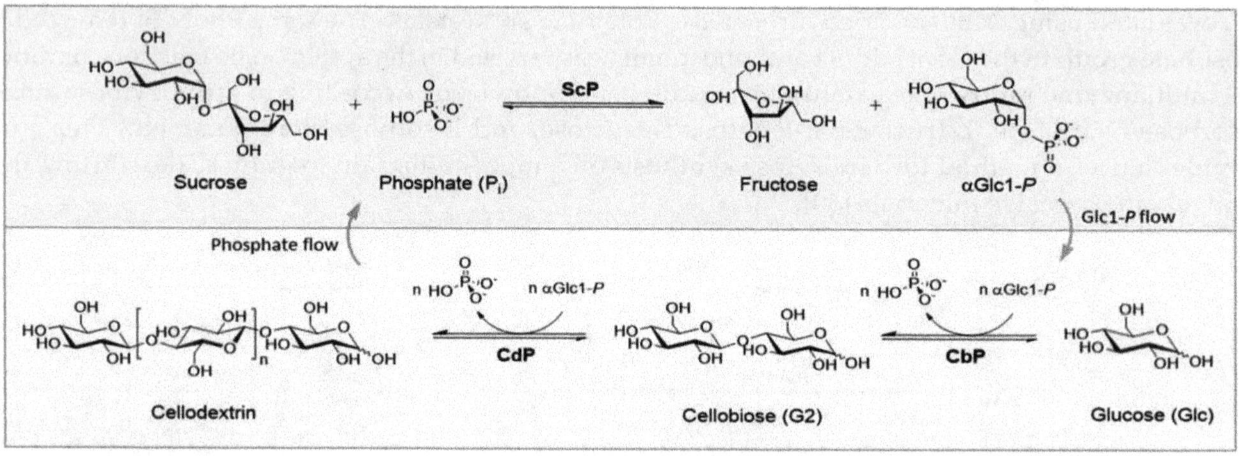

Figure 3. Synthesis of cellodextrin from sucrose by a three-enzyme cascade reaction [36].

The synthesis of cellodextrins was performed by two approaches, by two sequential steps and in simultaneous reaction. In the two-sequential steps, first in the sucrose ScP catalysis, a phosphate concentration equal to, or higher than, that of the sucrose was supplied to limit the phosphate/αGlc1-P shuttle to only one single operational cycle. In the second step, cellodextrin synthesis was carried out, and then the precipitation of the magnesium salt of the phosphate, released from αGlc1-P, allowed pulling the CbP-CdP phosphorylases reaction toward completion. In the case of simultaneous reaction, analysis of critical process factors for cellodextrin synthesis showed that in order to maximize soluble product formation, careful balancing of the individual phosphorylase activities is key for process efficiency. The increase of glucose concentration did not significantly affect sucrose utilization but benefited the formation of soluble cellodextrins. This study establishes a basis for the integrated production of soluble cellodextrins.

The application of a cascade of five glycoenzymes (sucrose synthase (SUS), UDP-glucose 4-epimerase (GalE), galactinol synthase (GS), raffinose synthase (RS), and stachyose synthase (STS)) has been performed in the synthesis of bioactive oligosaccharides, raffinose (Gal-Glu-Fru) and stachyose (Gal-Gal-Glu-Fru) [37], the major bioactive components of soybean oligosaccharides.

Another interesting property of glycoenzyme cascade has been focused on for the development of glycosylated natural products. This has been described as an interesting strategy to improve the solubility, stability, and bioactivity of bioactive molecules [38].

In this regards, nucleoside diphosphate (NDP)-sugar dependent Leloir glycosyltransferases (GT) have received considerable attention in recent years offering excellent control over the reactivity and selectivity of glycosylation reactions with unprotected carbohydrates [39]. Due to the high cost of NDP-sugars, cascade enzymatic systems for their regeneration have been developed in order to efficiently obtain natural and non-natural polyphenol saccharides [40–42].

Sohng and co-workers [42] have reported the glycosylation of naringenin, a flavonoid found in citrus fruits, with different bioactive properties such as being antioxidant, a free radical scavenger, and anti-inflammatory [43].

A cascade enzymatic system was used based on the combination of the glucosyltransferase (YjiC), for transferring glucose from UDP-α-D-glucose to naringenin, and the sucrose synthase AtSUS1, for regenerating UDP-α-D-glucose from sucrose (Figure 4) [42].

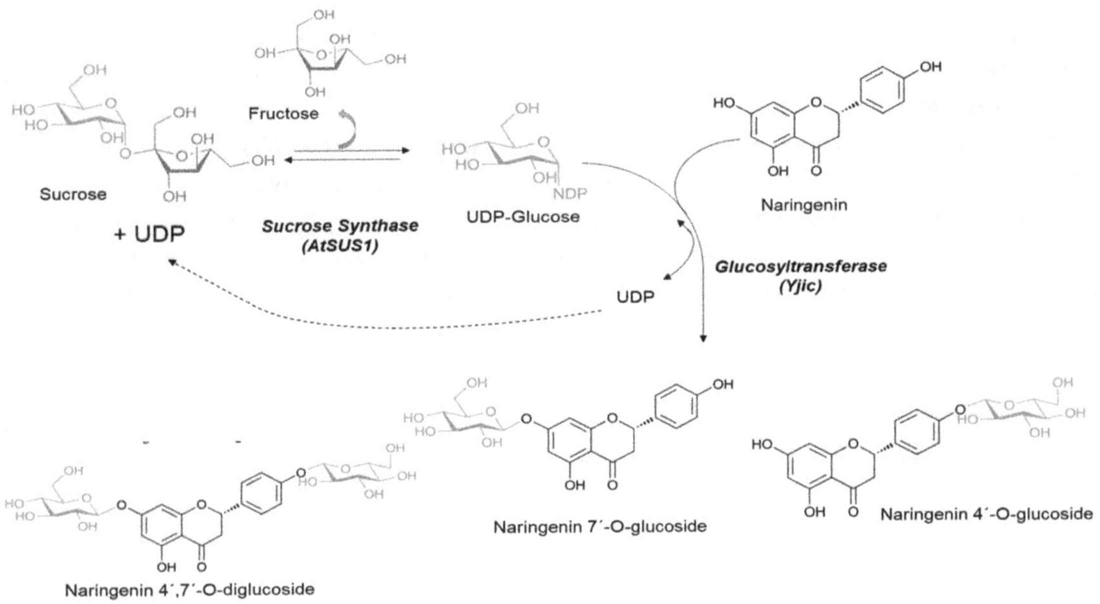

Figure 4. Synthesis of Naringenin-glycosylated flavonoids by multi-enzymatic cascade processes.

Optimization of the cascade processes was developed in order to produce a highly feasible system for large-scale production of different derivatives of naringenin. Three different naringenin glucosides were synthesized, 4'-O-β-D-glucoside, 7-O-β-D-glucoside, and 4',7-O-β-D-diglucoside derivatives (Figure 4).

A similar strategy was used for the synthesis of glycosylated quercetin derivatives [42]. In this case a three-enzyme cascade process was developed. Rhamnose synthase (AtRHM) was used to transform UDP-α-D-glucose to UDP-rhamnose, which was selectively incorporated on the 7-O position of quercetin by rhamnosyltransferase UGT89C1. The sucrose synthase was again used to regenerate UDP-glucose from sucrose.

The cascade reaction of glucosyltransferase (Yjic) and sucrose synthase was also applied in the synthesis of Ginsenoside Rh2 (3b,12b-Di-O-Glc-protopanaxadiol), with anticancer, anti-inflammatory, and antiallergic pharmaceutical activities [44], from protopanaxadiol (PPD) (Figure 5) [45].

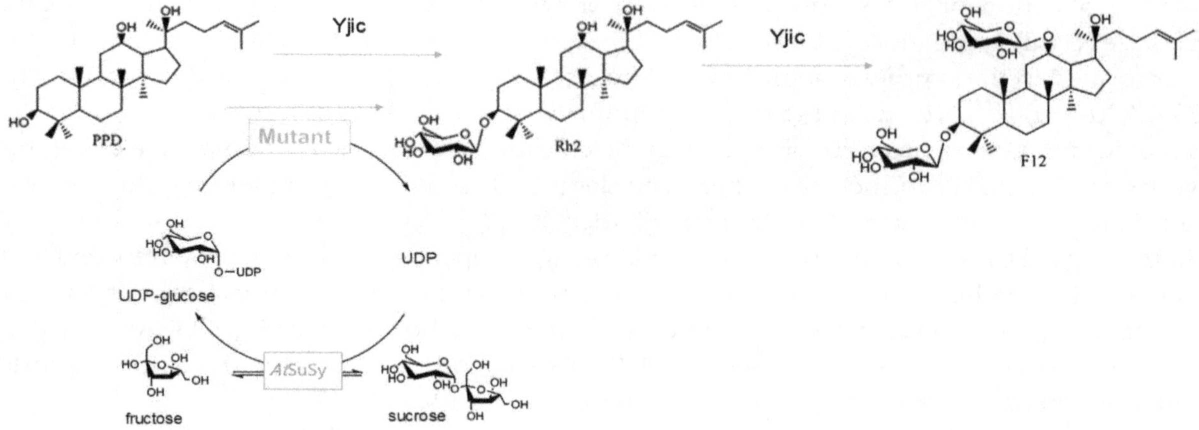

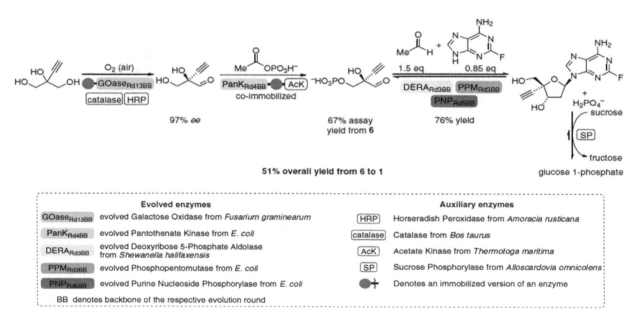

Figure 5. Cascade enzymatic process for Ginsenoside Rh2 synthesis. Figure was reprinted from Ref. [45].

A second glycosylation by Yjic was observed, producing finally, mainly F12 (3b,12b-di-*O*-Glc-protopanaxadiol) in a ratio of 7:3 compared to Rh2. Therefore, the authors used genetic engineering of the Yjic to improve the selectivity. Mutant M315F efficiently synthesized Rh2 (~99%) (Figure 5).

Following the idea described previously of preparation of sugars from small molecules, the ability to apply a multi enzymatic cascade for preparation of complex molecules, such as nucleoside derivatives has been recently described [46]. An outstanding method has been developed in the synthesis of islatravir, a drug for the treatment of HIV, by a multi-enzymatic cascade process [46]. Five of the nine enzymes utilized in this synthesis were developed through directed evolution in order to favor their action on unnatural substrates. Islatravir was synthetized with a 51% overall yield, using the alcohol ethinyl glycerol as the starting substrate (Figure 6). The atom economy far exceeded that of previous syntheses of this target, and the number of steps was less than half. The entire sequence takes place under mild conditions in a single aqueous solution, without the isolation of intermediates.

Figure 6. Multi-enzymatic synthesis of islatravir. Figure printed from Ref. [46]. Copyright © 2020, American Association for the Advancement of Science.

The application of evolved enzymes together with immobilization strategies, especially in the first two steps (Figure 6), selective oxidation of ethinyl glycerol by galactose oxidase, together with peroxidase and catalase, and posterior selective phosphorylation by the two kinases, makes the process feasible with a yield of 67% of the phosphate product at this point (Figure 6). The second step of combining an aldolase (DERA), a phosphopentomutase (PPM), and purine nucleoside phosphorylase (PNP) finally produced the bioactive compound.

2.2. Cascade Enzymes in Aminoacid Chemistry

Amino acids and their derivatives represent a sustainable approach for producing various high added value compounds [47]. Recent works in the area of amino acid synthesis through cascade reactions have focused mainly on the incorporation of lower cost starting materials and/or enzymes with genetically improved properties into the synthetic route [48].

By this premise, semi-scale synthesis of L-alanine was performed by a biocatalytic cascade process starting from glucose [49]. The process involved a multi-enzymatic cascade with an optimized combination of six biocatalysts (a glucose dehydrogenase, two dihydroxy acid dehydratases, a keto-deoxy-aldolase, an aldehyde dehydrogenase, and an L-alanine dehydrogenase) to produce the aminoacid with more than 95% yield. The production level of L-alanine was 0.17 $g·L^{-1}·h^{-1}$, which although being below the current industrial production values (13.4 $g·L^{-1}·h^{-1}$) [50], the proof of concept that has been developed is still of great interest, as it confirms the feasibility of carrying out multistep in vivo routes outside the cell.

Another interesting example is the efficient synthesis of L-tyrosine derivatives by a one-pot biocascade using L-lactate as a starting material (Figure 7) [51].

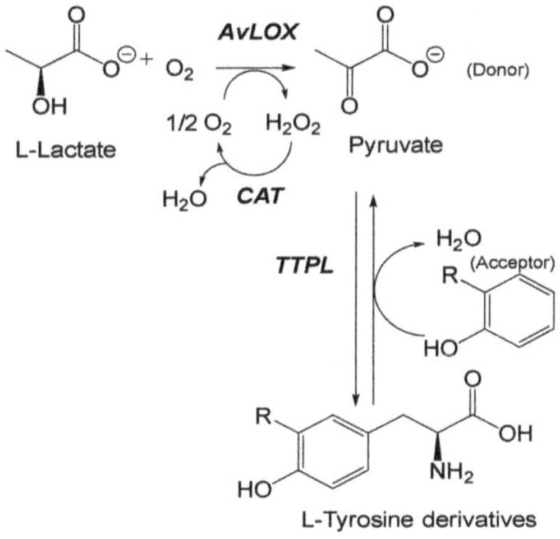

Figure 7. Enzymatic cascade synthesis of L-tyrosine derivatives.

The combination of a L-lactate oxidase (AvLOX), a catalase (CAT), and a tyrosine phenol-lyase (TTPL) permitted obtaining different tyrosine derivatives in high yields, between 81 and 98%, while the enantiomeric excess (e.e) was higher than 99% (Figure 7). A first step combined the use of a high active oxidase to produce pyruvate from lactate and catalase to degrade the formed hydrogen peroxide. Then, tyrosine phenol-lyase catalyzed the C–C bond reaction with different phenol derivatives as acceptor. This environmentally friendly procedure provides a real alternative to the conventional synthesis of L-tyrosine derivatives, since by using L-lactate as a starting material it is possible to avoid the high cost and easy decomposition of pyruvate.

β-Methyl-α-amino acids represent a very attractive synthetic challenge because of their high potential as chiral building blocks [52]; however, limitations in stereoselective synthesis still remain.

Very recently, Seebeck and coworkers [53] described a four-enzyme cascade system for the efficient asymmetric synthesis of different β-Methyl amino acids, in high yields (>80%) and stereoselectivity (>90%) (Figure 8).

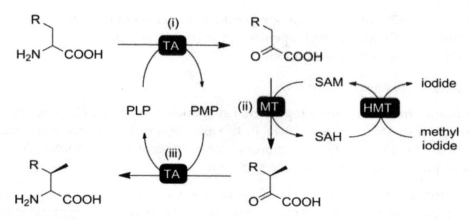

Figure 8. Enzymatic cascade synthesis of β-methyl-α-amino acids. TA: transaminase, MT: methyltransferase, HMT: halide methyltransferase.

The cascade was composed of a PLP-dependent L-amino acid or D-amino acid transaminase from Escherichia coli *(E. coli)* (IlvE or D-TA) that catalyzes the oxidation of the α-amino acid substrate to the corresponding α-keto acid. The α-ketoacid is methylated by an S-adenosylmethionine (SAM) dependent α-ketoacid methyltransferase from *Streptomyces griseoviridis* (SgvM). The methylated R-α-ketoacid obtained recovers its amino group from pyridoxamine (PMP) contained in the enzymes L-TA (IlvE) or D-TA to form L-β-Me-α-amino acids or D-β-Me-α-amino acids, respectively. The third enzyme in the cascade is a halide methyltransferase from *Burkholderia xenovorans* (HMT), used for SAM regeneration by stereoselective S-methylation of S-adenosylhomocysteine (SAH), through the use of methyl iodide as a methyl donor (Figure 8).

Moreover, strategies for enzymatic production of high added value alcohols from aminoacids have also been developed [54,55].

3. Chemo-Enzymatic Cascades

Chemoenzymatic cascades enable reactions with a high productivity of chemocatalysts and a high selectivity of enzymes. Nevertheless, combining chemo- and biocatalysts is extremely challenging, since the chosen chemocatalysts need to tolerate the presence of the enzyme generally in aqueous media [56].

To combine the advantages of both catalysts, chemoenzymatic C–C coupling reactions are often combined with enzymatic oxidations or reduction reactions [57].

One interesting example, where enzymes and organometallic complexes have been successfully combined, was the synthesis of cycloalkanes and cycloalkenes, not being reported via enzymatic pathways. Ward and coworkers [58] described a multi-enzymatic cascade system combined with (Hoveyda)-Grubbs ruthenium (II) catalysts. They combined a ruthenium-catalyzed ring closing metathesis reaction as the last step of one-pot chemo-enzymatic cascades that included (up to) nine enzymatic steps for production of cyclopentene, cyclohexene, and cycloheptene from olive oil derived intermediates, hosted by whole cells of *E. coli* in a single reaction vessel (Figure 9). Different routes were purposed where, in one, the enzymatic cascade was involved in the selective decarboxylation of oleic acid using *E. coli* decarboxylases (UndB), previously obtained from hydrolysis of olive oil using *E. coli Thermomyces lanuginosus* lipase *(TLL)*, whereas a final ring closed-metathesis (RCM) catalyzed by the Ru catalyst gave the final cycloheptene product in 44% yield. Another route went through an enzymatic cascade *E. coli* (C9) (by hydroxylation, oxidation, hydrolysis) to finally produce the corresponding dicarboxylic acid. Then successive chemoenzymatic cascades by decarboxylases, in *E. coli* UndB and Ru catalysis, produced cyclohexane and cyclopentane, in 22% or 65% yield, respectively (Figure 9).

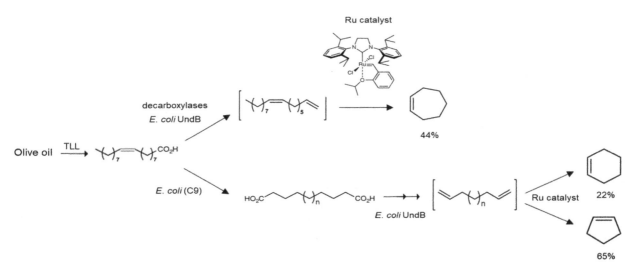

Figure 9. Chemo-enzymatic cascades to produce cycloalkenes from olive oil.

Catalytic selective oxidative C–H activation of hydrocarbon substrates remains a challenge today, especially due to the low regioselectivity of the catalysts and the overoxidation proneness of the substrate [59].

In order to prevent diminishing yields due to these phenomena, a vast number of enzymes have been characterized that do not suffer these shortcomings [60]. Among these are unspecific peroxygenases (UPOs), and extracellular fungal heme-thiolate enzymes that only use hydrogen peroxide (H_2O_2), as both oxygen donor and final acceptor of electrons, to generate the activated oxygen species at the active site [61]. In addition, they are relatively stable with high efficiency toward H_2O_2, and do not require ancillary flavoproteins and cofactors that need regeneration [62].

However, these enzymes are not exempt from limitations, because they can be easily deactivated by modest concentrations of H_2O_2, requiring a constant supply at low concentrations [61].

In this regard, a successfully chemo-enzymatic strategy has been developed for the selective hydroxylation of cyclohexane.

Freakley and coworkers [63] developed a tandem catalytic system, combining the use of peroxygenases for the C–H activation with $AuPd/TiO_2$ for the in situ supply of H_2O_2 at mild conditions (Figure 10).

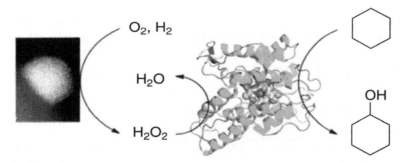

Figure 10. Chemoenzymatic cascade for selective oxidation of cyclohexane using in situ produced H_2O_2. Figure adapted from Ref. [63].

This cascade system produced cyclohexanol with high yields (87%) and total turnover numbers (TTN) of 25,300, with a minimal over oxidation to cyclohexanone, meaning that the bimetallic AuPd catalyst was not facilitating further overoxidation reactions.

In order to expand this chemoenzymatic strategy, the enzyme–metal tandem was applied in the C–H activation of ethylbenzene and isophorone [63]. In the case of using ethylbenzene as a substrate, selective production of enantiopure R-phenyl ethanol (>99%) with a TTN value of 25,900 was obtained.

The hydroxylation of isophorone resulted in two products, 4-hydroxyisophorone and 7-hydroxyisophorone, at 75% yield in roughly equal amounts.

Another interesting application of chemoenzymatic cascades has been described in the synthesis of benzofuran derivatives, which are known to be pharmacologically active against breast cancer cell lines, and which show antimicrobial activity against diverse pathogenic fungi and bacteria [64].

From the different approaches, one interesting strategy has focused on the development of these bioactive molecules from short starting materials, combining C–C bonding and oxidation reactions. For that purpose, the application of Pd free Sonogashira coupling with regio- and stereoselective enzymatic catalysis (C–H activation, dehydration) properties described by cytochrome P450 BM3 [19], was attempted [65].

Schwaneberg and coworkers [65] described a one-pot two-step catalytic cascade reaction for the synthesis of a bis (2-substituted benzofuran) derivative, combining an organometallic Cu catalyst and P450 BM3 (Figure 11).

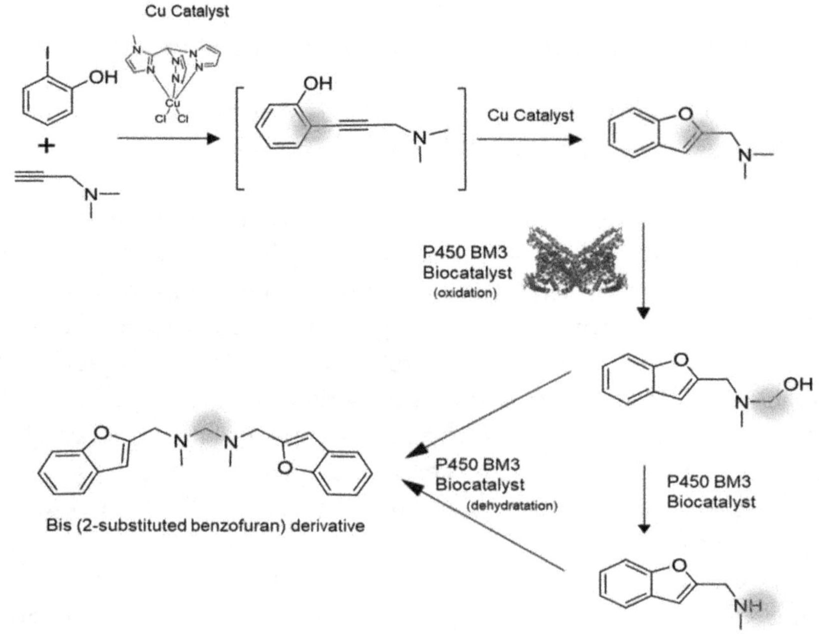

Figure 11. Chemoenzymatic cascade synthesis of a bis (2-substituted benzofuran) derivative [65].

In the first reaction step, a copper scorpionate catalyst (bis(pyrazol) (methyl imidazolyl)-methane copper (II) chloride complex) Cu scorpionate catalyst was used to synthesize a 2-substituted benzofuran from 2-iodophenol to perform a cascade Sonogashira coupling/cyclation reaction. In the next step, enzymatic hydroxylation via P450 BM3, followed by formaldehyde elimination of the benzofuran produced a novel bis (2-substituted benzofuran) derivative in a 33% yield.

This seems to be a moderate yield, considering that the first step after Cu catalyst, the yield of the benzofuran was 88%.

Therefore, conditions were optimized, an excess of EDTA was supplemented to reactions with the crude reaction mixture after completion of the first cascade step, acting as a chelating agent and capturing the complex metal ions. Using this simple additive, final yield of the product increased up to 84% yield.

This is a further promising approach, since it is a sequential two-step catalytic cascade reaction and moreover, it was the first time in this area that a Pd-free Sonogashira coupling could be combined with an enzyme.

Chemo-enzymatic one-pot syntheses offer the replenishment of the versatility of traditional chemical catalysis with the unbeatable selectivity and mild reaction conditions of enzymes. However, they also pose several challenges, like the compatibility of the considerable differences of ideal reaction conditions for the individual process steps [66]. Alternative solvents, such as deep eutectic solvents (DESs) can help to overcome this obstacle of solvent compatibility [67], and also, can give alternative solutions for issues such as low substrate solubility, enzyme activity, and stability. On the other hand, DESs bring along some drawbacks, including occasional toxicity and high viscosity, depending on the starting materials.

Following this line, Grabner, Kourist, and Gruber-Woelfler and coworkers [68] have developed a chemoenzymatic process by a fully integrated two-step flow setup for the asymmetric synthesis of stilbenes in DESs [68].

The process consists of the production of bio-based stilbenes from p-coumaric acid and iodobenzene (low price and easily available substrates) by combining a first enzymatic decarboxylation with a phenolic acid decarboxylase, followed by a Heck reaction catalyzed by a Pd-substituted Ce–Sn-oxide (Figure 12).

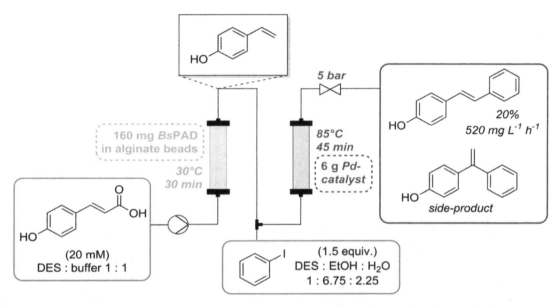

Figure 12. General scheme for asymmetric chemoenzymatic synthesis of stilbenes [68].

For the development of the flow system, both enzymatic and metallic catalysts were previously immobilized on different supports and optimized at the best conditions. The decarboxylated product from coumaric was obtained in a 90% yield in DESs: buffer mixture using immobilized decarboxylase, whereas a full conversion in Heck reaction producing (E)-4-hydroxy stilbene was possible using immobilized Pd catalyst in this solvent mixture, with ethanol: water 1:1 (v/v) containing iodobenzene. The residence time in the reactors was determined to be 30 min for the decarboxylation, and 45 min for the Heck coupling resulting in a space–time yield of $4.8\ \mathrm{g \cdot L^{-1} \cdot h^{-1}}$ for decarboxylation, and $0.52\ \mathrm{g \cdot L^{-1} \cdot h^{-1}}$ for Heck coupling (Figure 13).

4. Solid-Phase Multi-Chemoenzymatic Cascade Reactions

In the development of cascade processes combining enzymatic and chemo reactions, nanomaterials mimicking enzymatic activity (nanozymes) have emerged in the last years. They show interesting advantages, such as high stability, long-term storage, ease of modification, and also a multimodal platform, interfacing complex biologic environments [69–72].

Benefiting from the combination of enzymes and nanozymes, the activity of the cascade catalysis and the stability of the enzyme can be significantly improved. However, the overall performance of multi-enzyme systems is affected not only by specific interactions between catalysts, substrates, and the reactor, but also by other factors, such as substrate channeling, kinetic coincidence, and spatial distribution in the reactor [73]. Therefore, the immobilization on supports of the multienzyme system is the key to efficient cascade catalysis and high stability.

From the different approaches described in the literature, recently the application of graphene or derivatives has gained special attention, because of their unique properties; they can be used as material for selective immobilization of enzymes but also show enzyme-like activity, acting at the same time as nanozymes, being a very interesting system for chemo-enzymatic cascade processes [74–77].

The conversion of CO_2 in high-added value compounds is very attractive, particularly in terms of sustainable chemical processes [78].

From the different strategies, the enzymatic conversion has been described as a very interesting alternative. Formate dehydrogenase (FateDH) catalyzes CO_2 reduction into formic acid in the presence of nicotinamide adenine dinucleotide (NADH) as electron and proton donor [79]. This enzyme, in combination with formaldehyde dehydrogenase (FaldDH), and alcohol dehydrogenase (ADH), has been described as an excellent enzymatic cascade to transform CO_2 into methanol [80]. However, although the NADH cofactor is a highly efficient proton and electron donor for CO_2 reduction, it is irreversibly oxidized to NAD^+ (i.e., it is sacrificial), limiting its potential [81].

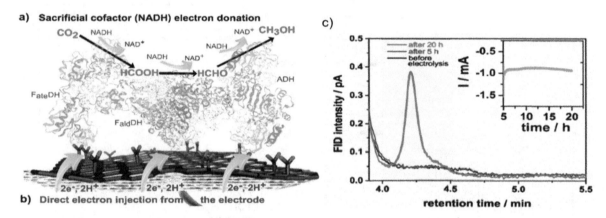

Figure 13. Multi-enzymatic cascade reduction of CO_2 to methanol. (**a**) NADH sacrificial cofactor approach; (**b**) NADH-free approach; (**c**) Faradaic efficiency of the cascade. Figure adapted from Ref. [82].

In this regard, Sariciftci and coworkers [82] have developed a novel heterogeneous cascade system formed by three dehydrogenases (FateDH, FaldDH, ADH) homogeneously covalent immobilized on carboxyl-functionalized grapheme (~15% degree of functionalization) as three graphene-based dual-function biocatalysts for evaluation of the conversion of CO_2 to methanol in two approaches: chemical reduction using NADH as cofactor (Figure 13a), and the NADH-free cascade electroreduction (Figure 13b).

The reduction of CO_2 to methanol by this cascade resulted in about 20 ppm of methanol production (by GC), corresponding to a Faradaic efficiency of 12% (Figure 13c). As a control, electrolysis of the modified electrode was carried out under N_2 saturated conditions. This experiment did not reveal methanol formation after 20 h of electrolysis, which confirms the conversion of CO_2 to methanol through the catalysts. The direct electron injection mechanism for the reduction of CO_2 to methanol, without immobilization of the enzymes, showed a concentration of 0.1 ppm (corresponding to around 10% faradaic efficiency). In the present case, graphene modified with three dehydrogenases showed absolute currents of one order of magnitude higher delivered to the reaction sites, suggesting a much more efficient transport of electrons from the electrode to the active sites of the enzymes, through the conductive graphene support, as well as a higher production rate (around 0.6 $\mu mol \cdot h^{-1}$).

Another example of cascade reaction of graphene-supported enzymes is a simple and effective strategy for the detection of organophosphate (OP) pesticides using an enzyme-based assay; such as acetylcholinesterase (AChE) that was inhibited by OP, whose concentration could be measured colorimetrically by enzymatic activity through a decreased color reaction, described by Chu et al. [83].

This method shows a fast response, but has the problems of instability and sensitivity, because enzymes are susceptible to environmental influences. Recently, advances in nanotechnology have improved this analysis. In most cases, AChE was immobilized onto nanomaterials for developing biosensors with a low detection limit, a rapid response, and long storage stability [84]. For instance, the AChE/TiO$_2$/reduced GO-based electrode increased the catalytic activity of AChE and could detect OPs at nanomolar levels with high reliability [85].

This method required a chemoenzymatic cascade combining two enzymes, AChE, choline oxidase (CHO) and graphene derivative (GO) acting in sequence, in which AChE/CHO catalyzed acetylcholine (ACh) to betaine aldehyde (BA), forming H$_2$O$_2$ that was detected by the peroxidase-like activity of GO, by converting 3,3,5,5-tetramethylbenzidine (TMB) into the oxidized blue product (Figure 14a).

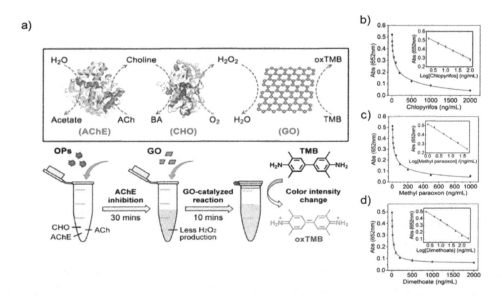

Figure 14. Chemoenzymatic cascade process as a pesticide biosensor. (**a**) General scheme of the strategy. (**b–d**) Dose–response curve and semilogarithmic plot (inset) for OP detection: chlorpyrifos, methyl paraoxon and dimethoate. Figure adapted from Ref. [83].

This chemoenzymatic cascade was applied for the detection of different pesticides, dimethoate, methyl paraoxon, and chlorpyrifos at low levels (Figure 14b–d). As a result, the inhibition of AChE by OPs resulted in a decrease in color intensity, which is linearly related to the concentration of OPs. The corresponding limit of detection (LOD) was measured to be lower than 2 ppb. In addition, the detection could be completed within 40 min, and provided a linear response for OP concentration levels from 1–200 ng mL^{-1}, revealing its great potential for quantitative analysis of pesticide residues.

Recently, metal-organic frameworks (MOFs) with regular pores or channels have garnered considerable research interest due to their excellent potential as a universal platform for the immobilization of various functional materials, including enzymes and nanozymes [86,87]. High flexibility and tunability of MOFs, in terms of their pore size and shape allow the encapsulation of catalysts with different sizes and functions in specific spaces, for efficient cascade reaction.

Thus, Zheng et al. [88] have developed a novel sensitive and stable electrochemical biosensor for glucose detection, using a chemo-enzymatic cascade system (Figure 15). In this multienzyme system, glucose oxidase (GOx) was immobilized within Cu-based MOF layers grown on the surface of a three-dimensional (3D) porous conducting copper foam (CF) electrode (GOx@Cu-MOF/CF) via a facile electrochemical assisted biomimetic mineralization (Figure 15a). This system was compared with the simple GOx immobilized on copper foam electrode.

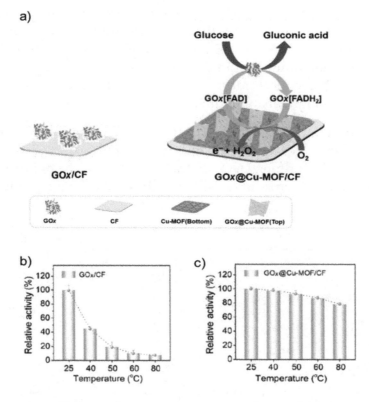

Figure 15. Chemoenzymatic cascade process as a new glucose biosensor. (**a**) Scheme of the bio-electrocatalytic cascade reaction; (**b**) Thermostability GOx/CF; (**c**) Thermostability GOx@MOF/CF. Figure adapted from Ref. [88].

The chemoenzymatic cascade in the detection process of glucose was based on, first, enzymatic glucose oxidation to gluconic acid (producing H_2O_2) catalyzed by GOx embedded within the Cu-MOF layer, and then, the electro-catalyzed reduction of H_2O_2 immediately by the Cu-MOF/CF (Figure 15a).

This system presents the advantage over a normal immobilization of the enzyme on the electrode, that the open and porous 3D feature of the copper foam provides enhanced specific surface area for enzyme immobilization, and facilitates the substrate transport and charge collection. Another advantage of the architecture is that the encapsulation structure provides good protection for GOx. Due to the bio-electrocatalytic cascade reaction mechanism, this well-designed GOx@Cu-MOF modified electrode exhibited superior catalytic activity and thermal stability for glucose sensing.

Notably, this chemoenzymatic system GOx@Cu-MOF/CF retained more than 80% activity after being incubated at 80 °C (Figure 15b). In sharp contrast, the activity of the unprotected electrode was reduced to the original 10% after the same treatment (Figure 15c).

Therefore, this work offers an interesting cascade system with potential application for efficient photo-thermal therapy, and other platforms subjected to harsh operating conditions.

5. Concluding Remarks

Recent enzyme-based cascade processes have been discussed. In this article we have approached the cascade system in terms of applicability of the enzyme as a synthetic tool, with advantages for a possible industrial implementation. Three different approaches have been proven as excellent tools, and as advanced technology in enzyme applications.

First, multi-enzyme cascades, from two to nine different enzymes working in one-pot to obtain a final product in a more effective way. These systems have been successfully tested for different important sugars, oligosaccharides, glycosylated natural products, amino acid derivatives, or very importantly, demonstrating the value of these systems in the synthesis of an anti-HIV drug.

Furthermore, in many cases the combination of enzymes with organometallic complexes has made it possible to solve some basic problems formed by enzymes. For example, chemoenzymatic success has been achieved in the synthesis of new biofuels, the regeneration of green oxidants, and the synthesis of complex molecules with high added value.

As a last point, examples have been developed of the application of a cascade system that involves the use of combined hybrid systems, in solid phase, where the support material of the enzymes acts in the process, acting as catalyst. In this case, these advantageous systems that use graphene or derivatives have made it possible to design new pesticide biosensors, or to solve environmental problems (such as the transformation of CO_2 into high added value compounds). Solid phase metal enzyme systems have also been described as novel glucose biosensors.

Therefore, these systems open the possibility of combining catalytically supported materials, enzymes and metals, with new trends such as flow (micro) reactors, as a novel approach to develop new cascade routes for biopharmaceuticals, but also in the development of environmentally friendly biomedical or electronic (new-bio batteries) devices.

Author Contributions: Z.C. and P.U. writing Section 2: enzymatic cascade processes. C.G.-S. and A.A. writing Section 3: chemo-enzymatic cascades. N.L.-G. writing Section 4: Solid-phase multi-chemoenzymatic cascade reactions and coordinating Sections 3 and 4. J.M.P. for conceiving topic of review article, planning content of review, coordinating overall content and writing, writing the introduction, abstract and conclusions, revising text and content. All authors have read and agreed to the published version of the manuscript.

References

1. Carlson, J.C.; Li, S.; Gunatilleke, S.S.; Anzai, Y.; Burr, D.A.; Podust, L.M.; Sherman, D.H. Tirandamycin biosynthesis is mediated by co-dependent oxidative enzymes. *Nat. Chem.* **2011**, *3*, 628–633. [CrossRef] [PubMed]

2. Galanie, S.; Entwistle, D.; Lalonde, J. Engineering biosynthetic enzymes for industrial natural product synthesis. *Nat. Prod. Rep.* **2020**, *37*, 1122–1143. [CrossRef] [PubMed]

3. Reetz, M.T. Biocatalysis in organic chemistry and biotechnology: Past, present, and future. *J. Am. Chem. Soc.* **2013**, *135*, 12480–12496. [CrossRef] [PubMed]

4. Shiroodi, R.K.; Gevorgyan, V. Metal-catalyzed double migratory cascade reactions of propargylic esters and phosphates. *Chem. Soc. Rev.* **2013**, *42*, 4991–5001. [CrossRef] [PubMed]

5. Zhao, C.; Lercher, J.A. Upgrading pyrolysis oil over Ni/HZSM-5 by cascade reactions. *Angew. Chem. Int. Ed.* **2012**, *51*, 5935–5940. [CrossRef] [PubMed]

6. Grondal, C.; Jeanty, M.; Enders, D. Organocatalytic cascade reactions as a new tool in total synthesis. *Nat. Chem.* **2010**, *2*, 167–178. [CrossRef] [PubMed]

7. Arnold, F.H. Innovation by evolution: Bringing new chemistry to life (Nobel lecture). *Angew. Chem. Int. Ed.* **2019**, *58*, 14420–14426. [CrossRef] [PubMed]

8. Shylesh, S.; Wagener, A.; Seifert, A.; Ernst, S.; Thiel, W.R. Mesoporous organosilicas with acidic frameworks and basic sites in the pores: An approach to cooperative catalytic reactions. *Angew. Chem. Int. Ed.* **2010**, *49*, 184–187. [CrossRef]

9. Palo-Nieto, C.; Afewerki, S.; Anderson, M.; Tai, C.W.; Berglund, P.; Córdova, A. Integrated heterogeneous metal/enzymatic multiple relay catalysis for eco-friendly and asymmetric synthesis. *ACS Catal.* **2016**, *6*, 3932–3940. [CrossRef]

10. Zhou, Y.B.; Liu, B.W.; Yang, R.H.; Liu, J.W. Filling in the gaps between nanozymes and enzymes: Challenges and opportunities. *Bioconjugate Chem.* **2017**, *28*, 2903–2909. [CrossRef]

11. Singh, S. Nanomaterials Exhibiting Enzyme-Like Properties (Nanozymes): Current Advances and Future Perspectives. *Front. Chem.* **2019**, *7*, 46. [CrossRef]

12. Ricca, E.; Brucher, B.; Schrittwieser, J.H. Multi-enzymatic cascade reactions: Overview and perspectives. *Adv. Synth. Catal.* **2011**, *353*, 2239–2262. [CrossRef]

13. Wang, Y.; Ren, H.; Zhao, H. Expanding the boundary of biocatalysis: Design and optimization of in vitro tandem catalytic reactions for biochemical production. *Crit. Rev. Biochem. Mol. Biol.* **2018**, *53*, 115–129. [CrossRef] [PubMed]

14. Muschiol, J.; Peters, C.; Oberleitner, N.; Mihovilovic, M.D.; Bornscheuer, U.T.; Rudroff, F. Cascade catalysis-strategies and challenges en route to preparative synthetic biology. *Chem. Commun.* **2015**, *51*, 5798–5811. [CrossRef] [PubMed]

15. Wilding, K.M.; Schinn, S.M.; Long, E.A.; Bundy, B.C. The emerging impact of cell-free chemical biosynthesis. *Curr. Opin. Biotech.* **2018**, *53*, 115–121. [CrossRef] [PubMed]

16. Wiltschi, B.; Cernava, T.; Dennig, A.; Casas, M.G.; Geier, M.; Gruber, S.; Haberbauer, M.; Heidinger, P.; Herrero Acero, E.; Kratzer, R.; et al. Enzymes revolutionize the bioproduction of value-added compounds: From enzyme discovery to special applications. *Biotechnol. Adv.* **2020**, *40*, 107520. [CrossRef] [PubMed]

17. Arana-Peña, S.; Carballares, D.; Morellon-Sterlling, R.; Berenguer-Murcia, Á.; Alcántara, A.R.; Rodrigues, R.C.; Fernandez-Lafuente, R. Enzyme co-immobilization: Always the biocatalyst designers' choice ... or not? *Biotechnol. Adv.* **2020**, 107584. [CrossRef]

18. Hwang, E.T.; Lee, S. Multienzymatic Cascade Reactions via Enzyme Complex by Immobilization. *ACS Catal.* **2019**, *9*, 4402–4425. [CrossRef]

19. Cutlan, R.; De Rose, S.; Isupov, M.N.; Littlechild, J.A.; Harmer, N.J. Using enzyme cascades in biocatalysis: Highlight on transaminases and carboxylic acid reductases. *Biochim. Biophys. Acta. Proteins. Proteom.* **2020**, *1868*, 140322. [CrossRef]

20. Perl, J.M.; Sieber, V. Multienzyme Cascade Reactions—Status and Recent Advances. *ACS Catal.* **2018**, *8*, 2385–2396.

21. Cheng, C.; Zuo, X.; Tu, D.; Wan, B.; Zhang, Y. Synthesis of 3,4-Fused Tricyclic Indoles through Cascade Carbopalladation and C-H Amination: Development and Total Synthesis of Rucaparib. *Org. Lett.* **2020**, *22*, 4985–4989. [CrossRef]

22. Mekheimer, R.A.; Al-Sheikh, M.A.; Medrasi, H.Y.; Sadek, K.U. Advancements in the synthesis of fused tetracyclic quinoline derivatives. *RSC Adv.* **2020**, *10*, 19867–19935. [CrossRef]

23. Schmaltz, R.M.; Hanson, S.R.; Wong, C.-H. Enzymes in the Synthesis of Glycoconjugates. *Chem. Rev.* **2011**, *111*, 4259–4307. [CrossRef] [PubMed]

24. Wen, L.; Edmunds, G.; Gibbons, C.; Zhang, J.; Gadi, M.R.; Zhu, H.; Fang, J.; Liu, X.; Kong, Y.; Wang, P.G. Toward Automated Enzymatic Synthesis of Oligosaccharides. *Chem. Rev.* **2018**, *118*, 8151–8187. [CrossRef]

25. Krasnova, L.; Wong, C.-H. Oligosaccharide Synthesis and Translational Innovation. *J. Am. Chem. Soc.* **2019**, *141*, 3735–3754. [CrossRef] [PubMed]

26. Yang, J.; Zhang, T.; Tian, C.; Zhu, Y.; Zeng, Y.; Men, Y.; Chen, P.; Sun, Y.; Ma, Y. Multi-enzyme systems and recombinant cells for synthesis of valuable saccharides: Advances and perspectives. *Biotechnol. Adv.* **2019**, *37*, 107406. [CrossRef]

27. Li, Z.; Cai, L.; Qi, Q.; Wang, P.G. Enzymatic synthesis of D-sorbose and D-psicose with aldolase RhaD: Effect of acceptor configuration on enzyme stereoselectivity. *Bioorg. Med. Chem. Lett.* **2011**, *21*, 7081–7084. [CrossRef]

28. Fessner, W.-D.; Sinerius, G. Synthesis of Dihydroxyacetone Phosphate (and Isosteric Analogues) by Enzymatic Oxidation; Sugars from Glycerol. *Angew. Chem. Int. Ed.* **1994**, *33*, 209–212. [CrossRef]

29. Burek, B.O.; Bormann, S.; Hollmann, F.; Bloh, J.Z.; Holtmann, D. Hydrogen peroxide driven biocatalysis. *Green Chem.* **2019**, *21*, 3232–3249. [CrossRef]

30. Li, Z.; Li, F.; Cai, L.; Chen, Z.; Qin, L.; Gao, X.-D. One-Pot Multienzyme Synthesis of Rare Ketoses from Glycerol. *J. Agric. Food Chem.* **2020**, *68*, 1347–1353. [CrossRef]

31. Zhu, P.; Zeng, Y.; Chen, P.; Men, Y.; Yang, J.; Yue, X.; Zhang, J.; Zhu, Y.; Sun, Y. A one-pot two-enzyme system on the production of high value-added D-allulose from Jerusalem artichoke tubers. *Process Biochem.* **2020**, *88*, 90–96. [CrossRef]

32. Lorillière, M.; Dumoulin, R.; L'Enfant, M.; Rambourdin, A.; Thery, V.; Nauton, L.; Fessner, W.-D.; Charmantray, F.; Hecquet, L. Evolved Thermostable Transketolase for Stereoselective Two-Carbon Elongation of Non-Phosphorylated Aldoses to Naturally Rare Ketoses. *ACS Catal.* **2019**, *9*, 4754–4763. [CrossRef]

33. Lorillière, M.; Guérard-Hélaine, C.; Gefflaut, T.; Fessner, W.D.; Clapés, P.; Charmantray, F.; Hecquet, L. Convergent in situ Generation of Both Transketolase Substrates via Transaminase and Aldolase Reactions for Sequential One-Pot, Three-Step Cascade Synthesis of Ketoses. *ChemCatChem* **2020**, *12*, 812–817. [CrossRef]

34. Sybesma, W.; Kort, R.; Lee, Y.K. Locally sourced probiotics, the next opportunity for developing countries? *Trends Biotechnol.* **2015**, *33*, 197–200. [CrossRef] [PubMed]

35. Chen, P.; Shrotri, A.; Fukuoka, A. Soluble Cello-Oligosaccharides Produced by Carbon-Catalyzed Hydrolysis of Cellulose. *ChemSusChem* **2019**, *12*, 2576–2580. [CrossRef] [PubMed]

36. Zhong, C.; Nidetzky, B. Three-Enzyme Phosphorylase Cascade for Integrated Production of Short-Chain Cellodextrins. *Biotechnol. J.* **2020**, *15*, 1900349–1900358. [CrossRef]

37. Tian, C.; Yang, J.; Zeng, Y.; Zhang, T.; Zhou, Y.; Men, Y.; You, C.; Zhu, Y.; Sun, Y. Biosynthesis of Raffinose and Stachyose from Sucrose via an In Vitro Multienzyme System. *Appl. Environ. Microbiol.* **2019**, *85*, 02306-18. [CrossRef]

38. Hofer, B. Recent developments in the enzymatic O-glycosylation of flavonoids. *Appl. Microbiol. Biotechnol.* **2016**, *100*, 4269–4281. [CrossRef]

39. Tegl, G.; Nidetzky, B. Leloir glycosyltransferases of natural product C-glycosylation: Structure, mechanism and specificity. *Biochem. Soc. Trans.* **2020**, *48*, 1583–1598. [CrossRef]

40. Hu, Y.; Min, J.; Qu, Y.; Zhang, X.; Zhang, J.; Yu, X.; Dai, L. Biocatalytic Synthesis of Calycosin-7-O-β-D-Glucoside with Uridine Diphosphate–Glucose Regeneration System. *Catalysts* **2020**, *10*, 258–269. [CrossRef]

41. Huang, F.-C.; Hinkelmann, J.; Hermenau, A.; Schwab, W. Enhanced production of β-glucosides by in-situ UDP-glucose regeneration. *J. Biotechnol.* **2016**, *224*, 35–44. [CrossRef] [PubMed]

42. Thapa, S.B.; Pandey, R.P.; Bashyal, P.; Yamaguchi, T.; Sohng, J.K. Cascade biocatalysis systems for bioactive naringenin glucosides and quercetin rhamnoside production from sucrose. *Appl. Microbiol. Biotechnol.* **2019**, *103*, 7953–7969. [CrossRef] [PubMed]

43. Salehi, B.; Fokou, P.V.T.; Sharifi-Rad, M.; Zucca, P.; Pezzani, R.; Martins, N.; Sharifi-Rad, J. The therapeutic potential of naringenin: A review of clinical trials. *Pharmaceuticals* **2019**, *12*, 11. [CrossRef] [PubMed]

44. Kim, Y.-J.; Zhang, D.; Yang, D.-C. Biosynthesis and biotechnological production of ginsenosides. *Biotechnol. Adv.* **2015**, *33*, 717–735. [CrossRef]

45. Ma, W.; Zhao, L.; Ma, Y.; Li, Y.; Qin, S.; He, B. Oriented efficient biosynthesis of rare ginsenoside Rh2 from PPD by compiling UGT-Yjic mutant with sucrose synthase. *Int. J. Biol. Macromol.* **2020**, *146*, 853–859. [CrossRef]

46. Huffman, M.A.; Fryszkowska, A.; Alvizo, O.; Borra-Garske, M.; Campos, K.R.; Canada, K.A.; Devine, P.N.; Duan, D.; Forstater, J.H.; Grosser, S.T.; et al. Design of an in vitro biocatalytic cascade for the manufacture of islatravir. *Science* **2020**, *368*, 1255–1259. [CrossRef]

47. Song, W.; Chen, X.; Wu, J.; Xu, J.; Zhang, W.; Liu, J.; Chen, J.; Liu, L. Biocatalytic derivatization of proteinogenic amino acids for fine chemicals. *Biotechnol. Adv.* **2020**, *40*, 107496. [CrossRef]

48. Laurent, V.; Gourbeyre, L.; Uzel, A.; Hélaine, V.; Nauton, L.; Traïkia, M.; Salanoubat, M.; Gefflaut, T.; Lemaire, M.; Guérard-Hélaine, C. Pyruvate Aldolases Catalyze Cross-Aldol Reactions between Ketones: Highly Selective Access to Multi-Functionalized Tertiary Alcohols. *ACS Catal.* **2020**, *10*, 2538–2543. [CrossRef]

49. Gmelch, T.J.; Sperl, J.M.; Sieber, V. Optimization of a reduced enzymatic reaction cascade for the production of L-alanine. *Sci. Rep.* **2019**, *9*, 1–9. [CrossRef]

50. Takamatsu, S.; Tosa, T.; Chibata, I. Production of L-alanine from ammonium fumarate using two microbial cells immobilized with κ-carrageenan. *J. Chem. Eng. Jpn.* **1985**, *18*, 66–70. [CrossRef]

51. Li, G.; Lian, J.; Xue, H.; Jiang, Y.; Ju, S.; Wu, M.; Lin, J.; Yang, L. Biocascade Synthesis of L-Tyrosine Derivatives by Coupling a Thermophilic Tyrosine Phenol-Lyase and L-Lactate Oxidase. *Eur. J. Org. Chem.* **2020**, *2020*, 1050–1054. [CrossRef]

52. Fernández-Tejada, A.; Corzana, F.; Busto, J.H.; Avenoza, A.; Peregrina, J.M. Conformational effects of the non-natural α-methylserine on small peptides and glycopeptides. *J. Org. Chem.* **2009**, *74*, 9305–9313. [CrossRef] [PubMed]

53. Liao, C.; Seebeck, F.P. Asymmetric β-Methylation of l- and d-α-Amino Acids by a Self-Contained Enzyme Cascade. *Angew. Chem. Int. Ed.* **2020**, *59*, 7184–7187. [CrossRef] [PubMed]

54. Contente, M.L.; Paradisi, F. Self-sustaining closed-loop multienzyme-mediated conversion of amines into alcohols in continuous reactions. *Nat. Catal.* **2018**, *1*, 452–459. [CrossRef]

55.   Burgener, S.; Cortina, N.S.; Erb, T.J. Oxalyl-CoA Decarboxylase Enables Nucleophilic One-Carbon Extension of Aldehydes to Chiral α-Hydroxy Acids. *Angew. Chem. Int. Ed.* **2020**, *59*, 5526–5530. [CrossRef]

56.   Groeger, H.; Hummel, W. Combining the 'two worlds' of chemocatalysis and biocatalysis towards multi-step one-pot processes in aqueous media. *Curr. Opin. Chem. Biol.* **2014**, *19*, 171–179. [CrossRef]

57.   Ríos-Lombardía, N.; García-Álvarez, J.; González-Sabín, J. One-pot combination of metal-and bio-catalysis in water for the synthesis of chiral molecules. *Catalysts* **2018**, *8*, 75. [CrossRef]

58.   Wu, S.; Zhou, Y.; Gerngross, D.; Jeschek, M.; Ward, T.R. Chemo-enzymatic cascades to produce cycloalkenes from bio-based resources. *Nat. Commun.* **2019**, *10*, 1–10. [CrossRef]

59.   Schuchardt, U.; Cardoso, D.; Seercheli, R.; Pereira, R.; da Cruz, R.S.; Guerreiro, M.C.; Mandelli, D.; Spinacé, E.V.; Pires, E.L. Cyclohexane oxidation continues to be a challenge. *Appl. Catal. A Gen.* **2001**, *97*, 1–17. [CrossRef]

60.   Soussan, L.; Pen, N.; Belleville, M.P.; Marcano, J.S.; Paolucci-Jeanjean, D. Alkane biohydroxylation: Interests, constraints and future developments. *J. Biotechnol.* **2016**, *2022*, 117–142. [CrossRef]

61.   Hrycay, E.G.; Bandiera, S.M. The monooxygenase, peroxidase, and peroxygenase properties of cytochrome P450. *Arch. Biochem. Biophys.* **2012**, *522*, 71–89. [CrossRef]

62.   Wang, Y.; Lan, D.; Durrani, R.; Hollmann, F. Peroxygenases en route to becoming dream catalysts. What are the opportunities and challenges? *Curr. Opin. Chem. Biol.* **2017**, *37*, 1–9. [CrossRef]

63.   Freakley, S.J.; Kochius, S.; Van Marwijk, J.; Fenner, C.; Lewis, R.J.; Baldenius, K.; Marais, S.S.; Opperman, D.J.; Harrison, S.T.L.; Alcalde, M.; et al. A chemo-enzymatic oxidation cascade to activate C–H bonds with in situ generated H_2O_2. *Nat. Commun.* **2019**, *10*, 1–8. [CrossRef] [PubMed]

64.   Khanam, H. Bioactive Benzofuran derivatives: A review. *Eur. J. Med. Chem.* **2015**, *97*, 483–504. [CrossRef] [PubMed]

65.   Mertens, M.S.; Thomas, F.; Nöth, M.; Moegling, J.; El-Awaad, I.; Sauer, D.F.; Dhoke, G.V.; Xu, W.; Pich, A.; Schwaneberg, U. One-Pot Two-Step Chemoenzymatic Cascade for the Synthesis of a Bis-benzofuran Derivative. *Eur. J. Org. Chem.* **2019**, *2019*, 6341–6346. [CrossRef]

66.   Schmidt, S.; Castiglione, K.; Kourist, R. Frontispiece: Overcoming the Incompatibility Challenge in Chemoenzymatic and Multi-Catalytic Cascade Reactions. *Chem. Eur. J.* **2018**, *24*. [CrossRef]

67.   Cicco, L.; Ríos-Lombardía, N.; Rodríguez-Álvarez, M.J.; Morís, F.; Perna, F.M.; Capriati, V.; García-Álvarez, J.; González-Sabín, J. Programming cascade reactions interfacing biocatalysis with transition-metal catalysis in Deep Eutectic Solvents as biorenewable reaction media. *Green Chem.* **2018**, *20*, 3468–3475. [CrossRef]

68.   Grabner, B.; Schweiger, A.K.; Gavric, K.; Kourist, R.; Gruber-Woelfler, H. A chemo-enzymatic tandem reaction in a mixture of deep eutectic solvent and water in continuous flow. *React. Chem. Eng.* **2020**, *5*, 263–269. [CrossRef]

69.   Geng, P.B.; Zheng, S.S.; Tang, H.; Zhu, R.M.; Zhang, L.; Cao, S.; Xue, H.G.; Pang, H. Transition metal sulfides based on graphene for electrochemical energy storage. *Adv. Energy Mater.* **2018**, *8*, 1703259. [CrossRef]

70.   Losada-García, N.; Rodríguez-Otero, A.; Palomo, J.M. Tailorable synthesis of heterogeneous enzyme–copper nanobiohybrids and their application in the selective oxidation of benzene to phenol. *Catal. Sci. Technol.* **2020**, *10*, 196–206. [CrossRef]

71.   Palomo, J.M. Nanobiohybrids: A new concept for metal nanoparticles synthesis. *Chem. Commun.* **2019**, *55*, 9583–9589. [CrossRef]

72.   Wei, H.; Wang, E.K. Nanomaterials with enzyme-like characteristics (nanozymes): Next-generation artificial enzymes. *Chem. Soc. Rev.* **2013**, *42*, 6060–6093. [CrossRef]

73.   Zhang, Y.F.; Ge, J.; Liu, Z. Enhanced activity of immobilized or chemically modified enzymes. *ACS Catal.* **2015**, *5*, 4503–4513. [CrossRef]

74.   Wang, Q.; Zhang, X.; Huang, L.; Zhang, Z.; Dong, S. One-pot synthesis of Fe_3O_4 nanoparticle loaded 3D porous graphene nanocomposites with enhanced nanozyme activity for glucose detection. *ACS Appl. Mater. Interfaces* **2017**, *9*, 7465–7471. [CrossRef]

75.   Ruan, X.; Liu, D.; Niu, X.; Wang, Y.; Simpson, C.D.; Cheng, N.; Du, D.; Lin, Y. 2D Graphene oxide/Fe-MOF nanozyme nest with superior peroxidase-like activity and its application for detection of woodsmoke exposure biomarker. *Anal. Chem.* **2019**, *91*, 13847–13854. [CrossRef] [PubMed]

76.   Losada-Garcia, N.; Berenguer-Murcia, A.; Cazorla-Amorós, D.; Palomo, J.M. Efficient production of multi-layer graphene from graphite flakes in water by lipase-graphene sheets conjugation. *Nanomaterials* **2020**, *10*, 7. [CrossRef]

77. Losada-Garcia, N.; Rodriguez-Oliva, I.; Simovic, M.; Bezbradica, D.I.; Palomo, J.M. New Advances in Fabrication of Graphene Glyconanomaterials for Application in Therapy and Diagnosis. *ACS Omega* **2020**, *5*, 4362–4369. [CrossRef]

78. Malkhandi, S.; Yeo, B.S. Electrochemical conversion of carbon dioxide to high value chemicals using gas-diffusion electrodes. *Curr. Opin. Chem. Eng.* **2019**, *26*, 112–121. [CrossRef]

79. Seelbach, K.; Riebel, B.; Hummel, W.; Kula, M.-R.; Tishkov, V.I.; Egorov, A.M.; Wandrey, C.; Kragl, U. A Novel, Efficient Regenerating Method of NADPH Using a New Formate Dehydrogenase. *Tetrahedron Lett.* **1996**, *37*, 1377–1380. [CrossRef]

80. Alissandratos, A.; Easton, C.J. Biocatalysis for the application of CO_2 as a chemical feedstock. *Beilstein J. Org. Chem.* **2015**, *11*, 2370–2387. [CrossRef]

81. Schlager, S.; Neugebauer, H.; Haberbauer, M.; Hinterberger, G.; Sariciftci, N.S. Direct Electrochemical Addressing of Immobilized Alcohol Dehydrogenase for the Heterogeneous Bioelectrocatalytic Reduction of Butyraldehyde to Butanol. *ChemCatChem* **2015**, *7*, 967–971. [CrossRef]

82. Seelajaroen, H.; Bakandritsos, A.; Otyepka, M.; Zbořil, R.; Sariciftci, N.S. Immobilized Enzymes on Graphene as Nanobiocatalyst. *ACS Appl. Mater. Interfaces* **2019**, *12*, 250–259. [CrossRef]

83. Chu, S.; Huang, W.; Shen, F.; Li, T.; Li, S.; Xu, W.; Lv, C.; Luo, Q.; Liu, J. Graphene oxide-based colorimetric detection of organophosphorus pesticides via a multi-enzyme cascade reaction. *Nanoscale* **2020**, *12*, 5829–5833. [CrossRef]

84. Periasamy, A.P.; Umasankar, Y.; Chen, S.M. Nanomaterials-acetylcholinesterase enzyme matrices for organophosphorus pesticides electrochemical sensors: A review. *Sensors* **2009**, *9*, 4034. [CrossRef] [PubMed]

85. Cui, H.F.; Wu, W.W.; Li, M.M.; Song, X.; Lv, Y.; Zhang, T.T. A highly stable acetylcholinesterase biosensor based on chitosan-TiO2-graphene nanocomposites for detection of organophosphate pesticides. *Biosens. Bioelectron.* **2018**, *99*, 223–229. [CrossRef] [PubMed]

86. Wang, Q.Q.; Zhang, X.P.; Huang, L.; Zhang, Z.Q.; Dong, S.J. GO*x*@ZIF-8 (NiPd) nanoflower: An artificial enzyme system for tandem catalysis. *Angew. Chem. Int. Ed.* **2017**, *56*, 16082–16085. [CrossRef] [PubMed]

87. Chen, G.S.; Huang, S.M.; Kou, X.X.; Wei, S.B.; Huang, S.Y.; Jiang, S.Q.; Shen, J.; Zhu, F.; Ouyang, G.F. A convenient and versatile aminoacid- boosted biomimetic strategy for the nondestructive encapsulation of biomacromolecules within metal−organic frameworks. *Angew. Chem. Int. Ed.* **2019**, *58*, 1463–1467. [CrossRef] [PubMed]

88. Cheng, X.; Zhou, J.; Chen, J.; Xie, Z.; Kuang, Q.; Zheng, L. One-step synthesis of thermally stable artificial multienzyme cascade system for efficient enzymatic electrochemical detection. *Nano Res.* **2019**, *12*, 3031–3036. [CrossRef]

4

Polymer-Assisted Biocatalysis: Polyamide 4 Microparticles as Promising Carriers of Enzymatic Function

Nadya Dencheva [1], **Joana Braz** [1], **Dieter Scheibel** [2], **Marc Malfois** [3], **Zlatan Denchev** [1,*] and **Ivan Gitsov** [2,4,*]

1   IPC-Institute for Polymers and Composites, University of Minho, 4800-056 Guimarães, Portugal; nadiad@dep.uminho.pt (N.D.); joanabraz@dep.uminho.pt (J.B.)
2   Department of Chemistry, State University of New York-ESF, Syracuse, NY 13210, USA; dmscheib@syr.edu
3   ALBA Synchrotron Facility, Cerdanyola del Valés, 0890 Barcelona, Spain; mmalfois@cells.es
4   The Michael M. Szwarc Polymer Research Institute, Syracuse, NY 13210, USA
*   Correspondence: denchev@dep.uminho.pt (Z.D.); igivanov@syr.edu (I.G.)

Abstract: This study reports a new strategy for enzyme immobilization based on passive immobilization in neat and magnetically responsive polyamide 4 (PA4) highly porous particles. The microsized particulate supports were synthesized by low-temperature activated anionic ring-opening polymerization. The enzyme of choice was laccase from *Trametes versicolor* and was immobilized by either adsorption on prefabricated PA4 microparticles (PA4@iL) or by physical in situ entrapment during the PA4 synthesis (PA4@eL). The surface topography of all PA4 particulate supports and laccase conjugates, as well as their chemical and physical structure, were studied by microscopic, spectral, thermal, and synchrotron WAXS/SAXS methods. The laccase content and activity in each conjugate were determined by complementary spectral and enzyme activity measurements. PA4@eL samples displayed >93% enzyme retention after five incubation cycles in an aqueous medium, and the PA4@iL series retained ca. 60% of the laccase. The newly synthesized PA4-laccase complexes were successfully used in dyestuff decolorization aiming at potential applications in effluent remediation. All of them displayed excellent decolorization of positively charged dyestuffs reaching ~100% in 15 min. With negative dyes after 24 h the decolorization reached 55% for PA4@iL and 85% for PA4@eL. A second consecutive decolorization test revealed only a 5–10% decrease in effectiveness indicating the reusability potential of the laccase-PA4 conjugates.

Keywords: laccase; polyamide 4; enzyme immobilization; magnetic enzyme supports; enzyme decolorization

1. Introduction

Biocatalysts (extracellular enzymes or whole-cells) constitute an alternative to traditional catalytic systems, being able to cause transformation of natural or synthetic substrates under mild reaction conditions, with high substance-, stereo-, and regiospecificity, and lack of toxic byproducts [1,2]. With the advances in the production of relatively inexpensive free enzymes, the steady increase of their extensive implementation as effective catalysts in many industrial and biomedical applications is beyond any doubt [3].

The wide use of industrial-scale processes with enzymatic catalysts is restricted by the free enzymes' inactivation by different denaturant agents and by their difficult or even impossible recovery from the reaction medium after use. These problems can be resolved by immobilization (conjugation) of the enzymes to prefabricated matrices of various nature, geometry, and topography. Immobilized enzymes, apart from their application as reusable heterogeneous biocatalysts, can serve as appropriate platforms for the development of biosensors, biofuel cells, or microscale devices for controlled release of protein drugs [4].

The most frequently used immobilization techniques can be sorted into five main groups, each with its associated advantages and disadvantages [5]: (i) non-covalent adsorption or deposition; (ii) immobilization via ionic interactions; (iii) entrapment in a polymeric gel or capsule (encapsulation); (iv) covalent attachment to the support (tethering); and (v) cross-linking of the enzyme. In the first three cases (i–iii), the enzyme immobilization is related to physical phenomena, i.e., entropy effects, electrostatic interactions, or hydrogen bonding. The distinct advantage of these methods is that neither the enzyme, nor the support, needs to be chemically modified or pretreated, which contributes to a cleaner and cost-effective biocatalyst. Moreover, the enzyme retains mobility, thus avoiding unfavorable screening of the active site. However, conjugates based on physical interactions often suffer from enzyme leaching, especially if used in aqueous media. To avoid leakage, immobilization protocols from groups (iv–v) are used. In them, covalent bonds across the enzyme–support interface are produced or cross-linking of the polypeptide moieties of the enzyme is caused by appropriate chemical reagents. These chemical modifications should be carefully directed so as not to cause deactivation of the biomacromolecule. As a rule, the immobilization method is selected on a case-by-case basis, by trial and error, although attempts to optimize this process by computer-aided analysis and molecular modeling have also been communicated [3,6].

General requirements for polymer supports are that they should not deactivate the immobilized enzyme, and for many important applications they have to be biocompatible. Therefore, traces of solvents, emulsifiers, unreacted monomers, or other accompanying substances that could be toxic or potentially deactivate the immobilized enzyme must be totally excluded. More specifically, immobilization by enzyme entrapment requires formation of the polymer support in situ, e.g., by low-temperature processes in the presence of the respective enzyme, thus avoiding denaturation. Furthermore, the monomer, the polymer support, and the enzyme should possess suitable functional groups in order to enable covalent or non-covalent interactions before, during, or after the polymerization. Finally, the polymer support should have appropriate physical structure and be formable into controllable shapes and textures at various length scales [7].

A large number of supports have been employed for enzyme immobilization in the form of beads, porous particles, membranes, and micro- or nanosized fibers made of inorganic materials [8] or organic polymers such as polyacrylates, cross-linked styrene-based copolymers, smart polymers, and modified polysaccharides [3]. Smart biocatalysts for industrial and biomedical applications produced by conjugation of enzymes to stimuli-responsive polymer supports have attracted a growing interest [9]. The activity of the enzyme can be favorably modified by complexation with supramolecular linear-dendritic block copolymers [10] or amphiphilic block copolymers self-assembling into micelles or physical networks [11].

Among the big variety of industrially relevant enzymes, the group of laccases (EC 1.10.3.2, benzenediol:oxygen oxidoreductases) have generated significant biotechnological interest since they require molecular oxygen as the final electron acceptor and release only water as a by-product [12]. Due to their broad specificity laccases are being used in different industrial fields for very diverse purposes, from food additives and beverage processing to biomedical diagnosis, pulp delignification, wood fiber modification, chemical or medicinal synthesis, or in the production of biofuels [13]. They show great promise as 'green' synthesis and polymerization mediators [14–16] and catalysts for Bisphenol A removal [17]. Also, laccases have great potential to biotechnological applications of industrial effluent treatment including textile dye decolorization and degradation [18,19]. Enzymes from the laccase group have also been studied intensely in nanobiotechnology for the development of implantable biosensors and biofuel cells, as well as wastewater and soil remediation [20].

Synthetic polyamides are among the known supports for laccase immobilization since they are cheap and can be easily produced in different forms. Thus, Fatarella et al. [21] explored the preparation of polyamide 6 (PA6) film or electrospun nanofiber carriers for covalently bonded laccase from *Trametes versicolor,* reaching enzyme loadings of ca. 60% and 71%, respectively. As reported, the resultant PA6/laccase conjugates are useful in industrial biocatalyst applications. Furthermore, a nanofibrous membrane for 3,3'-dimethoxybenzidine detoxification was developed by Jasni et al. [22] on the basis of PA66/laccase/Fe^{3+} conjugate produced by covalent bonding of the enzyme to the support by glutaraldehyde crosslinking. The immobilized enzyme displayed enhanced storage stability and a good potential for emerging pollutant detoxification. In a study of Silva et al. [23], laccase from *Trametes hirsuta* was covalently immobilized on woven PA66 supports employing glutaraldehyde and a 1,6-hexanediamine spacer. The PA66-laccase conjugate obtained under optimized conditions displayed higher half-life time than the free enzyme and potential for application in membrane reactors for continuous decolorization of effluents.

To the best of our knowledge, at this point of time, no attempt has been reported on laccase immobilization by polyamide particulate supports. Very recently, Dencheva et al. [24] synthesized neat and magnetic responsive PA4 [(poly-2-pyrrolidone), PPD] microparticles in good yields, with controlled size, shape, and porosity using activated anionic ring-opening polymerization (AAROP) of 2-pyrrolidone (2PD) carried out at 40 °C. This low-temperature and solventless process produces magnetic PA4 microparticles with controllable size, shape, and structure. There are certain advantageous factors in this technique that can open promising routes towards novel recyclable and efficient biocatalysts: PA4 is biodegradable in various environments [25,26], its monomer 2-pyrrolidone can be produced by sustainable biosynthesis [27], and the polyamide support is synthesized directly in the form of porous microparticles without the need of any additional treatment. Furthermore, the chemical structure of PA4 with its dense amide linkages and final functional groups resembles that of the proteins, thus enhancing the H-bonding between the support and the immobilized enzyme. Evidently, the neat and magnetic responsive PA4 microparticles combine most of the requirements for effective supports of biomolecules. Moreover, they have already been found to be good supports for model proteins [24] or to be used in molecular imprinting [28].

The present work investigates the potential of neat and magnetic-responsive PA4 microparticles obtained by low-temperature AAROP to serve as supports for immobilization of laccase. Two immobilization strategies were employed: (i) physical adsorption of laccase upon preformed PA4 microparticles and (ii) in situ encapsulation (entrapment) of laccase during the AAROP of PA4 synthesis. The morphology and the crystalline structure of the PA4-laccase conjugates obtained by these two methods were analyzed by microscopy, spectral and X-ray scattering techniques, and the respective enzyme activities toward 2,2'-azino-bis (3-ethylbenzothiazoline-6-sulphonic acid (ABTS) were compared. The biocatalytic capability of the synthesized PA4-laccase conjugates, as well as their reusability, were tested in two reactions of environmental interest related to the treatment of effluents from the textile industry. The activity of the immobilized laccase was tested in decolorization reactions of two dyes, widely used as textile coloring agents (malachite green and bromophenol blue).

2. Results and Discussion

Three microparticulate PA4 supports, without and with magnetic susceptibility, designated as PA4, PA4-Fe, and PA4-Fe_3O_4, were synthesized by low-temperature AAROP. They are in the form of fine powders with white, grey or brownish color, respectively. The scheme for the AAROP representing the chemical structure of all substances involved is presented in Figure S1 of the Supporting Information. When the polymerization is carried out in the presence of magnetic micro/nanoparticles, the PA4 MP become susceptible to external magnetic fields [24].

The 2PD monomer is a cyclic analog of the amino acids and the chemical composition of PA4 is comparable to that of proteins and enzymes. This will expectedly enable intense H-bond formation between the laccase and the PA4 supports: neat PA4, PA4-Fe, and PA4-Fe_3O_4. The fact that AAROP was possible at 40 °C allowed the entrapment of active laccase into the shell of the forming PA4 MP during the polymerization, resulting in three PA4@eL samples. Separately, the enzyme was immobilized by physical adsorption on prefabricated PA4 supports, thus preparing the three PA4@iL samples. This work presents a comparative discussion on the morphology, structure, and the catalytic activity of the three conjugates immobilized by adsorption PA4@iL and the three conjugates immobilized by entrapment PA4@eL.

2.1. Synthesis and Morphology of Empty PA4, PA4-Fe, and PA4-Fe$_3$O$_4$ Microparticles

Some basic characteristics of the synthesized neat and magnetic PA4 microparticles (MP) are listed in Table 1. The yields of polymer were in the 59–63 wt % range meaning that the presence of iron micro- and iron (II, III) oxide nanosized particles in the reaction mixture did not affect the activity of DL/C20 catalytic system and the kinetics of the AAROP reaction. Intrinsic viscosity (η) (see the Supporting Information) of 0.926 dL/g was found for the neat PA4 MP, which is similar to the values of PA4 microspheres obtained by a different method involving AAROP of 2PD [29].

Table 1. Designation and some characteristics of the PA4 MP empty supports.

Sample	PA4 Yield, % (a)	% of Oligo-mers	Real Fe Content R$_L$, % (b)	η, dL·g^{-1}	d_{max}, μm	d_{max}/d_{min}
PA4	59.4	4.4	-	0.926	8-15	1.1–1.2
PA4-Fe	61.2	6.9	1.7	-	10-20	1.2–1.3 3.6–4.1
PA4-Fe$_3$O$_4$	62.7	4.7	1.2	-	10-15	1.2–1.3 3.4–4.0

(a) In relation to the 2PD monomer; (b) Determined by thermogravimetric analysis (TGA) according to Equation (2) (see Materials and Methods—Section 3.2).

The average maximum size of the particles d_{max} in all PA4 empty supports based on optical microscopy with image-processing was found to be between 8–20 μm (Table 1, Figure S2 of the Supporting Information). The d_{max}, of PA4 MP without magnetic load is the smallest, ranging between 8 and 15 μm. Similar values of 10–15 μm are registered for the support with nanosized Fe$_3$O$_4$ loads, while in the presence of Fe the upper limit of d_{max} grows above 20 μm. The Fe- and Fe$_3$O$_4$ containing PA4 microparticles become less spherical, as seen from the roundness parameter d_{max}/d_{min} (Table 1, Figure S2). This means that the PA4-Fe MP would contain in their core up to 4–5 iron microparticles whose proper size is 3–5 μm, while the PA4-Fe$_3$O$_4$ core would expectedly contain iron oxide particles in the nanometer length scale [24,28]. The d_{max}/d_{min} in the range of 3–4 found in both magnetic empty supports are attributable to self-assembly of magnetized Fe and Fe$_3$O$_4$ to higher aspect ratio aggregates with their subsequent coating with PA4 during the AAROP.

All PA4 supports have porous structure as demonstrated by BET analysis (Table S1 of the Supporting Information). The neat PA4 microparticles showed the largest specific surface area, S_{BET} = 2.766 m^2/g with a total pore volume V_{total} = 0.012 cm^3/g, followed by PA4-Fe$_3$O$_4$ (2.542 m^2/g; 0.008 cm^3/g) and PA4-Fe (2.052 m^2/g; 0.004 cm^3/g) samples. The largest average pore size σ_{ave} = 85 Å was found for the PA4 MP, the values of PA4-Fe$_3$O$_4$ and PA4-Fe being 42 and 32 Å, respectively. These changes in σ_{ave}, V_{total}, and S_{BET} should be attributed to the presence of dense magnetic loads in the microparticles core of the last two samples.

More details on the morphology of the empty supports' particles can be obtained by SEM (Figure 1).

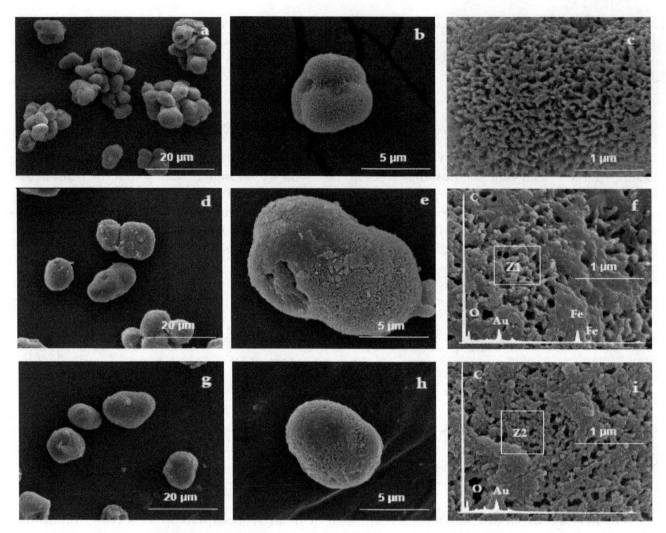

Figure 1. SEM images of empty PA4 particulate supports: (**a–c**) PA4; (**d–f**) PA4-Fe; (**g–i**) PA4-Fe$_3$O$_4$.

The micrographs of the neat PA4 support (Figure 1a–c) display spheroidal particles with sizes of the individual entity between 5–8 µm, the latter forming also aggregates with average sizes close to 20 µm. At larger magnifications, the PA4 particles display porous, scaffold-like topology with visible pore diameters in the range of 100–150 nm. The use of Fe and Fe$_3$O$_4$ particles (Figure 1d,e,g–i, respectively), seem to decrease the number of surface pores changing their distribution but with no significant change of the pore size. Moreover, as seen from some selected EDX traces in Figure 1f,i, the FeKα and FeKβ peaks only appear if the electron beam is directed into the pores (location Z1), rather than if it hits a non-porous spot of the particles surface (location Z2). This means that most of the magnetic particulate fillers (especially the much larger Fe particles) are embedded in the core of the respective MP, thus confirming the supposed precipitation–crystallization mechanism of the polyamide MP formation [24].

2.2. Immobilization of Laccase on PA4 MP Supports

2.2.1. Immobilization by Physical Adsorption

As previously mentioned, owing to the analogous chemical structure of PA4 and the peptide-containing biomolecules, a high capacity toward H-bonding formation is expected between the laccase and the PA4 supports during immobilization by adsorption. Since the laccase isoelectric point is in the range of 3–4, if the adsorption is carried out at pH > 4 the enzyme will be negatively charged. From the Z-potential measurements (see the Supporting Information, Table S2) it can be seen that under such conditions the three PA4 supports will also be negatively charged with values of −35 eV at pH 7 and between −10 ÷ −12 eV at pH 5. Nevertheless, preliminary adsorption tests confirmed that larger amounts of laccase are adsorbed in DDW at neutral pH than in PB, pH 5. This means that the immobilization by adsorption is not upset by possible electrostatic repulsion between the enzyme and the support, therefore all adsorption experiments were performed at pH 7.

Table 2 shows that, as expected, the particles of the three PA4@iL laccase-adsorbed samples do not change their average size and roundness. Judging from the SEM images in Figure 2, the topography of all particulate PA4@iL conjugates is noticeably smoother as compared to the starting PA4 support particles, which is better expressed in the case of the PA4-iL sample (Figure 2c).

Table 2. Designation and some characteristics of the PA4 supports with adsorbed laccase (PA4@iL samples).

Sample	Real Fe Content, R_L, % [a]	d_{max}, μm	d_{max}/d_{min}
PA4-iL	-	8–15	1.1–1.2
PA4-Fe-iL	1.4	12–20	1.2–1.3 3.6–4.1
PA4-Fe$_3$O$_4$ -iL	1.3	10–15	1.2–1.3 3.4–4.0

[a] Determined by TGA, according to Equation (2) (see Materials and Methods—Section 3.2).

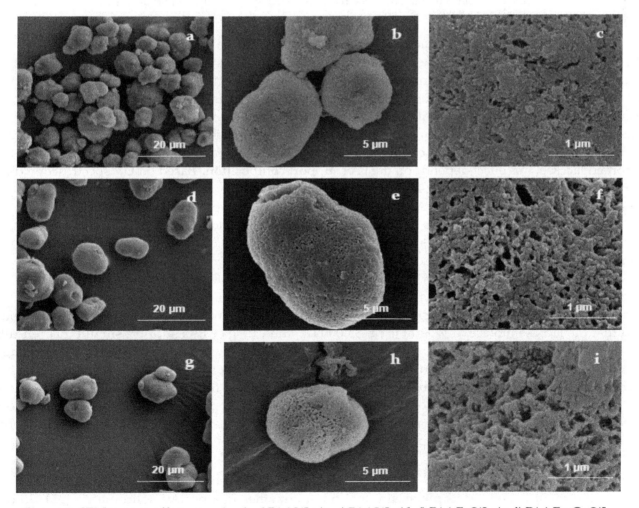

Figure 2. SEM images of laccase-adsorbed PA4@iL: (**a–c**) PA4@iL; (**d–f**) PA4-Fe@iL; (**g–i**) PA4-Fe$_3$O$_4$@iL.

The next step in the PA4@iL preparation was to quantify the amount of the adsorbed enzyme. Using the common method for determination of immobilized laccase, after adsorption completion, the residual supernatants were subjected to activity test toward ABTS. All three samples showed no activity, leading to the conclusion that there is no residual laccase in the supernatant, therefore almost 100% of the enzyme was adsorbed upon the PA4 supports (Table 3). This was a superior immobilization efficiency as compared to previous studies disclosing laccase immobilization by covalent bonding on PA6 [21] or carbon nanotube supports [30] with immobilization effectiveness (IE) values of 59–71% or 60–90%, respectively. However, the lack of activity of the residual supernatant means that it is free of active laccase but could contain some amounts of inactive enzyme. Therefore, a second method to verify the laccase content in the supernatant was used.

Table 3. Enzyme quantification in the PA4@iL samples.

Sample	Residual Laccase Activity, $\mu kat \cdot mL^{-1}$ [a]	Immobilized Laccase, $mg \cdot mL^{-1}$ [b]	Immobilized Laccase, $mg \cdot mL^{-1}$ [c]	Laccase on Support, mg	Laccase, $mg \cdot g^{-1}$ Support	Immobilization Efficiency, IE, % [d]
PA4-iL	0.0002	1.999	1.265	6.32	31.62	63.3
PA4-Fe-iL	0.0009	1.999	1.612	8.06	40.29	80.6
PA4-Fe$_3$O$_4$-iL	0.0002	1.999	1.072	5.36	26.81	53.6

[a] Activity of the residual laccase in the supernatant after immobilization; [b] Immobilized laccase calculated by standard curve based on the activation test toward ABTS; [c] Determined by UV analysis of the supernatant after immobilization using a standard curve based on the absorbance at $\lambda = 286$ nm; [d] Immobilization effectiveness (IE) is the ratio between the amounts of immobilized laccase and starting laccase used in the adsorption process (see Equation (3), Materials and Methods—Section 3.3.2).

An alternative method for laccase quantification can be based on its UV absorbance in the 280–290 nm range [31,32] due to the presence of aromatic amino acid residues: tryptophan, tyrosine, or phenylalanine. Thus, all supernatants were studied for residual laccase by UV spectroscopy measuring the intensity of the band appearing at 286 nm. As indicated in Table 3, 19–46% of residual laccase was found in the supernatants, which showed no activity in the previous test with ABTS. A possible reason for this behavior is that during the multiple adsorption/desorption processes in the PA4@iL sample preparation, a structural change occurs leading to lowering of the redox potential of the desorbed laccase, this effect depends on the PA4 support composition. Based on UV spectroscopy data, the maximum amount of laccase (40.3 mg/g or 81% from the enzyme amount) was adsorbed on the PA4-Fe MP support, whereas the PA4-Fe$_3$O$_4$ MP displayed the lowest IE of 54%. These IE values were considered more reliable and were therefore used in all further calculations.

Employing the approach of Qui et al. [33], the data in Table 3 can provide information about the mechanism of laccase adsorption on the PA4 supports of this study. Thus, since the size of laccase macromolecule is found to be $6.5 \times 5.5 \times 4.5$ nm [34], its largest footprint on a surface will be ca. 35.8 nm^2. The surface areas of the three PA4@MP samples determined by BET (Table S1) were in the range of 2.05–2.78 m^2/g. Assuming that laccase molecules arrange in ideal monolayers when adsorbing and that the laccase molecular weight is ca. 85,000 Da, then 1.0 g of PA4 support should theoretically adsorb 8–11 mg enzyme. According to Table 3, however, the real values vary in the 27–40 mg range, i.e., about 3–4 times larger. Evidently, the laccase adsorption occurs in multilayers, implying lateral intra- and interlayer interactions between adsorption sites. This finding is in good agreement with our previous work on protein adsorption upon similar PA4 microparticles [28] proving that without special treatment of the PA4 support the adsorption data are consistent with the Freundlich isotherm, thus proving the multilayer adsorption model.

2.2.2. Immobilization by Entrapment

Entrapment is defined as physical retention of enzymes in a porous solid matrix [35]. In our case, the laccase entrapment occurs during AAROP. The enzyme is first suspended in the monomer solution, and a subsequent polymerization process keeps the biomolecule trapped, preventing direct contact with the environment. To obtain laccase-entrapped PA4 MP with or without magnetic particles, the same low temperature AAROP was used as for the preparation of the empty PA4 supports, however carried out in the presence of the enzyme. The final 2PD conversion to PA4 was in the 45–58 wt % range (Table 4), i.e., in average 5–10% lower than of the empty PA4 supports (Table 1). This is explained with the more rigorous washing/purification procedure to eliminate not entrapped enzyme and oligomers with lower conversion to PA4. It can therefore be concluded that the presence of laccase in the AAROP reaction mixture did not significantly upset the catalytic system permitting to use the same protocol for synthesis and purification as for the empty PA4 particulate supports. However, Table 4 indicates larger average particle sizes of 15–25 μm and lesser roundness values of 1.2–1.3 in the PA4@eL samples than in the respective empty and laccase-adsorbed supports (Table 2).

Table 4. Designation and some characteristics of PA4 supports with entrapped laccase (PA4@eL).

Sample	PA4-eL Yield, % [a]	Real Fe Content, R_L, % [a]	d_{max}, μm	d_{max}/d_{min}
PA4-eL	45.3	-	15–20	1.1–1.2
PA4-Fe-eL	48.4	1.7	15–25	1.2–1.3
PA4-Fe$_3$O$_4$ -eL	57.6	1.9	15–25	1.2–1.3

[a] In relation to the 2PD monomer; [b] Determined by TGA, according to Equation (2) (see Materials and Methods—Section 3.2).

The possibility to carry out AAROP to PA4 in the presence of laccase at 40 °C had two major advantages. First, no denaturation or other disruption of the enzyme's secondary or tertiary structure caused by temperature will occur during the polymerization process. Second, in AAROP of all lactams, the chain propagation is accompanied by PA4 crystallization that affixes the topology of the enzyme-loaded microparticles and results in their precipitation from the reaction medium, thus facilitating the entrapment.

As seen from Table 5 showing the laccase quantification in the PA4@eL series, the total amount of entrapped laccase was 45–58 mg, corresponding to 13–18 mg enzyme per gram of MP. These concentrations are up to 2.5 times lower than in the case of immobilization by adsorption (Table 3). As to the entrapment factor EF, it is dependent on the 2PD conversion to PA4. Thus, the highest degree of 2PD conversion of ca. 70% was achieved with the PA4-Fe$_3$O$_4$@eL sample, which accounted for the minimum amount of entrapped enzyme per gram support in this case.

Table 5. Enzyme quantification in PA4@eL samples.

Sample	Laccase in AAROP Mixture, mg	Yield of PA4 MP, % [a]	Laccase on Support, mg [b]	Laccase mg·g^{-1} Support	Entrapment Efficiency, EE, % [c]
PA4-eL	85.10	45.29	50.91	17.86	59.8
PA4-Fe-eL	85.10	48.30	52.00	16.72	61.1
PA4-Fe$_3$O$_4$-eL	85.10	57.58	59.25	13.47	69.6

[a] Based on the sample weight after removal of excessive laccase and oligomer extraction; [b] Calculated by UV–VIS detection of the laccase in the combined DDW supernatants of the threefold wash. For more details, see the Experimental part; [c] EE is the ratio between the quantity of the entrapped and the starting laccase used in the syntheses (see Equation (4), Materials and Methods—Section 3.3.3).

Table 4 indicates larger average particle sizes of 15–25 μm and lesser roundness values of 1.2–1.3 in the PA4@eL samples than in the respective empty and laccase-adsorbed supports. Figure 3 displays selected SEM images at different magnifications of the three PA4 supports carrying entrapped enzyme.

Figure 3. SEM images of laccase-entrapped PA4@eL: (**a–c**) PA4@eL; (**d–f**) PA4-Fe@eL; (**g–i**) PA4-Fe$_3$O$_4$@eL.

Unlike the three empty PA4 supports or the magnetic-loaded laccase conjugates PA4@iL, the particles' shapes in PA4@eL samples are far from being spherical or elliptical. Most probably, this is due to the fact that upon its dispersion in the 2PD monomer the lyophilized enzyme produced aggregates that were subsequently covered by PA4. The growing PA4 molecules wind up upon these aggregates and, after reaching a critical molecular weight, crystallize upon them interacting by H-bonds with the enzyme. As a result, the specific topography of the PA4@eL conjugates is produced with visible average pore diameters up to 300–350 nm, being almost twice as large as in the other samples and with quite distinct shape.

2.3. Structure Characterization of PA4 Empty Supports and Laccase Conjugates

The PA4 porous microparticles obtained by AAROP are used for the first time as enzyme supports. The laccase-entrapped PA4@eL samples are also synthesized for the first time. Therefore, some initial structural characterization of all samples of this study was necessary in order to be able to explain their catalytic activity and decide about their potential as enzyme supports.

2.3.1. FTIR Spectroscopy

A FT-IR spectra comparison between the empty PA4 microparticles and the laccase-adsorbed and entrapped samples are presented in Figure 4. In all samples, the bands at 3300 cm^{-1} were assigned to the valence stretching vibrations of hydrogen atoms in secondary NH groups. The shoulder with a maximum at 3450 cm^{-1} was attributed to stretching vibrations in primary amines corresponding to terminal amine groups. Also, the spectra show well defined peaks for Amide I at 1631.5 cm^{-1} and Amide II at 1535.0 cm^{-1} with almost identical intensities. This is a clear indication for fixation of the trans-conformation of the NH–CO group, being typical for high molecular weight polyamides and proteins.

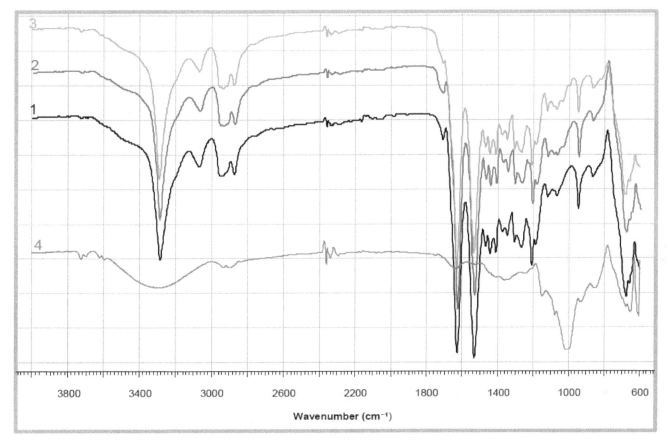

Figure 4. FTIR spectra with ATR of: 1—empty PA4 support; 2—PA4@iL; 3—PA4@eL; 4—native laccase.

The IR spectra in Figure 4 indicate undoubtedly that even at polymerization temperatures as low as 40 °C AAROP of 2DP led to high molecular weight polyamide in all samples studied. This was important to confirm for the PA4@eL and PA4@iL samples, in which the determination of (η) by viscosimetry is technically impossible. All curves display also a weak peak at 1710–1712 cm^{-1} attributable to terminal carboxyl groups. In the neat PA4 sample (curve 1) this narrow and well-resolved band belongs to the –COOH terminus of the polyamide macromolecule. In the enzyme-adsorbed and entrapped samples (curves 2 and 3) a broader composite peak appears due to superposition of the –COOH signal of PA4 with such belonging to laccase carboxyl groups. Curve 4 indicates that a weak and broad band centered at 1641.0 cm^{-1} really exists in the free laccase spectrum. However, FT-IR cannot provide more detailed quantitative or qualitative information about the entrapped or adsorbed laccase.

2.3.2. DSC and TGA

A typical shortcoming of PA4 is the intense degradation that precedes or accompanies the melting occurring normally in the broad range of 230–260 °C [24,36]. As shown previously, the presence of a model protein adsorbed onto PA4 microparticles can significantly increase the thermal resistance of the latter [28]. The TGA traces of all samples in this work displayed in Figure 5 confirm that the presence of laccase in the PA4@iL series can also considerably improve the thermal stability of the supports. Thus, the temperature of initial degradation T_d^{in} of PA4-iL is 255 °C, which is 38 °C higher than of the respective PA4 empty support particles. For the PA4-Fe-iL/PA4-Fe and PA4-Fe$_3$O$_4$-iL/PA4-Fe$_3$O$_4$ pairs $\Delta T_d^{in} = 20$ and 26 °C, respectively. Similar behavior was observed in PA4 obtained by AAROP of 2PD with subsequent conversion of the terminal –COOH into –NH$_2$ groups [37], which in our case may have occurred during the heating scan in the TGA.

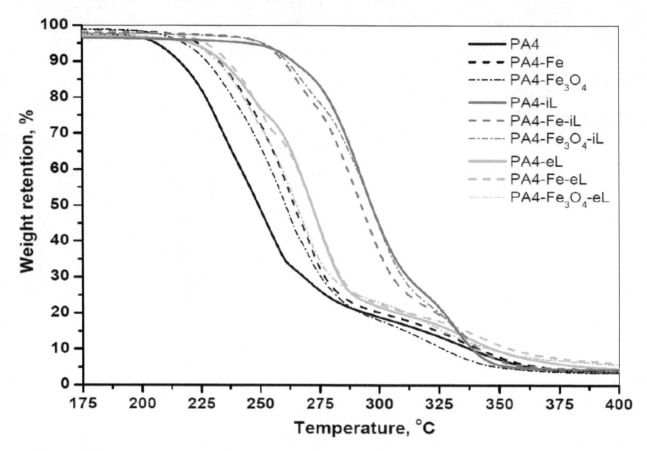

Figure 5. TGA curves at 10 deg/min of empty PA4 supports, laccase adsorbed (PA4@iL) conjugates, and laccase-entrapped (PA4@eL) conjugates.

It should be noted that the presence of magnetic particles alone in PA4 MP increases the T_d^{in} with 15–20 °C in comparison with the neat PA4 MP, most probably due to heat dissipation phenomena. At the same time, the ΔT_d^{in} for the entrapped PA4-eL/PA4 pair is only 12 °C. It can be hypothesized that in the case of the laccase-entrapped PA4@eL samples the said conversion of PA4 terminal groups is difficult or impossible. The TGA study permitted also to determine the carbonized residue at 600 °C of each sample of this study and to calculate on this basis the real content of Fe or Fe$_3$O$_4$ by means of Equation (2) (see Materials and Methods—Section 3.2).

The DSC curves of all samples presented in Figure 6 and the consolidated data extracted from them (Table S3 in the Supporting Information) display a trend of increasing the melting temperature T_m of PA4 in the presence of enzyme. After immobilization by adsorption in the PA4@iL series, the T_m of the three resulting conjugates goes over 260 °C irrespective of the support type. This is definitely higher than the T_m of the three empty supports, being in the range of 235–247 °C. As for the laccase-entrapped samples, the two of them i.e., PA4-eL and PA4-Fe-eL melt at slightly higher or similar T_m, whereas the PA4-Fe$_3$O$_4$-eL sample is the only one with a T_m with 11 °C lower than the respective empty support. The fact that both TGA and DSC display similar degradation behavior and melting temperatures for the empty supports and the laccase-entrapped samples means that the molecular weight of PA4 in these two sets should be similar and relatively high.

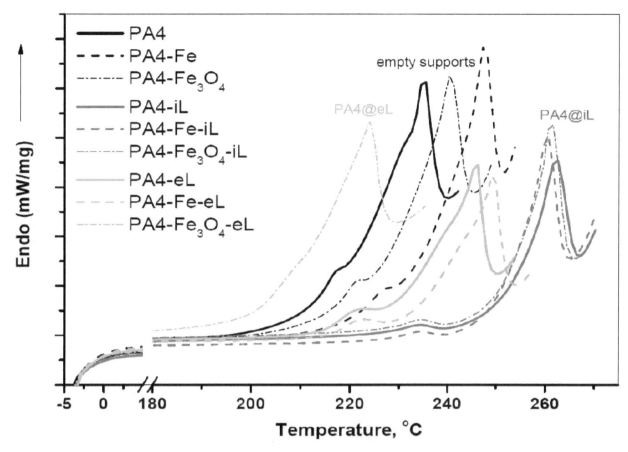

Figure 6. DSC (first scan, heating 10 deg/min) of empty PA4 supports, laccase adsorbed (PA4@iL) and laccase-entrapped (PA4@eL) conjugates.

2.3.3. Synchrotron WAXS

The activity of laccase physically adsorbed onto or entrapped into PA4 MP and the efficiency of the immobilization process in each case will directly depend on the nature and strength of the interactions at the enzyme/PA4 support interface. The postulated intense H-bond formation between the laccase and its structural analogue PA4 should have some influence on the enzyme configuration, or the crystalline structure of the polymeric support that can be probed by X-ray scattering techniques. It is important to know also whether or not the interior pores and channels of the PA4 microparticles that are impossible to access for direct SEM observation are filled with enzyme. In an attempt to evaluate these factors in the empty PA4 supports and in the respective adsorbed or entrapped PA4-laccase conjugates, synchrotron WAXS and SAXS were employed. Moreover, no structure studies about the PA4 crystalline structure by synchrotron X-ray were reported up to now.

Figure 7 shows a comparison between the linear WAXS patterns of samples representing the three empty supports, as well as the three PA4@iL immobilized and the three PA4@eL entrapped samples. All WAXS patterns display two strong reflections at $q \approx 14.5$ nm^{-1} and 17.0 nm^{-1} that, according to Bellinger et al. [38], should be ascribed to the monoclinic unit cell of the α-PA4 with $d_{\alpha[200]} = 4.33$ Å and $d_{\alpha[020]} = 3.69$ Å. Moreover, in accordance with the same study, it should be postulated that the PA4 chains are parallel to the lamellar normal and that an amide group is incorporated in the fold. Such incorporation does not take place in PA6 or PA66, however it is very typical for the β-bends in proteins [39]. Consequently, PA4 is really a closer structural analogue to all protein-containing biomolecules than other polyamides.

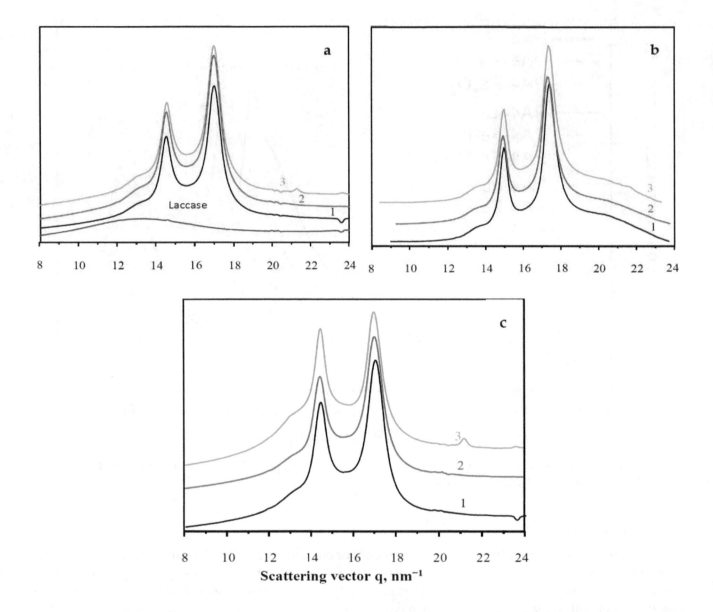

Figure 7. Background corrected linear WAXS patters of: **(a)** neat PA4 supports; **(b)** PA4@iL; **(c)** PA4@eL conjugates. 1—PA4; 2—PA4-Fe: 3—PA4 Fe_3O_4 supports. The curves are shifted along the vertical axis for better visibility.

The visual inspection of the WAXS patterns of the three PA4 supports in Figure 7a suggest that the presence of Fe or Fe_3O_4 fillers does not seem to change the PA4 crystalline structure leaving the angular position and the intensities of the two α-PA4 reflections unaffected. Figure 7a displays also the pattern of a free laccase producing a wide diffuse scattering peak (halo) typical of amorphous materials and centered at $q_a^{free} = 13$ nm^{-1}, i.e., the free laccase is amorphous at the length scale of various angstroms that is probed by WAXS.

The patterns of PA4@iL samples (Figure 7b) represent a superposition of the laccase and PA4 support scatterings. The two α-PA4 reflections apparently maintain their form and position, but the amorphous laccase halo appears centered at $q_a^{ads} \approx 20$ nm^{-1}. It is attributable to the adsorbed bulk laccase deposited on the surface and within the pores of the PA4 particulate support. The shift $\Delta q \approx 7$ nm^{-1} is significant and can be explained as follows. Since the amorphous halo in WAXS is related to intermolecular interactions [40], its position must be dependent on the degree of packing of the molecules (i.e., the density) of the respective amorphous phase. As pointed out by Alexander [41], the dependence of the amorphous halo angular position q_a on the intermolecular distance r_a is given as

$$r_a \sim \frac{\lambda}{2q_a}, \text{ nm} \tag{1}$$

which is the reciprocal dependence typical of all diffraction phenomena. Thus, the larger the scattering vector q_a, value is, the smaller the intermolecular distance and consequently the higher the density of this phase will be. This means that the adsorbed enzyme in the PA4@iL samples has a denser packing as compared to the free one, confirming the supposition for intensive interaction between the laccase and PA4 via multiple H-bonds.

The patterns of the entrapped PA4@eL samples (Figure 7c) do not show any difference in comparison to the empty PA4 supports in Figure 7a. This observation leads to two conclusions: (i) the enzyme arrested in the PA4 particles during AAROP cannot form a separate amorphous reflection and (ii) the entrapped enzyme does not upset the crystallization of the α-PA4 polymorph. It can be therefore hypothesized that, in the PA4@eL series, the enzyme macromolecules are distributed within the amorphous phase of the semi-crystalline PA4 support quite homogeneously with no significant interaction between one another.

Further information about the crystalline structure of the samples can be extracted after deconvolution of the WAXS patterns in Figure 8 by peak fitting. This procedure and the subsequent quantification of the α- and β-PA4 crystalline phases is made according to earlier publications, resolving the crystalline parameters and polymorph structure of PA4 [38] and PA6 [42]. All structural information from the fitted WAXS patterns is presented in Table 6.

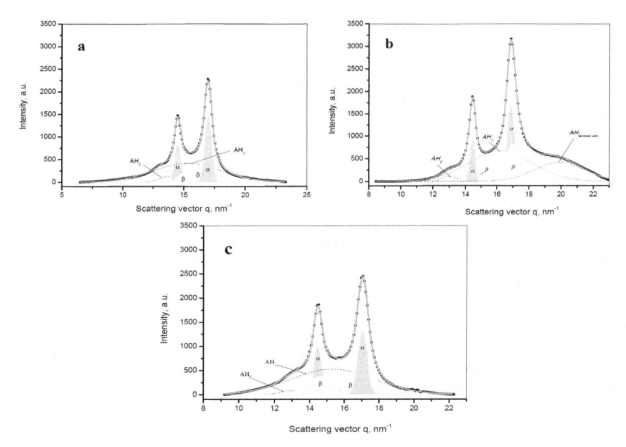

Figure 8. Examples for WAXS pattern deconvolution by peak fitting: (**a**) PA4 support; (**b**) PA4-iL; (**c**) PA4-eL.

Table 6. WAXS analysis of empty PA4 supports and PA4-laccase conjugates.

Sample	α, %	β, %	X_c^{WAXS}%	α/β	Δq_a, nm^{-1}	d$_{\alpha(200)}$ Å	d$_{\alpha(020)}$ Å	d$_{\beta(002)}$ Å	d$_{\beta(200)}$ Å
PA4	22.85	32.56	55.41	0.70	-	4.33	3.70	4.30	3.72
PA4-Fe	21.44	30.26	51.70	0.71	-	4.33	3.70	4.24	3.72
PA4-Fe$_3$O$_4$	23.71	28.12	51.83	0.84	-	4.33	3.70	4.27	3.73
PA4-iL	17.14 27.39	19.13 30.58	36.27 57.97	0.90 0.90	6.84	4.35	3.72	4.36	3.69
PA4-Fe-iL	18.43 21.84	20.96 24.84	39.39 46.68	0.88 0.88	7.21	4.36	3.73	4.34	3.71
PA4-Fe$_3$O$_4$-iL	17.09 20.77	20.80 25.28	37.90 46.06	0.82 0.82	6.50	4.34	3.72	4.35	3.69
PA4-eL	21.92	28.43	50.35	0.77	-	4.34	3.69	4.30	3.69
PA4-Fe-eL	21.54	26.62	48.16	0.81	-	4.36	3.69	4.15	3.71
PA4-Fe$_3$O$_4$-eL	21.91	30.60	45.25	0.72	-	4.36	3.68	4.30	3.70

Notes: For the d-spacings indexation presented the chain axis coincides with the b-axis [38]. The bolded values for the PA4@iL samples are determined excluding the adsorbed laccase amorphous reflection; $\Delta q_a = q_a^{ads} - q_a^{free}$; X_c^{WAXS} = WAXS crystallinity index. For more information, see the text.

All example deconvolutions in Figure 8 show that excellent fits with regression coefficients $R^2 \geq 0.99$ were only possible if along with the two peaks of $\alpha-$PA4 and the two amorphous halos AH_1 and AH_2 two more crystalline peaks were considered that should be assigned to an additional monoclinic phase designated by β-PA4. It was first described in the early work of Frederiks et al. [43] and later on shown to co-exist with α-PA4 [24].

Table 6 shows that the values of the long spacings of these two PA4 phases almost coincide. Thus, the reflections related to d$_{\beta(002)}$ and d$_{\beta(200)}$ are quite away from each other, which is typical of a monoclinic unit cell, contrary to the two γ-PA6 peaks that are often described as belonging to a pseudo-hexagonal unit cell [42,44].

Table 6 displays the structural data of all studied samples.

The α/β ratio in the empty PA4 supports and in the laccase-entrapped samples PA4@eL (Table 6) seems to be relatively constant varying between 0.70–0.85. The α/β ratio rises to ca. 0.90 upon physical adsorption of laccase in the PA4@iL samples, meaning that the PA4-enzyme interaction via hydrogen bonds may cause some slight $\beta-\alpha$ transition.

According to Table 6, the WAXS crystallinity indices X_c^{WAXS} of the empty PA4 supports vary in the narrow range of 52–55%, i.e., values relatively high for polyamides. Entrapping laccase by in situ AAROP drops the crystallinity with about 5%, being within the margin of the experimental error of the deconvolution method. More fluctuations in X_c^{WAXS} appear in the laccase-adsorbed samples. We are inclined to explain this with an increased error in the deconvolution due to interactions between the laccase halo $AH_{laccase}$ and AH_2 diffuse peak related to the PA4 own amorphous fraction, rather than to alteration of the crystallinity index.

As regards the Δq_a values displayed in Table 6 for all samples with laccase adsorption, it should be noted that this difference is larger with the PA4-Fe-iL, followed by the PA4-iL and the PA4-Fe$_3$O$_4$-iL samples. In accordance with Equation (1), in this sequence the density of the adsorbed bulk laccase is expected to decrease as a function of the polymer support composition.

2.3.4. Synchrotron SAXS

The use of synchrotron SAXS allows further clarification of the structure of the PA4 MP before and after laccase immobilization or entrapment. This method probes density periodicities with dimensions in the 20–250 angstroms range, which includes the sizes of the crystalline lamellae typically found in semi-crystalline polymers.

Figure 9 presents the background-subtracted and Lorentz-corrected SAXS linear profiles of the three empty PA4 supports (neat PA4 MP and such containing Fe or Fe$_3$O$_4$ fillers) and the respective laccase adsorbed (PA4@iL) and entrapped (PA4@eL) samples. To enable comparison, Figure 9b also contains the SAXS curve of the free laccase.

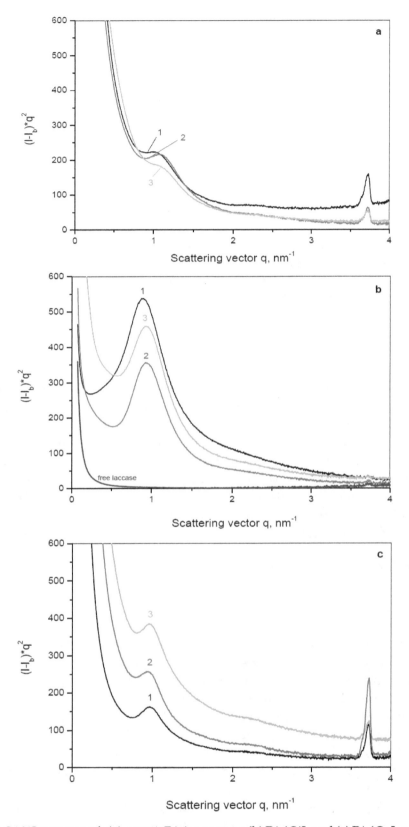

Figure 9. Linear SAXS patterns of: (**a**) empty PA4 supports; (**b**) PA4@iL and (**c**) PA4@eL samples. 1—PA4; 2—PA4-FE; 3—PA4-Fe$_3$O$_4$ supports. The laccase pattern is presented in Figure 9b for comparison.

It can be seen that the three empty PA4 supports (Figure 9a) show relatively well resolved Bragg peaks, indicating lamellar stack morphology. The neat PA4 displays a long spacing value L_B = 62 Å, i.e., close to the 63–66 Å established previously in PA4 bulk samples [38]. The presence of Fe or

Fe_3O_4 slightly reduces L_B to 57 Å. Having in mind that the total WAXS crystallinity index of the three empty supports is above 50% (Table 6), it should be concluded that it is the crystalline phase which is predominant in the lamellae. All the three PA4 empty supports display well-expressed and narrow SAXS peaks at $q_{max} \approx 3.75$ nm^{-1} corresponding to $L_B = 1.5$–1.7 nm that should be related to nanoporosity in the PA4 microparticles not observable directly by SEM but undoubtedly established by BET (Table S1). It is important to note that, as expected, this nanoporosity-related SAXS peak disappears if the empty support particles undergo melting and recrystallization (Figure S3 of the Supporting Information, the curves at 270 °C and at 30 °C after heating up to 270 °C).

As seen from Figure 9b, the free laccase does not show any periodicity in the SAXS q-range, which agrees with the WAXS results. Nevertheless, the laccase-adsorbed PA4@iL samples display very well resolved SAXS peaks with $L_B = 67$–70 Å, suggesting a much better phase contrast between the densities of the amorphous and crystalline regions, as compared to the respective neat PA4 supports (Figure 9a).

It should be noted that no significant changes in the crystalline structure of the PA4 supports could possibly occur during the physical adsorption of laccase. Therefore, the SAXS curves in Figure 9b allow the conclusion that after laccase immobilization most of the pores and channels of the empty supports get filled with enzyme whose density should be comparable to that of the amorphous PA4. This creates a clearer density gradient between the amorphous and crystalline fractions of the lamellar periodicity resulting in better resolved SAXS peaks of the PA4@iL samples. Notably, the narrow SAXS peaks at high q-values of the empty supports disappear completely after laccase adsorption, confirming its relation to the PA4 nanoporosity.

Considering the SAXS patterns in Figure 9c with their clear peaks with $L_B = 65$–67Å leads to the conclusion that a lamellar stack system similar to that of the empty PA4 supports is built after AAROP of 2PD in the presence of laccase that produces the PA4@eL samples. Let us note that the SAXS peaks with $q_{ma} \approx 3.75$ nm^{-1} remain present in all samples with enzyme entrapment, being the most intense in the PA4-Fe-eL sample. This finding can be logically explained if the enzyme is arrested within the amorphous phase of the PA4 support, leaving its nanoporosity unobstructed. At the same time, in the laccase adsorbed samples, all enzyme seems to be located within the cavities of the particulate support, completely obstructing its nanoporosity. This is an important structural conclusion that will be used in the explanation of the enzyme activities of the two types PA4-laccase conjugates.

2.4. Specific Activity of Laccase Immobilized by Adsorption and Entrapment

The catalytic activity of all PA4-laccase conjugates was studied using ABTS as a substrate and calculating its rate of oxidation in each case. As shown in preliminary studies, the free laccase displays highest activity at pH 5, so this condition was assumed for all activity tests (0.0125 M PB, 25 °C). Table 7 contains data on the total, specific, and relative activities of each PA4@iL and PA4@eL samples, taking the free laccase activity as 100%. As expected, the laccases immobilized by either adsorption or entrapment, was less active compared to the free laccase, the reduction of specific activity being different for the two immobilization strategies. The laccase-adsorbed samples (PA4@iL) were about 1.7–2.6 times less active, whereas the entrapped laccases displayed 2.4–6.2 times lower specific activity, as compared to the free laccase. Notably, the surface-immobilized laccase samples were from 1.4 to 3 times more active than the corresponding entrapped counterparts.

These results were expected having in mind the SEM and SAXS studies of the PA4@iL samples. After the adsorption, there is a large amount of 'exposed' laccase covering the surface of the PA4@iL and entering into their pores/cavities located near the surface. As seen directly from SEM and indirectly from the SAXS data, the topography of the entrapped samples (PA4@eL) is completely different. During the AAROP the laccase macromolecules are covered by a PA4 shell so there should be much less (or no) 'exposed' laccase in these samples. The porous shell makes more difficult the access of the ABTS substrate molecules to the active site of the entrapped enzyme. Moreover, the resulting oxidized cation-radical $ABTS^{+\cdot}$ should return to the aqueous medium where it is quantified by UV–VIS, which is also hampered by the PA4 MP shell.

Table 7. Specific activity of laccase immobilized by adsorption or entrapment.

Sample	Laccase Activity, $\mu kat \cdot mL^{-1}$	Laccase Activity, $\mu kat \cdot mL^{-1} \cdot g^{-1}$ Support	Specific Laccase Activity, $\mu kat \cdot mL^{-1} \cdot mg^{-1}$ Laccase	Relative Laccase Activity, %
Free laccase	0.1472	-	0.1472	100
PA4-i*L*	0.0508	2.5417	0.0804	54.60
PA4-Fe-i*L*	0.0714	3.5694	0.0886	60.18
PA4-Fe$_3$O$_4$-i*L*	0.0303	1.5139	0.0565	38.36
PA4-e*L*	0.0085	0.4246	0.0238	16.15
PA4-Fe-e*L*	0.0203	1.0139	0.0606	41.19
PA4-Fe$_3$O$_4$-e*L*	0.0075	0.3750	0.0278	18.91

Note: The substrate is 0.1 mL ABTS (5 mM in DDW). For more details, see the Materials and Methods, Section 3.3.4.

As seen from Table 7, the conjugate with the highest laccase activity is the PA4-Fe-iL sample, in which the laccase immobilization effectiveness is the highest (ca. 81%, Table 4). Logically, the enzyme densification measured by WAXS data (Table 6, Δq values) is the largest in this sample. Among the entrapped samples, it is also the Fe-containing PA4-Fe-eL that displays the highest activity. These data suggest some synergism between the laccase activity and the presence of Fe^0 in the PA4 support.

2.5. Laccase Retention Studies

Since both of our immobilization strategies count on H-bond formation between the PA4 support and laccase, leaching of enzyme will be an inevitable feature of PA4@iL and PA4@eL samples, as it is in all conjugates with no covalent bonding [33]. Figure 10 displays the laccase retention in each conjugate type as a function of the support composition, in five consecutive application cycles.

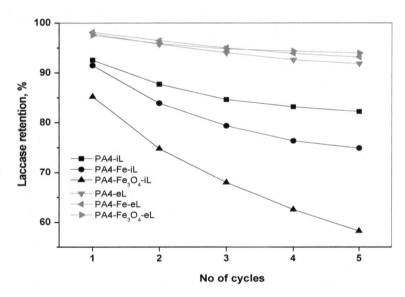

Figure 10. Laccase retention as a function of the immobilization strategy and cycles of application. Conditions: 50 mg of the respective PA4@iL or PA4@eL sample were dispersed in 5 mL PB, pH 0.0125M, pH 5, and shaken at 37 °C for 15 min. For more details, see the Experimental Part Section 3.3.4.

The amount of laccase leached was below the detection limit of the spectroscopy method measuring the absorbance at $\lambda = 286$ nm. Therefore, the supernatants after every application cycle were subjected to activity testing, in which the higher the rate of laccase-catalyzed oxidation of ABTS substrate, the higher the percentage of leached laccase (Section 2.3.4). The enzyme content in the starting conjugates (Tables 4 and 5) was considered 100%.

Figure 10 shows that the leaching from the PA4@eL samples is quite low and even after the fifth cycle the enzyme retention is close to 95%, not depending on the PA4 support. This can be explained with the steric hindrance of the enzyme entrapped deep in the PA4 shell. The laccase release from the PA4@iL samples is notably higher, whereby the strongest drop is observed during cycles 1–3. Apparently, the laccase that is closer to the surface leaches first at higher rates, and after that a stabilization is attained. Moreover, the leaching rate in this series clearly depends on the PA4 support composition. The leaching is the strongest in the $PA4-Fe_3O_4-iL$ sample ranging from 15% (cycle 1) to a total 42% (after cycle 5). The other two samples of this series display leaching percentages of 15–22% after the last cycle. In summary, the entrapment strategy affords better enzyme retention than immobilization by absorption.

2.6. Decolorization of Dyestuff Employing PA4-Laccase Conjugates

Malachite green (MG) is a cationic (i.e., positively charged) dyestuff widely used in the pigment industry and in agriculture. Bromophenol blue (BPB) is negatively charged with major applications as industrial or laboratory dyestuff and color marker. Both dyes contain polyaromatic hydrocarbon moieties (Figure S4 of Supporting Information) that have carcinogenic and mutagenic health effects. MG and BPB are frequently found in effluents from the textile industry and agriculture, so their neutralization is an important problem. As was proven in a previous report [45], MG and BPB can be successfully degraded by laccase oxidation. Therefore, the two dyes were selected in this study as substrates for decolorization with both PA4@iL and PA4@eL samples, produced by two different immobilization strategies. In both cases, the oxidizing activity of the conjugates was evaluated without any mediator.

Figure 11 illustrates the decolorization kinetics of MG and BPB by the PA4@iL and PA4@eL samples. For comparison, the action of the free laccase was also studied at the same conditions. All PA4-laccase conjugates showed 90–95% effectiveness of the MG decolorization after 15 min only, irrespective of the immobilization strategy, whereas about 300 min were necessary for the free laccase to reach similar values (Figure 11a,b). These yields are worth comparing to the data of Bagewadi et al. [46] who, in a similar experiment, reported 95–100% MG decolorization, but after 960 min using laccase from *Trichoderma harzianum* immobilized in a sol–gel matrix and 1-hydroxybenzotriazole (HBT) mediator.

The decolorization of BPB occurred quite differently with a clear dependence on the way of enzyme incorporation into the PA4 support microparticles. The adsorption-immobilized PA4@iL samples were less active than the free laccase, showing a lower decolorization effectiveness within the whole 24 h period that peaked only around 25–50% (Figure 11c,d). At the same time, the two laccase-entrapped samples PA4-eL and PA4-Fe-eL displayed values of 70–80% in 120 min, the free laccase value being 45% at this time. A final decolorization of 80–90% was achieved after 24 h against 100% of the free enzyme. The $PA4-Fe_3O_4-eL$ sample maintains an effectiveness of ca. 50% after 150 min that almost coincides with that of the free laccase. From this point on, the decolorization rate of the free enzyme continues to increase linearly, whereas that of the Fe_3O_4-containing conjugate saturates. Our results on the BPB decolorization can be favorably compared to the data of Forootanfar et al. [47] who performed decolorization studies of BPB with free laccase from *Trametes versicolor* and reached effectiveness values of 20% after 180 min without mediator and 32% with HBT mediator

Comparing the rates of decolorization of the two dyes by free laccase (Figure 11) suggests that the MG substrate is more susceptible to oxidative degradation than BPB. This experimental fact can be related to their different chemical structure. Thus, BPB contains bulky and heavy groups (four Br atoms and one SO_3 group) and its molecular weight is almost twice as high as that of MG, making BPB more difficult to degrade [46].

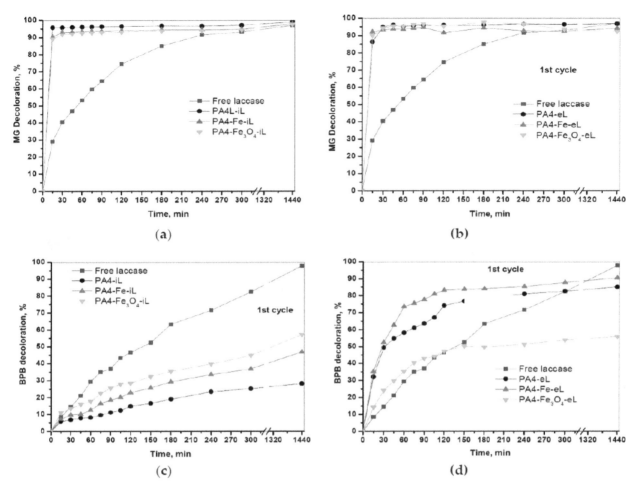

Figure 11. Decolorization of: (**a,b**)—malachite green (MG) and (**c,d**)—bromophenol blue (BPB) by PA4@iL or PA4@eL samples. The decolorization percentage was calculated according to Equation (5) (see Materials and Methods, Section 3.3.5).

When immobilized enzymes are used, theoretically, the disappearance of the color of the initial dye solution can be caused both by enzymatic action and by adsorption in the support, whereby the contribution of the latter could be quite significant and must be taken into account [48]. To assess the contribution of the physical adsorption in the above decolorization experiments, the empty PA4 supports were incubated in DDW MG and BPB solutions. Further treatment was exactly the same as indicated in Section 2.4. The results for MG and BPB are presented in Figure 12.

In the case of MG (Figure 12a), a ramp to 95% decolorization effectiveness was achieved within 15 min time with all three empty supports. This effect can only be due to the extremely high adsorption capacity of the PA4 microparticles (Z-potential of −36 eV at pH7) toward the MG cation. Clearly, this vigorous adsorption process is much faster than the enzymatic oxidative degradation of MG by free laccase. In the case of BPB, however (Figure 12b), the PA4 particles of the support adsorb only 2–3% dye (PA4 MP) or ca. 15% (Fe and Fe_3O_4-containing PA4 MP), these amounts being constant after 30 min exposure. The lower adsorption of BPB should be attributed to the electrostatic repulsion between the negatively charged dye molecules and PA4 microparticles. Evidently, the presence of Fe^0 or $Fe^{2+,3+}$ in the PA4 support slightly enhances the BPB removal by physical adsorption.

The results shown in Figure 12 help assess the contribution of the adsorption by the supports in the experiments in Figure 11. The instant decolorization of the MG substrate by PA4@iL or PA4@eL conjugates in Figure 11a,b should be explained by strong adsorption in the support particles, which influences the MG removal more than the oxidative degradation by laccase. In the case of BPB, the results in Figure 11c,d could be corrected by subtraction of the adsorption-caused dye removal thus allowing the evaluation of the sole enzymatic decolorization. The corrected data is presented in Figure 13.

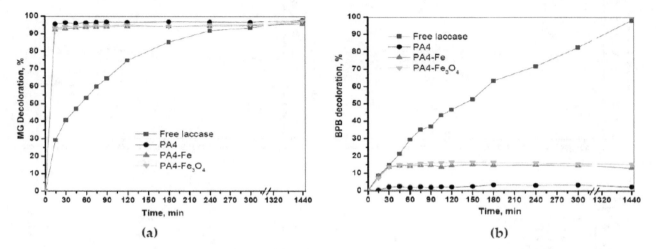

Figure 12. Decolorization of MG (**a**) and BPB (**b**) solutions by empty PA4 MP supports due to physical adsorption. The decolorization kinetics by free laccase are also given for comparison. The decolorization percentage was calculated according to Equation (5) (see Materials and Methods, Section 3.3.5). The structures of MG and BPB are presented in Figure S4 of the Supporting Information.

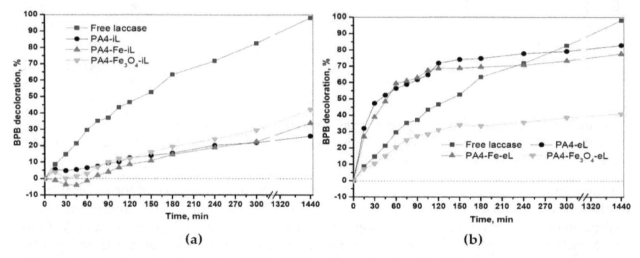

Figure 13. Resultant decolorization (RD) of BPB solution by: (**a**) PA4@iL samples and (**b**) PA4@eL samples. RD is obtained by subtracting the adsorption-caused decolorization (Figure 13) from the total decolorization (Figure 12).

The PA4@iL samples display slower decolorization kinetics than the free laccase in the whole 24 h interval studied. The Fe^0 and $Fe^{2+,3+}$ containing samples of this series display negative values of ΔA in the beginning of the experiment due to predominant adsorption-caused dye removal. As to the laccase entrapped samples, the PA4-eL and the PA4-Fe-eL samples show excellent decolorization kinetics, being faster than that of the free laccase during the first 3 h of the experiment. The maximum decolorization effectiveness of BPB due only to the enzyme action was between 65–75%. The $Fe^{2+,3+}$-containing support displayed a clear negative deviation from this behavior. Since the specific and relative laccase activities in the PA4-Fe$_3$O$_4$-eL sample are even slightly higher than those of the PA4-eL sample, the significantly lower decolorization kinetics in the former case cannot be attributed to lesser amount or lesser starting activity of the enzyme. Most probably, some laccase inactivation occurs by BPB functional groups during the decolorization process.

On the other hand, as seen from Table 7, the samples with entrapped laccase displayed lower activity toward the ABTS substrate, as compared to both free and PA4-immobilized laccases. The results in Figure 12 can be explained with some deactivation of the free laccase and that in the iL samples, while the catalytic activity of the eL samples remained constant. It should be noted that all samples were stored at exactly the same conditions and for the same time duration.

The potential of practical application of the PA4@iL and PA4@eL conjugates will be directly related to the possibility to remove them rapidly and completely from the reaction mixture and use them in several consecutive cycles. Figure S5 in the Supporting Information visualizes the removal of the PA4-Fe-eL sample from its suspension in DDW by means of a constant magnet. Figure 14 displays the second decolorization cycle of MG and BPB dyes by PA4@iL and PA4@eL conjugates.

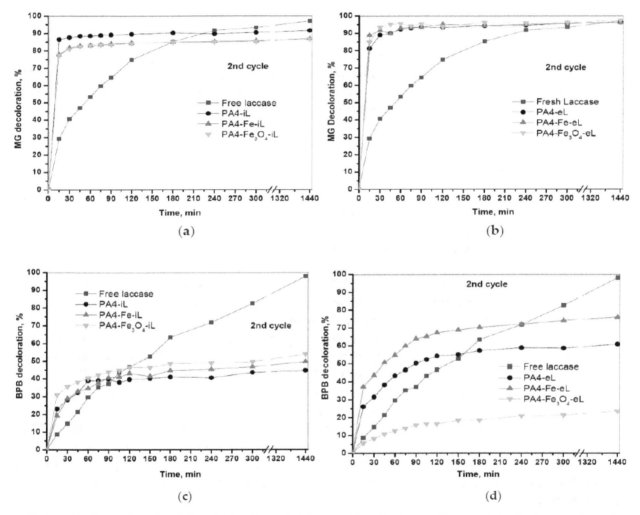

Figure 14. Second cycle of decolorization of: (**a,b**)—malachite green (MG) and (**c,d**)—bromophenol blue (BPB) by PA4@iL and PA4@eL samples. The decolorization percentage was calculated according to Equation (5) (see Materials and Methods, Section 3.3.5).

The comparison of Figures 11 and 14 show that in the case of MG both the PA4@iL and PA4@eL conjugates lose less than 5% of their decolorization capacity during the second cycle, which means they can be reused in many additional decolorizing cycles. However, in this case, the contribution of the enzymatic reaction cannot be evaluated due to the very rapid adsorption process. In regard to the BPB decolorization, where the contribution of the adsorption is negligible, the laccase-entrapped samples lose ~10% of their decolorization effectiveness. Supposing that this rate of loss is maintained in subsequent cycles, at least 3–5 more uses will be possible for PA4@iL and PA4@eL series, respectively, without adding free laccase.

3. Materials and Methods

3.1. Materials

The 2PD functional monomer for the PA4 synthesis and all solvents used in this work are of analytical grade and were supplied by Merck Life Science, Algés, Portugal. The laccase from *Trametes versicolor* ($\geq$0.5 U/mg) was also purchased from the same supplied and used without further purification. As activator of the anionic polymerization of 2PD the commercial product Brüggolen C20 from Brüggemann Chemical GmbH, Heilbronn, Germany was employed containing, according to the manufacturer, 80 wt % of aliphatic diisocyanate blocked in ε-caprolactam. The polymerization initiator sodium dicaprolactamato-bis-(2-methoxyethoxo)-aluminate (Dilactamate, DL), which is also a commercial product was supplied by Katchem, Prague, Czech Republic and used without further treatment. Soft, non-insulated iron particles (Fe content >99.8%), with average diameters of 3–5 μm were kindly donated by the manufacturer BASF, Ludwigshafen, Germany. The Fe_3O_4 magnetic particles are a product of Merck, Algés, Portugal with >99% purity and grain sizes of 50–80 nm. Diammonium 2,2'-azino-bis(3-ethylbenzothiazoline-6-sulfonate) (ABTS) with a purity of $\geq$98% (HPLC), as well as malachite green and bromophenol blue dyes, were purchased from the same supplier. All buffer solutions in this work were prepared with double-distilled water (DDW).

3.2. Instrumentation and Methods

Fourier-transform infra-red spectroscopy with attenuated total reflection (FTIR-ATR): The FTIR-ATR spectra were collected in a Perkin-Elmer Spectrum 100 apparatus (Waltham, MA, USA) using a horizontal ATR attachment with ZnSe crystal. Spectra were acquired between 4000 and 600 cm^{-1} accumulating up to 16 spectra with a resolution of 2 cm^{-1}. The PA4 samples were studied in the form of fine powders.

Ultraviolet–visible spectroscopy (UV–VIS): The UV–VIS analysis was performed on a Shimadzu UV-2501 PC spectrometer (Kyoto, Japan). The absorbance at λ = 286 nm of aqueous solutions placed in quartz cuvettes was measured to determine the protein content and calculate the immobilization efficiency (IE). The assessment of the laccase activity was performed using ABTS as substrate, measuring the absorbance at λ = 414 nm as previously indicated by Claus et al. [19]. One unit of laccase activity expressed in μkatals corresponds to the amount of enzyme transforming 1 μmol ABTS per second at pH 5.0 and 25 °C. The absorbance in decolorization experiments was measured at the respective wavelength of each dye that was λ = 616 nm for malachite green and λ = 591 nm for bromophenol blue.

Scanning electron microscopy (SEM): The SEM studies were performed in a NanoSEM-200 apparatus of FEI Nova (Hillsboro, OR, USA) using mixed secondary electron/back-scattered electron in-lens detection. The pulverulent samples were observed after sputter-coating with Au/Pd alloy in a 208 HR equipment of Cressington Scientific Instruments (Watford, UK) with high-resolution thickness control.

Thermo-gravimetric analysis (TGA): The real iron load, R_L, and the thermal stability of all neat PA4 micro-particulate supports and laccase@PA4 MP samples were established by means of thermo-gravimetric analysis (TGA) in a Q500 gravimetric balance (TA Instruments, New Castle, DE, USA), heating the samples in the 40–600 °C range at 20 °C/min in a nitrogen atmosphere. The R_L was calculated according to Equation (2)

$$R_L = (R_i - R_{PA4}) \times 100, \tag{2}$$

where R_{PA4} is the carbonized residue at 600 °C of the neat PA4 particles and R_i represents the carbonized residue of the respective Fe- or Fe_3O_4—containing PA4 MP or laccase@PA4 MP samples.

Differential scanning calorimetry (DSC): The DSC measurements were carried out in a 200 F3 equipment of Netzsch (Selb, Germany) at a heating/cooling rate of 10 °C/min under nitrogen purge.

The samples were heated to 290 °C, cooled down to 0 °C, and then heated back to 290 °C. The typical sample weights were in the 10–15 mg range.

Synchrotron X-ray studies: Synchrotron wide- (WAXS) and small-angle X-ray scattering (SAXS) measurements were performed in the NCD-SWEET beamline of the ALBA synchrotron facility in Barcelona, Spain [49]. Two-dimensional detectors were used, namely LH255-HS (Rayonix, Evanston, IL, USA) and Pilatus 1M (Dectris, Baden Daettwil, Switzerland) for registering the WAXS and SAXS patterns, respectively. The sample-to-detector distance was set to 111.7 mm for WAXS and 2700 mm for SAXS measurements, the λ of the incident beam being 0.1 nm and the beam size 0.35 × 0.38 mm (h × v). The 2D data were reduced to 1D data using pyFAI software [50]. For processing of the WAXS and SAXS patterns, the commercial package Peakfit 4.12 (2016) by SYSTAT (San Jose, CA, USA) was implemented.

3.3. Sample Preparation and Activity Testing

3.3.1. Synthesis of Empty or Magnetic PA4 MP

The low-temperature AAROP of 2PD to empty PA4 MP or to PA4 MP with magnetic response was described in detail in a previous publication [24]. Generally, 0.2 mol of 2PD were stirred with 1 wt % of Fe micro- or Fe_3O_4 nanoparticles at 25 °C for 30 min. Subsequently, the C20 activator and the DL initiator were added under stirring, in an inert atmosphere. Then, the temperature of the reaction mixture was set to 40 °C and the pressure to 50 mbar for the next 6 h. After AAROP completion, the resulting reaction mixture was dispersed in acetone and filtered, followed by a two-fold wash with methanol, thus removing most of the unreacted monomer. In order to eliminate the PA4 oligomers, the resulting fine powders were extracted with methanol in a Soxhlet for 4 h, dried in vacuum, and stored in a desiccator. For the synthesis of the empty PA4 MP, the procedure was the same but without addition of any magnetic nanoparticles. Altogether, three types of PA4 MP supports were synthesized: PA4, PA4-Fe, and $PA4-Fe_3O_4$. Some basic characteristics of these samples are presented in Table 1.

3.3.2. Immobilization of Laccase by Physical Adsorption

A typical immobilization was carried out by first preparing the laccase solution in DDW with concentration of 2 mg/mL. 200 mg of each PA4 MP sample were introduced into 5 mL of this laccase solution and the three sample tubes were incubated at room temperature for 24 h using a laboratory orbital shaker. Thereafter, the samples were centrifuged, the supernatant was decanted, and the laccase-immobilized PA4 MP samples were washed two times with distilled water to remove the non-adsorbed laccase. UV analysis was performed to determine the residual laccase and calculate the immobilization efficiency, IE, expressed as

$$IE = \frac{C_0 - C_s}{C_0} \times 100, \% \tag{3}$$

where C_0 is the starting laccase concentration and C_s is the laccase content in the resultant supernatant after completion of the immobilization process. Two methods were applied for the determination of C_s. In the first one, the absorbance at $\lambda = 286$ nm characteristic for the proteic part of the enzyme was measured and the laccase concentration was determined using a standard calibration plot. In the second method, the laccase activity toward ABTS was tested and C_s was assessed using an activity standard calibration plot. Altogether, three samples of laccase immobilized by adsorption on PA4 MP were prepared and designated as PA4-iL, PA4-Fe-iL, and $PA4-Fe_3O_4-iL$. They were stored in semi-dried conditions at 4 °C with basic characteristics listed in Table 2.

3.3.3. Immobilization of Laccase by Entrapment

The entrapment of laccase into the PA4 MP was carried out in situ during their synthesis by AAROP. Thus, 1.0 wt % of the enzyme in respect to the monomer was used, stirring their mixture for 30 min in an inert atmosphere at room temperature. Thereafter, the respective magnetic micro/nano particles (if necessary) and the DL/C20 catalytic system were introduced and the AAROP was conducted under the conditions described in Section 2.3.1. After 6 h of polymerization, the raw product was dispersed in acetone and filtered. The resulting laccase-entrapped PA4 MP samples were washed three times with DDW and stored at 4 °C. The water from each washing was stored separately and further subjected to UV analysis to determine the efficiency of the laccase entrapment (EE) using the equation

$$EE = \frac{(C_0 - \sum C_i)}{C_0} \times 100, \%$$ (4)

where C_0 is the starting laccase concentration in the reaction mixture and C_i ($i = 1$–3) is the laccase content in each DDW washing. To assess the content of the non-entrapped laccase, the two methods already described in Section 2.3.2. were applied. Thus, three PA4 MP samples with entrapped laccase were synthesized, namely: PA4-eL, PA4-Fe-eL, and PA4-Fe3O4-eL, with basic characteristics given in Table 3.

3.3.4. Laccase Activity Assay

Native laccase and all conjugates of laccase-immobilized PA4@-iL and entrapped PA4@-eL were assayed for their activities using ABTS as a color-generating substrate. The rate of the formation of ABTS-cation radical (ABTS$^{+\cdot}$) due to the catalytic action of laccase was proportional to the enzyme activity. In a typical assay of free laccase, 0.9 mL of the enzyme solution (1.0 mg/mL in 0.0125 M phosphate buffer, PB, pH 5) were introduced into the spectrometer cell, followed by 0.1 mL of freshly prepared 5 mM solution of ABTS in DDW. During the next 2 min the absorbance at 414 nm ($\varepsilon_{414} = 36{,}000$ M$^{-1}\cdot$cm^{-1}) was measured every 5 s. To determine the activity of the PA4@iL and PA4@eL conjugates, 20 mg of each sample were added to 1.0 mL of 0.0125 M PB, pH 5. The mixtures were stirred for 5 min followed by addition of 0.1 mL of ABTS solution. Then, the samples were centrifuged and 1 mL from the decanted supernatant was subjected to UV analysis to measure the absorbance at 414 nm for 10 min. In order to assay the amount and the activity of the laccase that possibly leached from both adsorption-immobilized and entrapped laccase samples, they were subjected to the following procedure: 50 mg of the respective PA4@iL or PA4@eL sample were dispersed in 5 mL PB, pH 0.0125M, pH 5, and shaken at 37 °C for 15 min. Thereafter, the system was centrifuged and the supernatant decanted and subjected to UV–VIS to determine the laccase concentration based on activity test calibration curve. Then, new portion of fresh PB was added and the above procedure was repeated up to five consecutive cycles.

In all tests the enzyme activities were expressed in microkatals. One microkatal of enzyme activity (μkat) was defined as the amount of enzyme that converted 1.0 μmol of ABTS to ABTS$^{+\cdot}$ per second, at pH 5 and 25 °C. The enzyme activity assay was always performed in duplicate, and the standard deviations in measurements were consistently below 3%. Throughout this study, enzyme activities are expressed in μkat per milliliter (μkat·mL^{-1}).

3.3.5. Application of Free, Adsorbed, and Entrapped Laccase

The enzyme-catalyzed decolorization of two structurally different dyes -malachite green ($\lambda_{max} = 616$ nm) and bromophenol blue ($\lambda_{max} = 591$ nm) were studied in the presence of free, immobilized and entrapped laccases without any mediator. The process was followed using UV–VIS spectroscopy by measuring the decrease in the λ_{max} absorbance of each dye. The reaction mixture for the decolorization contained a final concentration of 0.0153 mg/mL of individual dye in 0.0125 M PB, pH 5, and 1.0 mg free laccase (0.5 U) or 10–30 mg of adsorption immobilized or entrapped PA4@laccase conjugates (~0.5 U) in a total volume of 1.3 mL. All reaction mixtures were incubated at 30 °C for 24 h under shaking. Aliquots of 1 mL were withdrawn from each sample at certain intervals to measure the residual dye concentration. After the UV–VIS measurement the analyte was returned to the reaction mixture. After 24 h the immobilized or entrapped samples were reused for a second decolorization cycle. For that purpose, the liquid phase was drained, the immobilized or entrapped laccase conjugates were washed consecutively with 2×1 mL ethanol, DDW and 0.0125M PB, pH 5 to remove the adsorbed dyes, and then introduced into a fresh dye solution. All UV–VIS measurements were performed twice to determine the standard deviation. The percentage of decolorization ΔA was calculated as

$$\Delta A = \frac{A_i - A_t}{A_i} \times 100, \ \% \tag{5}$$

where, A_i is the initial absorbance of the dye at the respective wavelength, and A_t is the absorbance of the test sample.

4. Conclusions

This is the first study reporting on PA4 porous microparticles as effective supports for enzyme immobilization and describes PA4-laccase conjugates with all aspects of their preparation and characterization. The main objective was to investigate three new micron-sized particulate porous supports based on PA4 obtained by low-temperature AAROP, with and without magnetic susceptibility, establishing whether or not they are suitable for laccase immobilization. The structure and morphology were investigated by microscopic, spectral, thermal, and synchrotron WAXS/SAXS technics.

Two immobilization strategies—i.e., by adsorption or entrapment—were applied. Comparing the adsorption-immobilized (PA4@iL) and entrapped (PA4@eL) samples, an important difference in the synthetic procedure of the PA4@eL should be noted. There, the polymerization was carried out in the presence of the enzyme, whereby the PA4 is formed around the laccase molecules leading to their distribution within the shell of the porous microparticles. In the case of PA4@iL, the enzyme is adsorbed onto the surface and fills the pores of prefabricated PA4 microparticles. The SEM studies confirm these morphological differences. On the other hand, the WAXS/SAXS studies provide evidence for the different way of immobilization of the enzyme in PA4@eL and PA4@iL.

The adsorption strategy resulted in high content of immobilized laccase upon the PA4 supports (27–40 mg/g), the immobilization efficiency being in the range of 54–81%. The entrapment strategy produced 60–70% immobilization efficiency with enzyme content of 14–18 mg/g support. The laccase-adsorbed samples PA4@iL showed up to three times higher specific activity as compared to their entrapped laccase counterparts PA4@eL, whereby the highest specific activity of 0.09 $\mu kat \cdot mL^{-1}$ was registered with PA4-Fe-iL sample. The relative laccase activity of the PA4-based conjugates determined in relation to the free enzyme was between 16–41% for the PA4@eL samples and 38–60% for the PA4@iL series. The PA4@eL samples displayed >93% enzyme retention after five cycles of incubation, whereas for the PA4@iL series this value was ~60%. The potential application of the adsorption- and entrapment-immobilized PA4-laccase conjugates was tested in the enzymatic decolorization of two synthetic dyes. All laccase conjugates displayed excellent decolorization of malachite green dyestuff reaching ~100% in 15 min, which was mostly due to dye adsorption upon the PA4 support. The decolorization after 24 h of the bromophenol reached 55% by PA4@iL and 85% by PA4@eL samples. The reuse of the laccase-PA4 conjugates in a second consecutive decolorization test resulted in only up to 10% decrease in their effectiveness. Supposing that this will be the decrease after each cycle of utilization, it seems that the laccase-PA4 conjugates will be sufficiently active to use for 3–5 more consecutive cycles.

The present study justifies further investigations on PA4 microparticles as effective enzyme supports for biotechnological applications. Compared to the common polyamides as PA6, PA66, and PA12, PA4 has the advantage of biodegradability and possesses a crystalline structure, which is a closer structural analogue of proteic biomolecules. The possibility to apply different immobilization strategies and to use magnetic field as external stimulus could open the way to the use of PA4 microparticulate supports of this study in smart green catalytic systems.

Supplementary Materials:

Figure S1: Chemical reactions occurring during AAROP of 2PD to PA4 microparticles; Figure S2: Histograms of the average size (d_{max}) and roundness (d_{max}/d_{min}) distributions for PA4 MP supports based on optical microscopy; Figure S3: Disappearance of the scattering peak at $q = 3.73$ nm^{-1} in the SAXS patterns after melting and recrystallization of PA4 MP support; Figure S4. Structures of the dyes used in the discoloration studies with PA4@iL and PA4@eL laccase conjugates; Figure S5. PA4-Fe-eL sample fast removal from DDW suspension by means of a constant magnet; Table S1: Data from BET for the empty PA4 supports; Table S2: Z-potential values of empty particulate PA4-based supports; Table S3: Consolidated DSC data from an initial heating scan, subsequent cooling scan, and a second heating scan at 10 deg/min.

Author Contributions: Conceptualization, N.D., Z.D., and I.G.; Data curation, N.D., J.B., and D.S.; Formal analysis, N.D., J.B., and Z.D.; Funding acquisition, Z.D., N.D., and I.G.; Investigation, N.D., J.B., D.S., and M.M.; Methodology, N.D., Z.D., and I.G.; Software, M.M.; Supervision, Z.D. and I.G.; Validation, N.D., M.M., and Z.D.; Writing—original draft preparation, N.D. and Z.D.; Writing—review and editing, I.G. and Z.D. All authors have read and agreed to the published version of the manuscript.

References

1.   Bull, A.T.; Bunch, A.W.; Robinson, G.K. Biocatalysts for clean industrial products and processes. *Curr. Opin. Microbiol.* **1999**, *2*, 246–251. [CrossRef]

2.   Sheldon, R.A.; Rantwijk, F.V. Biocatalysis for sustainable organic synthesis. *Aust. J. Chem.* **2004**, *57*, 281–289. [CrossRef]

3.   Torres-Salas, P.; Monte-Martinez, A.; Cutiño-Avila, B.; Rodriguez-Colinas, B.; Alcalde, M.; Ballesteros, A.O.; Plou, F.J. Immobilized Biocatalysts–novel approaches and tools for binding enzymes to supports. *Adv. Mater.* **2011**, *23*, 5275–5282. [CrossRef]

4.   Ge, J.; Lu, D.; Liu, Z.; Liu, Z. Recent advances in nanostructured biocatalysts. *Biochem. Eng. J.* **2009**, *44*, 53–59. [CrossRef]

5.   Hanefeld, U.; Gardossi, L.; Magner, E. Understanding enzyme immobilization. *Chem. Soc. Rev.* **2009**, *38*, 453–468. [CrossRef] [PubMed]

6.   Hudson, S.; Magner, E.; Cooney, J.; Hodnett, B. Methodology for the immobilization of enzymes onto mesoporous materials. *J. Phys. Chem. B* **2005**, *109*, 19496–19506. [CrossRef]

7.   Dwevedi, A. Basics of enzyme immobilization. In *Enzyme Immobilization*; Springer: Cham, Switzerland, 2016; Chapter 2; pp. 21–44, ISBN 978-3-319-41416-4. [CrossRef]

8.   Sigurdardóttir, S.B.; Lehmann, J.; Ovtar, S.; Grivel, J.C.; Negra, M.; Kaiser, A.; Pinelo, M. Enzyme immobilization on inorganic surfaces for membrane reactor applications: Mass transfer challenges, enzyme leakage and reuse of materials. *Adv. Synth. Catal.* **2018**, *360*, 2578–2607. [CrossRef]

9.   Shakya, A.K.; Nandakumar, K.S. An update on smart biocatalysts for industrial and biomedical applications. *J. R. Soc. Interface* **2018**, *15*, 20180062. [CrossRef] [PubMed]

10.   Gitsov, I.; Hamzik, J.; Ryan, J.; Simonyan, A.; Nakas, J.P.; Omori, S.; Krastanov, A.; Cohen, T.; Tanenbaum, S.W. Enzymatic nanoreactors for environmentally benign biotransformations – 1. Formation and catalytic activity of supramolecular complexes of laccase and linear-dendritic blockcopolymers. *Biomacromolecules* **2008**, *9*, 804–811. [CrossRef]

11.   Scheibel, D.M.; Gitsov, I. Polymer-assisted biocatalysis: Effects of macromolecular architectures on the stability and catalytic activity of immobilized enzymes toward water-soluble and water-insoluble substrates. *ACS Omega* **2018**, *3*, 1700–1709. [CrossRef]

12.   Mate, D.M.; Alcalde, M. Laccase: A multi-purpose biocatalyst at the forefront of biotechnology. *Microb. Biotechnol.* **2017**, *10*, 1457–1467. [CrossRef] [PubMed]

13.   Couto, S.L.; Herrera, J.L. Lacasses in the textile industry. *Biotechnol. Mol. Biol. Rev.* **2006**, *1*, 115–120. Available online: https://academicjournals.org/journal/BMBR/article-full-text-pdf/1C6FD3040214 (accessed on 5 March 2020).

14.   Gitsov, I.; Simonyan, A.; Wang, L.; Krastanov, A.; Tanenbaum, S.W.; Kiemle, D. Polymer-assisted biocatalysis: Unprecedented enzymatic oxidation of fullerene in aqueous medium. *J. Polym. Sci. Part A Polym. Chem.* **2012**, *50*, 119–126. [CrossRef]

15. Gitsov, I.; Wang, L.; Vladimirov, N.; Simonyan, A.; Kiemle, D.J.; Schütz, A. "Green" synthesis of unnatural poly(amino acid)s with zwitterionic character and pH-responsive solution behavior, mediated by linear-dendritic laccase complexes. *Biomacromolecules* **2014** *15*, 4082–4095. [CrossRef] [PubMed]

16. Scheibel, D.M.; Gitsov, I. Unprecedented enzymatic synthesis of perfectly structured alternating copolymers via "green" reaction cocatalyzed by laccase and lipase compartmentalized within supramolecular complexes. *Biomacromolecules* **2019**, *20*, 927–936. [CrossRef] [PubMed]

17. Gitsov, I.; Simonyan, A. "Green" Synthesis of Bisphenol Polymers and Copolymers, Mediated by Supramolecular Complexes of Laccase and Linear-Dendritic Block Copolymers. In *Green Polymer Chemistry*; Cheng, R.A., Gross, H.N., Eds.; American Chemical Society: Washington, DC, USA, 2013.

18. Bilal, M.; Rasheed, T.; Nabeel, F.; Iqbal, H.M.N.; Zhao, Y. Hazardous contaminants in the environment and their laccase-assisted degradation–A review. *J. Environ. Manag.* **2019**, *234*, 253–264. [CrossRef] [PubMed]

19. Claus, H.; Faber, G.; König, H. Redox-mediated decolorization of synthetic dyes by fungal laccases. *Appl. Microbiol. Biotechnol.* **2002**, *59*, 672–678. [CrossRef] [PubMed]

20. Maloney, J.; Dong, C.; Campbell, A.S.; Dinu, C.Z. Emerging enzyme-based technologies for wastewater treatment. In Green Polymer Chemistry: Biobased Materials and Biocatalysis. *ACS Symp. Ser.* **2015**, *1192*, 73–75.

21. Fatarella, E.; Spinelli, D.; Ruzzante, M.; Pogni, R. Nylon 6 film and nanofiber carriers: Preparation and laccase immobilization performance. *J. Mol. Catal. B Enzym.* **2014**, *102*, 41–47. [CrossRef]

22. Jasni, M.J.F.; Sathishkumar, P.; Sornambikai, S.; Yusoff, A.R.M.; Ameen, F.; Buang, N.A.; Kadir, M.R.A.; Yusop, Z. Fabrication, characterization and application of laccase–nylon 6,6/Fe^{3+} composite nanofibrous membrane for 3,30-dimethoxybenzidine detoxification. *Bioprocess Biosyst. Eng.* **2017**, *40*, 191–200. Available online: https://link.springer.com/article/10.1007%2Fs00449-016-1686-6 (accessed on 20 February 2020). [CrossRef]

23. Silva, C.; Silva, C.J.; Zille, A.; Guebitz, G.M.; Cavaco-Paulo, A. Laccase immobilization on enzymatically functionalized polyamide 6,6 fibers. *Enzym. Microb. Technol.* **2007**, *41*, 867–875. [CrossRef]

24. Dencheva, N.; Braz, J.; Nunes, T.G.; Oliveira, F.D.; Denchev, Z. One-pot low temperature synthesis and characterization of hybrid poly(2-pyrrolidone) microparticles suitable for protein immobilization. *Polymer* **2018**, *145*, 402–4015. [CrossRef]

25. Yamano, N.; Nakayama, A.; Kawasaki, N.; Yamamoto, N.; Aiba, S. Mechanism and characterization of polyamide 4 degradation by Pseudomonas sp. *J. Polym. Environ.* **2008**, *16*, 141–146. Available online: https://link.springer.com/article/10.1007/s10924-008-0090-y (accessed on 10 March 2020). [CrossRef]

26. Tachibana, K.; Urano, Y.; Numata, K. Biodegradability of nylon 4 film in a marine environment. *Polym. Degrad. Stabil.* **2013**, *98*, 1847–1851. [CrossRef]

27. Park, S.J.; Kim, E.Y.; Noh, W.; Oh, Y.H.; Kim, H.Y.; Song, B.K.; Cho, K.M.; Hong, S.H.; Lee, S.H.; Jegal, J. Synthesis of nylon 4 from gamma-aminobutyrate (GABA) produced by recombinant Escherichia coli. *Bioproc. Biosyst. Eng.* **2013**, *36*, 885–892. Available online: https://link.springer.com/article/10.1007/s00449-012-0821-2 (accessed on 10 March 2020). [CrossRef] [PubMed]

28. Dencheva, N.V.; Oliveira, F.D.; Braz, J.F.; Denchev, Z. Bovine serum albumin-imprinted magnetic poly(2-pyrrolidone) microparticles for protein recognition. *Eur. Polym. J.* **2020**, *122*, 109375. [CrossRef]

29. Kim, N.; Kim, J.H.; Nam, S.W.; Jeon, B.S.; Kim, Y.J. Preparation of nylon 4 microspheres via heterogeneous polymerization of 2-pyrrolidone in a paraffin oil continuous phase. *J. Ind. Eng. Chem.* **2015**, *28*, 236–240. [CrossRef]

30. Costa, J.; Lima, M.J.; Sampaio, M.J.; Neves, M.C.; Faria, J.L.; Morales-Torres, S.; Tavares, A.P.M.; Silva, C.G. Enhanced biocatalytic sustainability of laccase by immobilization on functionalized carbon nanotubes/polysulfone membranes. *Chem. Eng. J.* **2019**, *355*, 974–985. [CrossRef]

31. Liers, C.; Ullrich, R.; Pecyna, M.; Schlosser, D.; Hofrichter, M. Production, purification and partial enzymatic and molecular characterization of a laccase from the wood-rotting ascomycete Xylaria polymorpha. *Enz. Microb. Technol.* **2007**, *41*, 785–793. [CrossRef]

32. Atalla, M.; Zeinab, H.; Eman, R.; Amani, A.; Abeer, A. Characterization and kinetic properties of the purified Trematosphaeria mangrovei laccase enzyme. *Saudi J. Biol. Sci.* **2013**, *20*, 373–381. [CrossRef]

33. Qiu, H.; Xu, C.; Huang, X.; Ding, Y.; Qu, Y.; Gao, P. Immobilization of laccase on nanoporous gold: Comparative studies on the immobilization strategies and the particle size effects. *J. Phys. Chem. C* **2009**, *113*, 2521–2525. [CrossRef]

34. Piontek, K.; Antorini, M.; Choinowski, T. Crystal structure of a laccase from Trametes versicolor at 1.90 Å resolution containing a full complement of coppers. *J. Biol. Chem.* **2002**, *277*, 37663–37669. [CrossRef] [PubMed]

35. Fernández-Fernández, M.; Sanromán, M.Á.; Moldes, D. Recent developments and applications of immobilized laccase. *Biotechnol. Adv.* **2013**, *31*, 1808–1825. [CrossRef] [PubMed]

36. Kawasaki, N.; Nakayama, A.; Yamano, N.; Takeda, S.; Kawata, Y.; Yamamoto, N.; Aiba, S. Synthesis, thermal and mechanical properties and biodegradation of branched PA4. *Polymer* **2005**, *46*, 9987–9993. [CrossRef]

37. Tachibana, K.; Hashimoto, K.; Tansho, N.; Okawa, H. Chemical modification of chain end in nylon 4 and improvement of its thermal stability. *J. Polym. Sci. Part A Polym. Chem.* **2011**, *49*, 2495–2503. [CrossRef]

38. Bellinger, M.A.; Waddon, A.J.; Atkins, E.D.T.; MacKnight, W.J. Structure and morphology of nylon 4 chain-folded lamellar crystals. *Macromolecules* **1994**, *27*, 2130–2135. [CrossRef]

39. Schulz, G.E.; Schirmer, R.H. *Principles of Protein Structure*; Springer Verlag: New York, NY, USA, 1987. [CrossRef]

40. Bartczak, Z.; Galeski, A.; Argon, A.S.; Cohen, R.E. On the plastic deformation of the amorphous component in semicrystalline polymers. *Polymer* **1996**, *37*, 2113–2123. [CrossRef]

41. Alexander, L.E. *X-ray Diffraction Methods in Polymer Science*; Wiley-Interscience: New York, NY, USA, 1969.

42. Dencheva, N.; Nunes, T.; Oliveira, M.J.; Denchev, Z. Microfibrillar composites based on polyamide/polyethylene blends. 1. Structure investigations in oriented and isotropic PA6. *Polymer* **2005**, *46*, 887–901. [CrossRef]

43. Fredericks, J.; Doyne, T.H.; Spague, R.S. Crystallographic studies of nylon 4. II. On the β and δ polymorphs of Nylon 4. *J. Polym. Sci. Polym. Phys.* **1966**, *4*, 913–922. [CrossRef]

44. Fornes, T.D.; Paul, D.R. Crystallization behavior of Nylon nanocomposites. *Polymer* **2003**, *44*, 3945–3961. [CrossRef]

45. Teerapatsakul, C.; Parra, C.R.; Keshavarz, T.; Chitradon, L. Repeated batch for dye degradation in an airlift bioreactor by laccase entrapped in copper alginate. *Int. Biodeterior. Biodegrad.* **2017**, *120*, 52–57. [CrossRef]

46. Bagewadi, Z.K.; Mulla, S.I.; Ninnekar, H.Z. Purification and immobilization of laccase from Trichoderma harzianum strain HZN10 and its application in dye decolorization. *J. Genet. Eng. Biotechnol.* **2017**, *15*, 139–150. [CrossRef] [PubMed]

47. Forootanfar, H.; Moezzi, A.; Khozani, M.; Mahmoudjanlou, Y.; Ameri, A.; Niknejad, F.; Faramarzi, M.A. Synthetic dye decolorization by three sources of fungal laccase. *Iran. J. Environ. Health Sci. Eng.* **2012**, *9*, 27–37. Available online: https://link.springer.com/article/10.1186%2F1735-2746-9-27 (accessed on 18 January 2020). [CrossRef]

48. Peralta-Zamora, P.; Pereira, C.M.; Tiburtius, E.; Moraes, S.G.; Rosa, M.A.; Minussi, R.C.; Durán, N. Decolorization of reactive dyes by immobilized laccase. *Appl. Catal. B Environ.* **2003**, *42*, 131–144. [CrossRef]

49. González, J.B.; González, N.; Colldelram, C.; Ribó, L.; Fonserè, A.; Manas, G.J.; Villanueva, J.; Llonch, M.; Peña, G.; Gevorgyan, A.; et al. NCD-SWEET Beamline Upgrade. In Proceedings of the 10th Mechanical Engineering Design Synchrotron Radiation Equipment Instruments, Paris, Franch, 25–29 June 2018; pp. 374–376. [CrossRef]

50. Ashiotis, G.; Deschildre, A.; Nawaz, Z.; Wright, J.P.; Karkoulis, D.; Picca, F.E.; Kieffer, J. The fast azimuthal integration Python library: pyFAI. *J. Appl. Crystallogr.* **2015**, *48*, 510–519. [CrossRef] [PubMed]

5

Novel Routes in Transformation of Lignocellulosic Biomass to Furan Platform Chemicals: From Pretreatment to Enzyme Catalysis

Grigorios Dedes [†], Anthi Karnaouri [†]◉ and Evangelos Topakas [*]◉

Industrial Biotechnology & Biocatalysis Group, School of Chemical Engineering, National Technical University of Athens, 9 Iroon Polytechniou Str., Zografou Campus, 15780 Athens, Greece; gdedes@chemeng.ntua.gr (G.D.); akarnaouri@chemeng.ntua.gr (A.K.)
* Correspondence: vtopakas@chemeng.ntua.gr
† These authors equally contributed to this work.

Abstract: The constant depletion of fossil fuels along with the increasing need for novel materials, necessitate the development of alternative routes for polymer synthesis. Lignocellulosic biomass, the most abundant carbon source on the planet, can serve as a renewable starting material for the design of environmentally-friendly processes for the synthesis of polyesters, polyamides and other polymers with significant value. The present review provides an overview of the main processes that have been reported throughout the literature for the production of bio-based monomers from lignocellulose, focusing on physicochemical procedures and biocatalysis. An extensive description of all different stages for the production of furans is presented, starting from physicochemical pretreatment of biomass and biocatalytic decomposition to monomeric sugars, coupled with isomerization by enzymes prior to chemical dehydration by acid Lewis catalysts. A summary of all biotransformations of furans carried out by enzymes is also described, focusing on galactose, glyoxal and aryl-alcohol oxidases, monooxygenases and transaminases for the production of oxidized derivatives and amines. The increased interest in these products in polymer chemistry can lead to a redirection of biomass valorization from second generation biofuels to chemical synthesis, by creating novel pathways to produce bio-based polymers.

Keywords: furan-based chemicals; biocatalysis; lignocellulose; 5-hydroxymethylfurfural; furfural; pretreatment

---

1. Introduction

The increasing interest in renewable, bio-based polymers has resulted in the quest for new synthesis routes for the production of polymer building blocks. This interest is further reinforced by the general concern regarding sustainability demands and environmental issues of modern-day societies. Polylactic acid, polyhydroxyalkanoate and other biomass-derived thermoplastic polyester products have been widely used to replace the traditional petroleum-based polyester products, thus paving the way for the industrial production of bio-based chemicals [1]. However, the monomers of these bio-based polymers are mainly synthesized from refined sugars derived from first generation feedstocks, consequently increasing the need for exploitation of non-food renewable biomass-based monomers to support the sustainable development of polymer industry. Similar to the "food versus fuel" conflict, lignocellulosic biomass offers a potential solution towards the exploitation of sugars in sources that would otherwise be discarded, compared to first generation sugars deriving from the food industry.

Lignocellulosic biomass is the most abundant renewable carbon source on the planet with an estimated annual production of 2×10^{11} tons [2]. Lignocellulosic biomass is mainly comprised of forest and agricultural residues and has traditionally been valorized as a feedstock for the production of second-generation biofuels. According to the Department of Energy, in the USA alone there are produced 1.3 billion tons of dry lignocellulosic biomass per year, of which agricultural residues contribute 933 million tons per year, while 368 million tons per year originate from forest wastes [3]. However, a sharp increase in the number of publications in the field of furan-based monomers research over the last five years indicates a possible shift of the biorefinery concept from the production of fuels towards the production of value-added chemicals [4].

Due to its recalcitrant nature, lignocellulosic materials are difficult to handle, therefore, a standard process has been established, in order to render these materials more amenable to downstream treatments. This process typically involves an initial pretreatment step, which holds a key role in the fractionation and separation of the different fractions of lignocellulosic biomass, namely lignin, hemicellulose and cellulose. The following step includes the chemical or enzymatic hydrolysis of cellulose and hemicellulose with the aim to release the monomeric sugars of each fraction [5]. The monomeric sugars can then be utilized as carbon sources for fermentation processes for the production of biofuels, such as ethanol and methane [6,7], or a repertoire of different valuable compounds, such as lactic acid, succinic acid and omega-3 fatty acids [8–10]. In addition, biomass-derived monosaccharides can serve as starting materials in chemical or enzymatic conversion processes towards the synthesis of advanced chemicals and compounds with different applications, such as in polymer industry, in drug synthesis or as nutraceuticals [11]. Among these chemicals, the most interesting section of chemical synthesis includes the platform chemicals, which can in turn lead to bio-based monomers and plastics. This part has recently received increased attention as an alternative to traditional petrochemically synthesized polymers, aiming to produce furans, an otherwise undesired product for microbial fermentation, as a building block for further use [12].

Numerous reviews concerning furan-based monomers published in the last few years underline the growing importance of these renewable sources. Uniquely, the furan ring is characterized by interesting peculiarities including its pronounced dienic character that renders it susceptible to Diels–Alder reactions towards polymerizations with suitable dienophiles. Furans, such as the *2,5-furandicarboxylic acid (FDCA)*, are moreover considered appropriate renewable substitutes for the corresponding benzene-based monomers (i.e., terephthalic acid) [13]. According to the report of the U.S. Department of Energy, FDCA is listed among the top 12 value-added chemicals from biomass nowadays and is considered a highly promising platform chemical with numerous applications in many fields [14]. Its structural analogy to terephthalic acid renders it a suitable candidate as a building block for the synthesis of polymers, such as polyethylene furanoate (PEF), which has been heralded as a green alternative of the petroleum-derived poly(ethylene terephthalate) (PET). PET has an annual market size of approx. 50 million tons and the use of FDCA represents a promising approach to obtain biomass-derived polyester, and fully or partially replace oil-based materials [4]. The application of FDCA is, however, not restricted to PEF and other polyesters, as it can be also used for preparation of polyamines and polyurethanes [15]. Apart from FDCA, *2-furancarboxylic acid (FCA)* also possesses significant potential for use as a monomer in polyesters, as it can be dimerized via a condensation reaction with aldehydes and ketones, and the resulting dimers bear a structural similarity with the bisphenol series, thus introducing novel properties to polyesters [16].

The valorization process of lignocellulosic biomass towards the production of furans is accomplished by means of chemical catalysis, thus involving the use of expensive metal catalysts or solvents that are also detrimental for the environment [16–19]. For this reason, over the last few years there has been an attempt to shift the valorization process towards the use of biocatalysts, as they require mild, less costly conditions and are environmentally friendly. In that context, much attention has been paid to utilizing furans as substrates for the production of chemicals as building blocks for the production of monomers, including FDCA and FCA, aiming to demonstrate environmentally-friendly bio-based polymer synthesis. However, the use of such high selectivity catalysts means a cascade of chemical reactions with numerous products, each presenting different benefits and possible pathways towards the end product [20]. In order to accomplish this, an arsenal of different enzymes including, among others, oxidases and peroxygenases, are required for the production of FDCA, while transaminases can also function as possible catalysts for the synthesis of amines. The aim of this review is to provide information on an overall process for the production of furans and furan-based derivatives from lignocellulosic biomass, by means of biocatalysis. The introduction of such a process could, therefore, function as a sustainable and eco-friendly approach towards the replacement of the traditionally petrochemically synthesized polymers.

2. Conversion of Lignocellulosic Feedstocks to Monomeric Sugars

Lignocellulosic biomass, as aforementioned, is the most abundant plant material on the planet and is mainly derived from perennial herbaceous plant species and woody crops, while other major sources are agricultural and forest residues [21]. However, the deconstruction process of the plant cell wall presents a bottleneck for the industrialization of lignocellulose biorefining, due to its recalcitrant nature. The term *"biomass recalcitrance"* refers to the rigid structure of the plant cell wall and the subsequent difficulties it poses to the enzymatic treatment of the lignocellulosic biomass [22]. In order to overcome the recalcitrant barriers of lignocellulosic biomass and to render these materials more vulnerable to subsequent chemical or enzymatic treatment, a procedure known as *pretreatment* is required. Pretreatment is the initial step of the biomass conversion process to monosaccharides, as depicted in Figure 1. Pretreatment aims to alter the rigid structure and chemical composition of the biomass feedstocks and, therefore, increase the digestibility of cellulose or hemicellulose, which are the main sources of utilizable sugars [2]. Another key role of pretreatment is to achieve efficient fractionation of three biomass constituents, namely lignin, cellulose and hemicellulose, thus yielding pure streams and facilitating the subsequent valorization processes [23]. However, the type of pretreatment can vary, depending on the need of each application. Therefore, the appropriate pretreatment can be designed taking into consideration whether a specific fraction of lignocellulose is of particular interest. Understanding how the chemical composition and physical structure of biomass contribute to its recalcitrance, as well as their effect on the enzymatic hydrolysis of lignocellulose, would greatly help to improve the current pretreatment technologies, and probably promote the development of novel pretreatment processes [22].

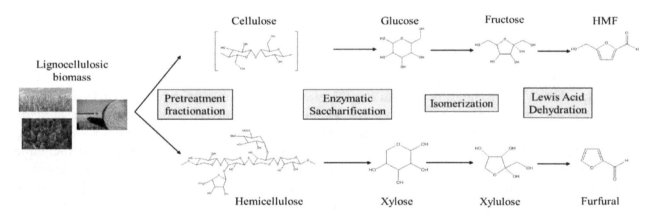

Figure 1. Overview of the process illustrating how furans can be produced from lignocellulosic biomass.

2.1. Plant Cell Wall Composition

Lignocellulose is composed of a complex structure of different compounds, in which cellulose, hemicellulose (polymeric carbohydrates) and lignin (an aromatic polymer) are intertwined together. The exact proportions of each fraction can vary depending on the type of the plant cell wall.

Lignocellulose also contains pectin, proteins, extractives and inorganic compounds in smaller amounts [24,25]. Different types of feedstock carry different amounts of each compound. Softwood, which is found at forest residues, consists of approximately 33–42% cellulose, 22–40% hemicellulose and 27–32% lignin [26]. The respective amounts in hardwood materials are 38–51% cellulose, 17–38% hemicellulose and 21–31% lignin [27]. Grasses contain 25–95% cellulose, 20–50% hemicellulose and 0–40% lignin [28,29]. The complex matrix constructed from the combination of these three fractions has a protective role for the plant against assault from the microbial or animal kingdoms, as it blocks the access of microorganisms or enzymes to the structural sugars of the cell wall and, therefore, their degradation. In terms of biomass valorization towards the production of furans, cellulose and hemicellulose fractions are those of main interest for the production of hexoses and pentoses, respectively. Hexoses can be subsequently utilized for the catalytic production of *5-hydroxymethylfurfural (HMF)*, while pentoses represent the starting material for the production of *2-furaldehyde (furfural, FA)*, as shown in Figure 1.

Cellulose is the largest and most abundant single component of lignocellulose with its percentage in the plant cell wall ranging between 35% and 50% depending on the type of the biomass. The cellulose fragment is composed of monomeric units of *D*-glucose linked together linearly with 1-4-β glucosidic bonds, thus enabling a stretched chain conformation with hydrogen bonds linking these chains into flat sheets. In addition, the packing of multiple cellulose strands leads to the creation of crystalline fibrils [30]. The crystalline nature of cellulose is partly responsible for the mechanical strength of plant tissues, as well as the difficulty in its disruption by enzymes. Cellulose exhibits a high degree of polymerization (higher than 10,000), which renders it less flexible and insoluble to water and most other solvents, even though this does not apply to glucose and its oligomers [26]. The enzymes responsible for the disruption of cellulose are cellulases and mainly involve four types of enzymes: endoglucanases, cellobiohydrolases, β-glucosidases and lytic polysaccharide monooxygenases, each providing a unique function for the conversion of cellulose to glucose. Cellulose has always been in the center of attention when it comes to second generation biofuels and especially bioethanol, since glucose produced from its disruption can function as a substrate for fermentative microorganisms for the production of ethanol. However, lately reports show that cellulose can be utilized also for other applications, such as the production of prebiotic cellooligosaccharides [31]. Production of HMF, not only from lignocellulosic biomass but also from other residual streams such as food wastes, is at the forefront in the synthesis of novel bio-based materials, as reviewed by Menegazzo et al. [32].

Hemicellulose is the second most abundant group of polysaccharides of lignocellulosic biomass. It is comprised mainly of pentose sugars, such as xylose and arabinose, exhibiting a degree of polymerization of 100 to 200, while hexose sugars, such as glucose, mannose and galactose are also present in small amounts [26]. The xylan backbone can be branched with its monomers decorated with acetyl and methyl groups or cinnamic, ferulic, glucuronic and galacturonic acids. These monomers, along with their decorations, can vary depending on the origin of each type of lignocellulose and function as the connections between lignin and hemicellulose [33]. There are indications that hemicellulose binds non-covalently to cellulose fibrils, thus resulting in a complex matrix that exhibits great resistance to enzymatic invasion and provides mechanical strength to the plant tissue. Due to its non-crystalline nature, hemicellulose is more susceptible to depolymerization than cellulose, an aspect of its behavior that is exploited by many deconstruction strategies, which include either a physico-chemical pretreatment, with the use of solvents and high temperature and pressure conditions or an enzymatic treatment upon addition of hemicellulose-degrading enzymes, or even both processes [26]. Indeed, many studies have demonstrated that hemicellulose removal leads to an increased yield of cellulose conversion [34,35]. Moreover, hemicellulose removal through efficient biomass fractionation further expands the potential of exploiting the pentose fraction for the production of, among others, furan-based valuable chemicals, such as FA, which has recently exhibited increased popularity, as reviewed by Luo et al. [36].

Lignin is the most abundant substance composed of aromatic moieties in nature [37] and it is composed of three basic monomeric units, namely *p*-hydroxyphenyl (H), guaicyl (G) and syringil (S) units, in different ratios depending on the origin of the plant [26]. Lignin represents the compound that provides structural reinforcement and resilience to the plant tissue, protecting it against invasion from pathogens and insects [38]. In the plant cell wall, lignin is connected to hemicellulose, and the subsequent complex is intertwined around cellulose. Therefore, lignin restricts the access of hydrolytic enzymes to cellulose and hemicellulose as a physical impediment. However, aside from its physical restrictions, it has been proved that lignin irreversibly absorbs cellulases, thus preventing them from reaching cellulose [39–41]. Hence, the effective enzymatic treatment of any material demands a delignification step, which is mainly accomplished by the pretreatment. Lignin is only soluble in organic solvents and not in water, and for that reason the most effective pretreatments for lignin removal are organosolv or alkali-catalyzed organosolv processes [34,42]. However, enzymatic digestibility is not exclusively dependent on lignin removal. For example, dilute acid pretreatment removes only a part of lignin, but the enzymatic digestibility of cellulose can also be increased due to hemicellulose removal, lignin redistribution and other structural changes induced [22]. Within the frame of a furan-based valorization process of lignocellulose, lignin remains as a side stream that can be utilized for numerous potential applications [43].

2.2. Fractionation Technologies

Fractionation is a term used to describe the effective separation of each fraction of lignocellulosic biomass with the highest possible purity. As described above, this separation is achieved by pretreatment, which aims to alter or remove the structural and compositional impediments for hydrolysis and increase the yield of monomeric hexoses or pentoses from cellulose or hemicellulose, respectively [2]. For developing an effective pretreatment technology within the frame of furan-based valorization of biomass, efficient fractionation holds the key role in the overall process. Separation of biomass into a cellulose-rich lignin-free solid pulp, which will be subsequently hydrolyzed to glucose, and a hemicellulose-rich liquor containing lignin that can easily precipitated and removed, will allow for obtaining clean fractions for the production of furans. Other parameters to be considered are the reduction in the biomass particle size, the increase in cellulose accessibility, as well as the absence of inhibitory products that can cause adverse effects in a subsequent step. Inhibitory compounds include carboxylic acids, furan derivatives and phenolic compounds that are produced as sugar degradation products during pretreatment [44,45]. Although HMF and FA are not considered undesired products in an integrated biorefinery for the production of furans, they might both affect the performance of isomerases, cellulases and/or hemicellulases, and reduce the sugar yield [46], while other compounds, such as formic, acetic or levulinic acid might also affect the dehydration step towards the production of furans. The type of pretreatment can vary depending on either the origin of the feedstock or the downstream process and the fraction that is of particular interest. Various pretreatment processes target mainly at breaking the linkages between the different biomass components through high temperature, pressure conditions and/or the use of chemical catalysts. In general, there are many types of pretreatment that can be roughly summed up in four categories: physical, chemical, physico-chemical and biological pretreatment.

Physical pretreatment typically involves the mechanical treatment of the lignocellulosic biomass, the objective of which is to reduce the particle size and increase the surface accessibility of biomass to hydrolytic enzymes. Low particle size along with effective stirring techniques are of utmost importance for maximizing the subsequent hydrolysis yield [47]. The most prominent types of mechanical treatment include chipping, grinding and milling, depending on the desirable particle size. Extrusion is also another promising process that subjects biomass to heating, mixing and shearing, which leads to physical and chemical modifications of lignocellulose during the passage through the extruder [6,48]. However, the power requirements of such processes are relatively high, especially for low particle size demands, thus limiting their industrial use.

Chemical pretreatment catalyzed by different compounds, such as alkali, acid, organic solvents and ionic liquids, has been utilized as an effective method of lignocellulosic biomass fractionation. The use of alkali is based on the ability to solubilize lignin and increase cellulose digestibility, while exhibiting low cellulose/hemicellulose solubilization [49]. Pretreatment with NaOH has been demonstrated to cause cellulose swelling and decrease its degree of polymerization and crystallinity, which further results in destabilization and dissolution of the lignin structure [50]. Another alkali showing great potential in lignocellulose pretreatment is $Ca(OH)_2$, also known as lime, which catalyzes lignin removal together with hemicellulose deacetylation and it can be easily recovered upon a post-treatment with CO_2 [51]. Alkali compounds are often used in low concentrations as bonus catalysts to physico-chemical pretreatments. On the contrary, the use of acid is based on the degradation and removal of hemicellulose fraction. Acid pretreatment can be performed by the use of either concentrated or diluted acid for the dissolution of hemicellulose. However, the implementation of concentrated acid encompasses equipment corrosion and acid recovery drawbacks, as well as high operational costs, making it unfavorable for commercial scale. On the other hand, the use of diluted acid not only solubilizes hemicellulose but can also hydrolyze it to its monomers, producing fermentable sugars. One of the drawbacks of acid pretreatments is the formation of furans, which are inhibitory substances for the growth of fermentative microorganisms and the production of ethanol [52]. However, these substances can be used efficiently for the increase in the overall yield of a furan-centered biorefinery concept. The most commonly used acid is H_2SO_4, while hydrochloric, phosphoric and nitric acid have also been used [53]. Finally, the use of fumaric and maleic acid as catalysts has been related to lesser amounts of FA produced compared to processes with H_2SO_4 [54].

Organosolv pretreatment employs the use of organic solvents, where the lignin fraction can be dissolved and, thus, removed in the liquid fraction. Upon solvent evaporation in the liquor, the water-insoluble lignin is precipitated and recovered in a relatively pure fraction [55]. Moreover, the use of organic solvent causes cellulose swelling and increases its degradability. Organosolv pretreatment is combined with other types of pretreatment for the efficient fractionation of biomass to its constituents. Organic solvents or their aqueous solutions that are commonly used involve methanol, ethanol, acetone, ethylene glycol and tetrahydrofurfuryl alcohol. By the use of a mixture of water and organic solvent, it is possible to collect both hemicellulose and lignin to each solvent, respectively, since hemicellulose is soluble in water and lignin is soluble in organic solvents. However, in organosolv procedures, effective removal of solvents through the appropriate separation techniques is required, as they may be inhibitory to enzymatic hydrolysis or sugar fermentation [56]. In addition, the increased cost for the industrial use of these solvents leads to the preference for easy to separate solvents with low molecular weights and low boiling points, such as methanol and ethanol. Organosolv is a very promising process within the frame of a furan-based biorefinery process, as it allows for obtaining lignin-free fractions, namely a cellulose-rich solid fraction and a hemicellulose-rich liquor that can be used for the production of HMF and FA, respectively.

Physicochemical pretreatment involves the simultaneous treatment of biomass with high temperature and pressure conditions upon the addition of chemicals as catalysts. This type of pretreatment combines the effect of chemical solvents on biomass structure, together with bond cleavage and fractionation through the harsh temperature and pressure conditions. The most widespread physicochemical pretreatment for lignocellulosic biomass is steam explosion. It is a hydrothermal process, where biomass is subjected to pressurized steam for several seconds or minutes followed by a sudden depressurization. The explosive decompression leads to a mechanical rupture of the plant cell wall polymers, thus enabling hemicellulose removal to the aqueous solvent and separation of the cellulose fibrils, rendering them more amenable to enzymatic hydrolysis. In addition, the harsh mechanical conditions affect the lignin fraction, leading to a temporary redistribution of its constituents [57]. During steam explosion pretreatment, the formation of acetic acid is very common due to the hemicellulose acetyl groups' exposure to high temperatures. This is the reason why many studies suggest the use of acid in low concentrations in order to increase the hemicellulose degradation. Further degradation of hemicellulose-derived sugars can lead to the formation of furans, such as FA, and HMF as well as formic and levulinic acid [51]. In general, steam explosion pretreatment represents a process that does not require the use of hazardous materials. Another process for biomass pretreatment is liquid hot water pretreatment, where biomass is exposed to high temperatures (160–230 °C), although without the immediate decompression step. This process leads to the solubilization and removal of hemicellulose, while slightly affecting the structure of lignin and cellulose [39].

Another interesting physicochemical pretreatment is the Ammonia Fiber Explosion (AFEX). This process follows the principle of steam explosion pretreatment, as the material is incubated with liquid anhydrous ammonia in temperatures ranging from 60 °C to 100 °C with a subsequent release in pressure. This leads to the rapid expansion of the ammonia, yielding a solid pulp in which the cellulose fibers are swollen and less crystalline. While this process removes only a small part of the lignin and hemicellulose fraction, it has been shown to disrupt the lignin–carbohydrate linkages and increase the enzymatic digestibility of the material. This is probably attributed to the partial removal of lignin that adsorbs the cellulolytic enzymes, which are responsible for cellulose degradation [58]. Another advantage of this pretreatment process lies in the limited formation of by-products that can inhibit the downstream biological processes, mainly phenolic fragments of lignin, as well as the easy separation of ammonia in spite of its high volatility [59]. Finally, wet oxidation is a pretreatment that subjects lignocellulosic biomass to high temperatures (170–200 °C) and pressures (10–12 bar O_2) for 10 to 15 min, employing oxygen or air as an oxidative catalyst. This process has been demonstrated to achieve effective fractionation of biomass and solubilization of hemicellulose and lignin, depending on the solvent used, for the subsequent production of ethanol [47,60]. However, even though wet oxidation is able to remove lignin efficiently, the cost of oxygen and chemical catalysts can be significant impediments for the industrialization and the development of such technologies.

The *biological treatment* of lignocellulosic biomass involves the use of microorganisms, mainly for the degradation of lignin and hemicellulose, leaving the cellulosic structure almost untouched. The key microorganisms involved in biological pretreatment are the brown and white-rot fungi that produce several lignin-degrading enzymes, such as laccases and peroxidases for the efficient degradation of lignin [61]. Similar to all pretreatment processes reported above, biological pretreatment can be combined with other treatment types for the production of a cellulose-rich material susceptible to enzyme digestibility. As an example, Itoh et al. used a combination of biological and organosolv pretreatment for the effective production of ethanol [62]. However, while biological pretreatment offers advantages such as low capital cost, low energy demands and no use for chemicals, all in mild environmental conditions, its use is limited by low hydrolysis rates and time requirements [56].

Summarizing all the above, pretreatment is a process that aims to efficiently fractionate the polymeric constituents of the plant cell wall of lignocellulosic biomass. Fractionation can either occur upon the use of extreme mechanical, physical or chemical environments that cause disruption to the linkages among lignin, hemicellulose and cellulose, however with a toll. The effective separation and recovery of each fraction can lead to high capital costs or increased formation of by-products. In order for the pretreatment to be effective and cost-competitive, there is a need for the design of custom processes depending on both the specifications of the raw materials and the targeted end-products.

2.3. Enzymatic Hydrolysis of Cellulose and Hemicellulose

As described in Section 2.1, cellulose consists of D-glucose units linked together with β-glucosidic bonds. Its structure exhibits a high crystallinity index, therefore hampering its disruption. The breakdown of cellulose towards the production of glucose can be achieved either by the use of chemicals (though the use of acids or ionic liquids) or enzymatically by the use of cellulases [50]. While the chemical hydrolysis of lignocellulosic biomass has been the norm in the past, the enzymatic approach offers a more targeted, effective and environmentally friendly way to produce glucose. However, the complete deconstruction of cellulose in real feedstocks is a hard procedure that requires an arsenal of enzymes, each with its unique role. Hydrolytic cellulases are divided into three classes: exo-1,4-β-D-glucanases (cellobiohydrolases, CBHs, EC 3.2.1.176), endo-1,4-β-D-glucanases (EGs, EC 3.2.1.4) and β-glucosidases (BGLs, EC 3.2.1.21). Exo-1,4-β-D-glucanases attach to the ends of the cellulose chains moving alongside the fiber releasing cellobiose units. Endo-1,4-β-D-glucanases hydrolyse internal glucosidic bonds randomly within the cellulose chain and, finally, β-glucosidases produce glucose either through the hydrolysis of cellobiose or through the cleavage of glucose units from cellooligosaccharides [63,64]. Lytic polysaccharide monooxygenases (LPMOs) are another class of enzymes with oxidative activity that have been shown to act synergistically with hydrolytic enzymes to boost the degradation of polysaccharidic substrates and increase the sugar yield [65]. Currently there is a great repertoire of commercially available cellulolytic cocktails for efficient biomass hydrolysis to monosaccharides, mostly produced by thermophilic filamentous fungi [66]. In addition, numerous monoenzymes have been cloned and expressed heterologously in eukaryotic hosts; this facilitates not only the study of mode of action of specific enzyme activities, but also the construction of tailor-made cocktails targeted for specific substrates [67–69]. Contrary to cellulose, hemicellulose is a heterogeneous polymer with a variety of substitutions that requires many enzymes for its deconstruction. Exo- and endo-xylanases cleave the xylan chain producing xylooligosaccharides, while β-xylosidases produce xylose from the cleaved xylooligosaccharides [70–73]. Mannanases and β-mannosidases function similarly in mannan. The side groups of hemicellulose can be removed by a variety of different enzymes, including α-galactosidases, α-L-arabinofuranosidases, α-glucoronidases, acetyl xylan esterases and feruloyl and p-coumaroyl acid esterases.

Inhibition of Enzymatic Hydrolysis

Cellulose deconstruction is a rigorous procedure that may be inhibited on many levels. The main factor inhibiting the enzymatic digestibility is the aforementioned complex structure of the plant cell wall. The matrix formed from the intertwined polymers of lignin and hemicellulose around cellulose, impedes the access of the cellulolytic enzymes to the cellulose fraction. Therefore, pretreatment is a key step for the overall yield of the process, as an effective fractionation will remove lignin and hemicellulose, the first and foremost obstacles. However, the cellulose hydrolysis holds itself as a process with many challenges.

Firstly, substrate composition is a key factor affecting the enzyme performance. It is known that many cellulases and hemicellulases attach themselves on the cellulose fibers through one of their domains, namely the carbohydrate-binding module (CBM) that promotes stabilization of the enzyme onto the carbohydrate substrate and coordinates the hydrolysis [74]. Non-productive adsorption of enzymes on lignin can occur, which can reduce the loading of active biocatalysts and, subsequently hamper the hydrolysis yields. Development of efficient pretreatment strategies that minimize the lignin content in cellulose and hemicellulose fractions is of pivotal importance in order to eliminate the adverse effects of residual lignin [23]. Substrate concentration also affects the overall saccharification rate, as operating at high solids conditions (above 12% wt.) can be detrimental due to the high viscosity of the mixture and the subsequent poor mass transfer conditions. In their work, Katsimpouras et al. portrayed the importance of the stirring technique to the yield of the hydrolysis, by comparing a traditional stirring system with flasks to an alternative free-fall mixer [47]. The sharp increase in cellulose conversion and glucose release that was observed in the free fall mixer under high initial solids content, confirms the advantages of effective mixing. These findings are very important for obtaining a high glucose syrup that is a prerequisite in order to produce a high-fructose syrup to be targeted for furans within the frame of a furan-based valorization process.

On another note, hydrolysis can be inhibited either by pretreatment degradation products (phenolic compounds, furans, carboxylic acids), as described above, or by the end-product of the process. During the enzymatic deconstruction of cellulose, released glucose shows a typical product inhibition to β-glucosidases [75], which leads to accumulation of cellobiose in the hydrolysate. Cellobiose, in turn, is also a strong inhibitor to cellobiohydrolases, impeding their function [76]. Therefore, in a typical hydrolysis process, reaching the full potential of glucose release is practically impossible, since the cellulolytic enzymes activity will be hampered due to accumulation of inhibitors. Hence, when fermentation is applied as a subsequent step of the process, in order to counteract against this problem, simultaneous saccharification and fermentation strategy (SSF) is preferred, when possible. In this process, the fermentative microorganism is added to the hydrolysate containing the cellulases, fermenting the existing glucose, while it is simultaneously produced. This way, the cellulolytic enzymes are relieved from their inhibitors, continuing the cellulose deconstruction, while the fermentative organism keeps consuming the produced glucose, increasing both the cellulose conversion and the end-product formation. It is profound that the effective degradation of cellulose is a multi-factor process. Considering all the aforementioned factors, as well as the possibility of variations in the pH or temperature optima of the enzymes or the microorganisms and the production of other inhibitors, such as formic acid [77], it is evident that cellulose hydrolysis is a matter of extensive research on many fronts. Therefore, in order to reach high yields, taking into account and combining multiple factors is of pivotal importance.

2.4. Isomerisation of Hexoses and Pentoses for the Production of Furan Derivatives

Throughout the literature, many reports describe different chemical routes for the transformation of glucose to HMF [17,18], and of pentoses to FA [78], by using metal catalysts, ionic liquids and other methods. However, the high cost of such materials, combined with the necessity of post-purification, raised the need for the development of a cost-effective, environmentally friendly process for the conversion of glucose to HMF. The production of furan derivatives from lignocellulose-derived sugars under less severe reactions requires as a first step the isomerization of glucose and xylose to fructose and xylulose, respectively. These ketoses can dehydrated more easily than aldoses in the presence of Lewis acid catalysts towards the corresponding furans, namely HMF and FA [79,80]. In order to reach high yields of furans, the development of an effective isomerization step is of utmost importance.

The enzymatic isomerization of glucose to fructose requires the activity of *glucose isomerase*, one of the largest-volume commercial enzymes typically used for the production of high-fructose syrups from starches [81]. Production of fructose from corn liquor glucose has been demonstrated on an industrial scale by the commercially available immobilized enzyme preparation Sweetzyme IT (Novozymes) from *Streptomyces murinus*. [82]. However, the isomerization reaction is restricted by a thermodynamic equilibrium, at around 50%, preventing glucose from being converted in high yields. One way to shift this equilibrium towards fructose is the addition of sodium tetraborate in the reaction. This occurs because borate ions react with glucose and fructose, attaching themselves onto the second carbon of the ring, forming ionized esters. These complexes shift the equilibrium of the reaction as they are not recognized by the isomerase and, therefore, do not participate in the reaction. In general, borate ions show higher affinity to ketoses than aldoses in the equilibrium between borate and borate esters. That is why the reaction equilibrium is shifted towards the direction of fructose and not glucose [83,84]. As a result, the default 50% conversion of glucose can increase to as much as 80% [85]. This way, it is possible to produce HMF at high yields by means of a method with high selectivity that is also environmentally friendly. In a recent study by Wand et al., it was verified that the addition of borate to a ratio of 0.5 to sugar could significantly increase isomerization of glucose- and xylose-rich streams from pretreatment of corn stover, achieving approximately 80% glucose and 90% xylose conversion [85].

Furthermore, hemicellulose constitutes a large part of the plant cell wall aside from cellulose, whose main representative in most plants is xylan. In this perspective, the aforementioned system for the production of HMF by means of glucose isomerization, could be employed also for the valorization of xylose towards the production of FA. Similar to glucose, xylose can be converted to FA with the use of metal catalysts. However, by using an isomerase, xylose can be transformed to xylulose; dehydration of xylulose to FA is easier upon the use of acid or base catalysts [86,87]. This indicates the possibility for the development of a system where both glucose and xylose produced from the hydrolysis of lignocellulose can undergo an isomerization step and a subsequent dehydration step for the production of HMF and FA, all in a one-pot process (Figure 1). This way, the biorefinery process can be further integrated to encompass the valorization of xylose in addition to glucose, preventing the hemicellulose side stream from being discarded entirely.

3. Enzymatic Conversion of Furans to Building Blocks for Polymer Synthesis

Furans have been in the spotlight of biotechnological research as either inhibitors in processes such as microbial fermentation or, more recently, as starting materials in the chemical synthesis of different valuable monomers, as depicted in Figure 2. Throughout the years, many sources report the use of microorganisms to metabolize furans [88]. However, the mechanisms employed by microorganisms to metabolize furans are complex and the conversion is accomplished with a number of different enzymes secreted. In addition, working with microbial cultures presents other problems, such as optimum pH and temperature conditions of the organism, as well as the need of downstream processes for purification and recovery of the final product. For these reasons, there is a need to seek catalysts that can selectively transform furans in a controlled manner, be easily removed from the reaction medium and require mild conditions to function. As a result, during the last few years, much attention has been given to discovering enzymes that can catalyze the conversion of furans. The most common biotransformation reactions of furans include oxidation, reductive amination and reduction of furan aldehydes to alcohols. These modifications are discussed below and some future perspectives in the field of biocatalysis are described.

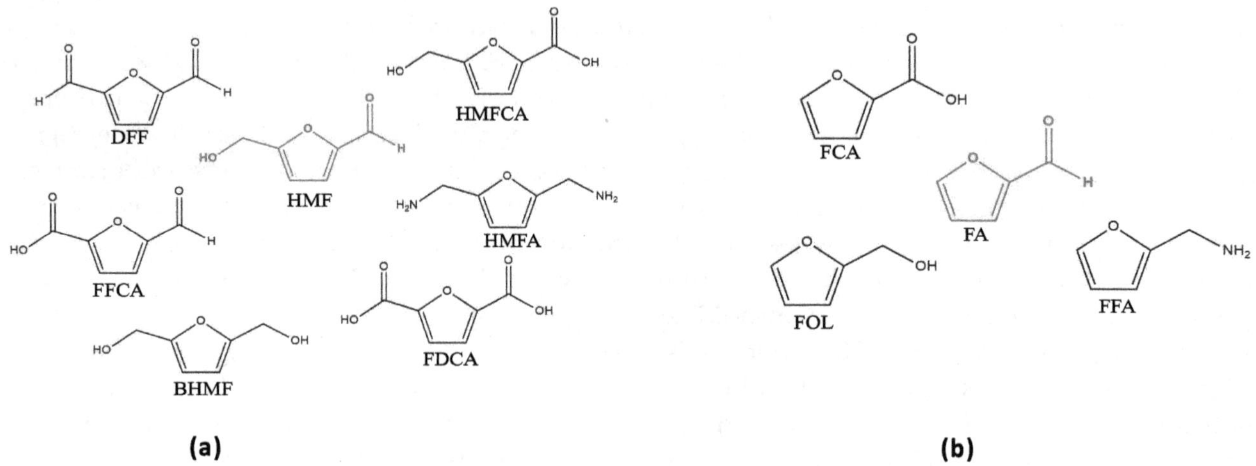

(a)   **(b)**

Figure 2.   Overview of different monomers originating from **(a)** HMF and **(b)** FA. **HMF**: 5-hydroxymethylfurfural, **DFF**: furan-2,5-dicarbaldehyde, **HMFCA**: 5-hydroxymethyl-2-furancarboxylic acid, **FFCA**: 2,5-formylfurancarboxylic acid, **FDCA**: 2,5-furandicarboxylic acid, **HMFA**: 5-(hydroxymethyl)furfurylamine, **FA**: furfural, 2-furaldehyde, **BHMF**: 2,5-bis(hydroxymethyl)furan, **FCA**: 2-furancarboxylic acid or furoic acid, **FFA**: 2-furfurylamine, **FOL**: 2-(hydroxymethyl)furan.

3.1. Oxidation Routes of Furans

3.1.1. Oxidative Reactions of HMF

As mentioned above, the most interesting compound that can be derived from furans is FDCA, which is a substitute for the petrochemically synthesized terephthalic acid. FDCA can be produced from HMF by following two routes (Figure 3). In the first route, HMF is converted to *5-hydroxymethyl-2-furancarboxylic acid (HMFCA)*, while in the second route HMF is converted to *furan-2,5-dicarbaldehyde (DFF)*. Both routes transform each respective chemical to *2,5-formylfurancarboxylic acid (FFCA)* which can, in turn, lead to FDCA. All steps in the above process and oxidation reactions require catalysts that can perform oxidation in alcohols and aldehydes. Due to the promiscuity in the mode of action of these oxidative enzymes, different activities have been reported to successfully participate in this cascade, as reviewed by [20]. The main enzyme activities implied include alcohol oxidases, galactose oxidases (GOs), 5-hydroxymethylfurfural oxidases, laccases, as well as catalases and peroxygenases.

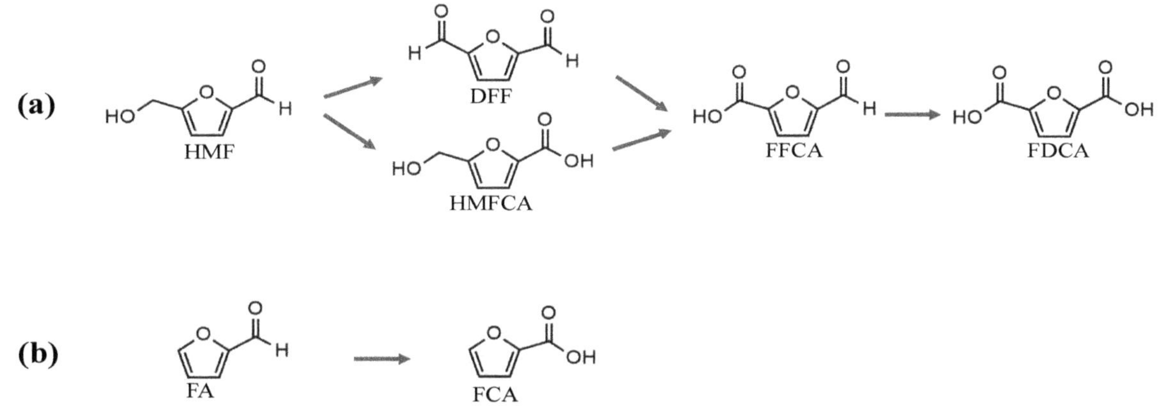

Figure 3. Schematic representation of oxidation of furan-derivatives that have been reported in the literature (**a**) HMF to FDCA, (**b**) FA to FCA. **HMF**: 5-hydroxymethylfurfural, **DFF**: furan-2,5-dicarbaldehyde, **HMFCA**: 5-hydroxymethyl-2-furancarboxylic acid, **FFCA**: 2,5- formylfurancarboxylic acid, **FDCA**: 2,5-furandicarboxylic acid, **FA**: furfural, 2-furaldehyde, **FCA**: 2-furancarboxylic acid or furoic acid.

In 2013, Krystof et al. [89] reported the use of a one-pot system of 2, 2, 6, 6-tetramethylpiperidine-1-oxyl (TEMPO) and lipase for the conversion of HMF to FDCA. This system includes a chemical catalyst (TEMPO) for the oxidation of HMF to DFF, combined with the activity of a lipase for the formation of peracid that catalyzes further oxidation to FDCA. In 2015, Carro et al. [90] tried to produce FDCA by means of a fully enzymatic cascade reaction. In that study, an aryl-alcohol oxidase (AOO) was employed, with the aim to perform all the oxidative steps required. However, while AAO successfully catalyzed the production of HMFCA in high yields, further oxidation was inhibited from the hydrogen peroxide produced in the reaction. For that reason, the authors performed the same reaction by adding an unidentified peroxygenase (UPO) as a subsequent step after the action of AAO, eventually producing FDCA in a yield of 91%. Serrano et al. [91] also investigated the effect of an AAO from *Pleurotus eryngii* on HMF for FDCA production. While AAO was demonstrated to efficiently oxidize HMF to HMFCA, further oxidation to FDCA was inhibited. However, by using HMFCA as an initial substrate to the AAO, the enzyme could successfully transform the substrate to FDCA. H_2O_2 produced during the reaction has an inhibitory effect on the second oxidation step, therefore a catalase was added, leading to complete transformation of HMF to FDCA after 12 days.

All the above indicate the ability of AAO enzymes belonging to the AA3 CAZy family (EC 1.1.3.7) to perform the oxidation of HMF. However, due to their limited ability to perform all three oxidation steps to FDCA, some report the oxidative ability of other enzymes of the AA3 family, such as glucose oxidases (EC 1.1.3.4), alcohol oxidases (EC 1.1.3.13) and pyranose oxidases (EC 1.1.3.10). Dijkman et al. characterized an HMF/FA oxidoreductase HmfH on a genetic level [92]. Following this, due to the fact that it was not possible to obtain the enzyme by means of expression [93], they sought for homolog enzymes that led to an HMF oxidase (HMFO) from *Methylovorus* sp., strain MP688. The HMFO again produced FFCA in high amounts, however the latter's low degree of hydration did not permit its further hydration to HMF. For that reason, by solving the crystal structure and introducing a double mutation, a 1000-fold increase in the catalytic efficiency of the enzyme on FFCA was achieved, finally producing FDCA in higher yields. Viña-Gonzalez et al. [94] tried to follow a similar mutational pattern on the amino sequence of an AAO in order to render it able to perform all three oxidation steps required for FDCA production. Indeed, while the wild type AAO initially strongly impeded the linkage of FFCA in the catalytic site of the enzyme, by introducing a tryptophan residue to a mutant carrying a H91N mutation that resembles the active site and structure of the HMFO, a 6-fold increase in the production of FDCA from HMF was achieved. Although the actual FDCA production yield was relatively low

(3%), that study reported for the first time an enzyme catalyzing complete HMF conversion to FDCA, paving the way for novel biocatalysts for furan valorization.

Enzymes of the Recently Revisited AA5 Family

The members of the broad AA5 family of CAZy database (http://www.cazy.org/AA5.html) have recently attracted significant interest regarding their potential in catalyzing oxidative modifications of furan derivatives. This group encompasses oxidases that require oxygen as a receptor (EC 1.1.3.-), including GOs (EC 1.1.3.9), glyoxal oxidases (GlOs) (EC 1.2.3.15), alcohol oxidases (EC 1.1.3.13) and raffinose oxidases (EC 1.1.3.-). Among other oxidative enzymes, GOs and GlOs have been reported to catalyze the conversion of furans, similar to AAOs and HMFO described above. Over the last years, there has been an increasing number of studies examining the potential of these enzymes in cascade processes on HMF conversion.

The first report of an AA5 enzyme acting on HMF was from the group of McKenna et al. [95], where the ability of a custom-developed variant of a GO from *Fusarium graminearum* to produce FDCA was examined. GO was able to convert HMF to DFF, but not to further oxidize it to FDCA. For that reason, an *Escherichia coli* periplasmic aldehyde oxidase (PaoABC) that was shown to be active on HMF, DFF and FFCA was employed to study whether the production of FDCA was possible. The key advantage of PaoABC in the process lied in the fact that it did not require the hydrate form of FFCA, contrary to other catalysts used in the process. Indeed, by adding both GO and PaoABC to HMF, production of FDCA was observed, although incomplete, due to the fact that PaoABC additionally produced HMFCA from HMF, which is a poor substrate for GO. Hence, in a subsequent experiment, GO was allowed to completely oxidize HMF to DFF prior to the addition of PaoABC to the system, thus accomplishing complete transformation to FDCA. Catalase was also added in the system to relieve the oxidative enzymes from the H_2O_2 produced throughout the process and provide the O_2 needed. A one-pot process was designed based on these findings [96]. A horseradish peroxidase (HRP) was added in the HMF-GO-PaoABC-catalase system and was found to significantly boost the activity of GO towards DFF production. This way, the HMFCA production was shifted, leading to 100% production yield of FDCA.

In 2015 Qin et al. examined the oxidation of HMF using an array of enzymes [97]. Firstly, a xanthine oxidase (XO) from *E. coli* producing HMFCA at a yield of 94% after 7 days, without progressing in further oxidation steps, was used. Secondly, three laccases were employed to boost the production of FFCA using TEMPO as a mediator. The highest conversion achieved was 82% yield, while lower yields of 4% and 10% were observed for DFF and FDCA, respectively. Finally, an enzyme cascade of GO, HRP and catalase to produce DFF from HMF was developed. The final step included the use of a lipase B (CAL-B) to convert DFF to FDCA with a yield of 88%.

Recently, Karich et al. attempted to design an enzyme cocktail to benefit from both routes of the FDCA synthesis process, including a UPO, a GO and an AAO [98]. In particular, it was supported that GO's and AAO's role was to oxidize the HMF towards HMFCA and DFF, respectively, and produce H_2O_2, while UPO's role was double; it was employed as oxidative factor to HMF and its oxidized derivatives, as well as relief agent of the H_2O_2 that was produced throughout the reaction. With this experiment, it was proved that the production of FFCA was rapid, while the FFCA to FDCA conversion was the most time-demanding step of the process. However, after 24 h, an 80% FDCA yield in the reaction was successful, while HMF was almost entirely converted to its oxidized derivatives.

Daou et al. studied the oxidation of HMF with three GlOs from the *Pycnoporus cinnabarus* [99]. Even though none of these enzymes exhibited HMF conversion yield over 40%, PciGlO2 and PciGlO3 activity was doubled upon the addition of catalase. The main oxidized product of HMF conversion was HMFCA. In order to bypass the inability to oxidize further HMFCA, an AAO was utilized in tandem with GlOs. Eventually, after AAO oxidized HMF to FFCA, GlO was added to the reaction in presence of catalase, producing small, but considerable, amounts of FDCA.

Mathieu et al. recently discovered a novel AAO from *Colletotrichum graminicola* that is able to efficiently catalyze the oxidation of aryl alcohols but demonstrates weak activity towards carbohydrates [100]. As such, AAO (EC 1.1.3.7) was reported as the first member of the AA5 family, whose activity is typically similar to the flavin-dependent glucose-methanol-choline (GMC) oxidoreductase family AA3. In the same work, it was demonstrated that AAO was able to oxidize HMF more efficiently than other biocatalysts previously reported; however, the enzyme could efficiently oxidize HMF towards DFF, but could only partially convert HMFCA in FFCA, with a yield of 54%.

While most enzymes require external co-factors to perform the oxidation reaction, such as (NADPH/NADP+)-dependent oxidases, this does not seem to be the case for enzymes belonging to AA5 family. The majority of members of this family represent copper-dependent enzymes which are easy to produce and purify, while they can function under no other requirement to perform the oxidation [101]. This is the reason why these enzymes can contribute significantly to the furan valorization procedure. However, in any case, it is evident that a cascade of enzymes is required in order to either reach the final steps of the sequential oxidations or achieve high yields in the final step. In most cases, the use of biocatalysts, such as catalases or peroxygenases, is required in order to alleviate the system from the adverse effects of the H_2O_2 produced, especially in case of the AA5 family members. Nevertheless, it is a promising observation that through protein engineering, mutants of a single enzyme can perform the entire oxidative pathway, even at low yields. As a result, further investigation can potentially transfer research from the current whole-cell to a single-enzyme strategy, thus significantly boosting the bioconversion process.

3.1.2. Oxidative Reactions for FA

Similar to HMF, FA also displays a potential for oxidative conversion, leading to the formation of *2-furancarboxylic acid (furoic acid, FCA)*. Early reports mention this process as a bio-detoxification process in fermentation, thus utilizing whole-cell catalysis for the reduction of FA [94,102,103]. However, further studies have been conducted in order to identify the specific enzymes involved in FA degradation.

A TEMPO-lipase system that was developed for HMF transformation was shown to effectively oxidize FA to FCA [89]. Kumar et al. tested the ability of 15 different Bayer-Villinger monoxygenases (BVMOs) to oxidize FA [104]. Among them, a phenylacetone monooxygenase (PAMO, EC 1.14.13.92) was shown to convert FA to FCA as the main product, achieving a 60% conversion yield after 12 h. However, a small amount of another by-product of oxidation reaction, which was identified as the formyl ester, was observed. This ester represents an intermediate compound that can be further transformed into the corresponding alcohol and formic acid [104]. The above results verify the possibility of effectively valorizing FA towards the production of its oxidized derivatives. Since FA can be formed by the hemicellulose-derived sugars, its valorization is a prerequisite towards lignocellulose biorefineries integration. For that reason, the promiscuity of utilizing oxidative enzymes on FA as a substrate can potentially contribute to achieve this goal.

3.2. Reductive Amination

Furfulylamines represent an interesting group of compounds that can be produced from the reductive amination of furans. They have numerous applications as starting materials in the synthesis of polyamine biopolymers and pharmaceuticals [105]. *2-Furfurylamine (FFA)*, a primary amine that can be produced from FA, is often used as a modifier in high performance thermosets due to its crosslinkable furan ring, where the introduction of furan groups can increase the crosslinking densities of the resulting polymer, and further enhance its thermal stability due to the formation of furfurylamine bridges during polymerization [106]. *5-(Hydroxymethyl)furfurylamine (HMFA)*, the amine occurring from reductive amination of HMF, has been patented as a curing agent in epoxy resins, besides its potential application in pharmaceutical industry as an intermediate compound in drug synthesis [107–109]. A high-titer synthesis of furfurylamine on an industrial scale is an attractive alternative to benzylamine and derivatives for the production of bio-based polymers with high-quality features, especially for biomedical applications [110]. The current chemical process for the synthesis of furfurylamines occurs upon the addition of a hydrogenation catalyst under high pressure via multiple steps and leads to by-product formation and other issues related to the sensitivity of the furan ring to reductive conditions. The equipment costs, together with concerns from the ecological and safety point of view, still hamper its production, therefore, novel synthesis routes need to be explored. Reductive amination has been recently demonstrated in the literature by employing electrocatalysis, thus providing novel insights into chemical catalysts-free amination [111]. However, biocatalysis is an attractive approach and offers the great advantage of enantioselectivity in mild reaction conditions. One-pot selective synthesis of optically pure amines from furan bio-based derivatives is a prerequisite for establishing novel enzymatic routes with significant commercial value and eliminating the production costs in this highly challenging area of research.

Transaminases (EC 2.6.1.18) are highly attractive versatile enzymes that catalyze the transformation of aldehydes and ketones towards the synthesis of chiral amines. These biocatalysts can provide a sustainable, high-yield, selective route to amines under mild aqueous conditions, by transfer of an amino group from a donor substrate to an acceptor compound. They have been used for the amination of FA and derivatives to produce furfurylamines [112], while the high-potential of this family of enzymes in industrial biotransformation processes is demonstrated by Merck's biosynthesis of the antidiabetic drug Sitagliptin [113]. Reductive amination of furans is possible both for HMF and FA as starting materials. Both aldehyde groups can be substituted with primary amines to generate amides with the use of transaminases. Transaminases have been reported to hold a primary role in the one-step biotransformation of furans (Figure 4) towards the production of furfurylamines with multiple applications.

There has been a limited amount of research on transaminases and furan derivatives, although highly promising and challenging. In the pioneering work by [112], the authors demonstrate the potential of different bacterial transaminases with varying enantioselectivities to convert furans to their corresponding amines, utilizing isopropylamine as a low-cost amine donor, at ambient temperature (30 °C). The results of this study show that enzymes from *Chromobacterium violaceum*, *Arthrobacter* sp. and *Mycobacterium vanbaalenii* were able to transfer amino groups not only to FA, HMF, FFCA and DFF, but also to various furan compounds and halogenated FAs, producing mono- and di-amines with yields that reach >90% for HMF, FA and FFCA and 70% for DFF. Apart from isopropylamine, alanine can also serve as an amine donor. In another study, a transaminase from *Vibrio fluvialis* was employed for the reduction amination of HMF with alanine as an amine donor, while an alanine dehydrogenase was added in order to shift the equilibrium [107]. A salt-tolerant ω-transaminase capable of catalyzing the amination of FA in salt water was studied [114]. The enzyme, produced by a halophilic bacterium *Halomonas* sp. CSM-2, exhibited 53.6% conversion of FA with (S)-methylbenzylamine as amino donor in a pH 9.0 seawater system; that yield was higher than its performance in freshwater. These results are quite promising, since there is a tendency for biorefineries in coastal areas to utilize seawater in order to cope with freshwater depletion. Transaminases have also been used together with lipases towards the production of optically pure substituted furfurylamines as functionalized building blocks [115].

Despite the high conversion yield, transaminases often suffer from low stability, which leads to many adverse effects when it comes to an industrial process. Immobilization has been used as an attractive option, as immobilized transaminases have shown increased thermal stability [116]. Moreover, immobilization offers fast recovery and reuse of the enzyme, as well as a simpler and less costly final product purification. In a work by [117], HMFA was synthesized by amination transfer reaction of HMF, which was catalysed by immobilized transaminase enzymes. Authors performed screening of a set of commercially available transaminases and achieved >99% conversion and reuse of the enzyme. Computational approaches constitute another attractive method in order to predict and design mutations of enzyme amino acids towards the stabilization of the subunit interface [118]. Transaminases, which are typically homodimers, are likely to be subjected to inactivation by subunit dissociation and local unfolding. This can lead to loss of the enzyme cofactor (pyridoxal 5'-phosphate), which serves as a molecular shuttle to transfer an amino group from a donor (amine, amino acid) to an acceptor (ketone, aldehyde, keto acid). Folding energy calculations together with molecular dynamics simulation have been used to design surface and interface modifications; the produced mutants have shown higher activity, enhanced co-solvent compatibility and overall increased stability at their new elevated optimal temperature of action [118].

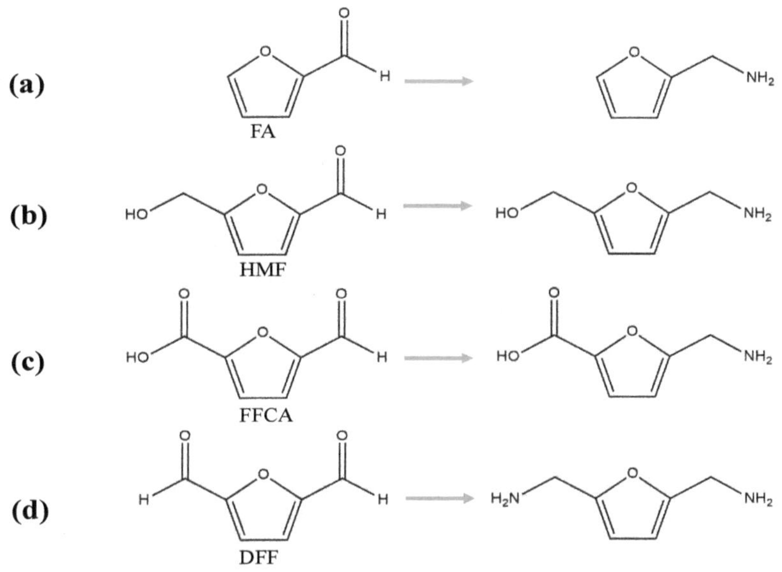

Figure 4. Schematic representation of furan-derivatives that have been reported in the literature to be transformed by aminotransferases to their corresponding amines [112,113,117]. **(a)** HMF to HMFA, **(b)** FA to FFA, **(c)** FFCA and **(d)** DFF to monoamine and diamine respectively. **HMF:** 5-hydroxymethylfurfural, **DFF:** furan-2,5-dicarbaldehyde, **FFCA:** 2,5-formylfurancarboxylic acid, **FA:** furfural, 2-furaldehyde.

A rising class of enzymes for amine synthesis are amine dehydrogenases (EC 1.4.99.3) [119]. Similar to transaminases, they can efficiently catalyze asymmetrically the reductive amination of chiral aldehydes and ketones, reaching up to 100% yield, albeit exhibiting a different mode of action than transaminases. Amine dehydrogenases are not strongly inhibited by neither the substrate nor the final product, such as transaminases, and they require a cheap nitrogen donor (NH_3) for their action. However, they exhibit lower enantioselectivity than transaminases, while cofactor regeneration is required. An amine dehydrogenase (a variant of the wild-type enzyme of *Bacillus badius*) was used for the efficient amination of a range of diverse aromatic and aliphatic ketones and aldehydes [120]. The enzyme showed high stereoselectivity in the amination of prochiral ketones, reaching amines with >99% optical purity, underlining the potential of these enzymes for amination reactions. Although amine dehydrogenases are promising candidates, currently there are no reports in the literature demonstrating the action of amine dehydrogenases on furans.

3.3. Reduction to Furan Alcohols

Furan alcohols, such as *2-(hydroxymethyl)furan (furfuryl alcohol, FOL)* and *2,5-bis(hydroxymethyl)furan (BHMF)* that originate from FA and HMF, respectively, (Figure 5), are used as starting materials in the synthesis of biopolymers [121]. BHMF is a versatile building block in the synthesis of polymers, fuels, and macrocycle polyethers; this molecule is further converted to *5,5'-dihydroxymethyl-furoin (DHMF)*, which undergoes enzymatic polycodensation reactions catalyzed by lipase towards aliphatic–aromatic oligoesters [122].

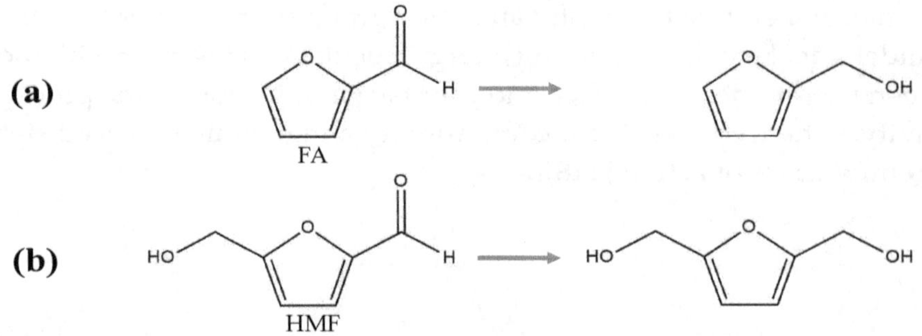

Figure 5. Schematic representation of furan-alcohol derivatives from (a) FA and (b) HMF that are have been reported in the literature and are catalyzed by alcohol dehydrogenases. **HMF:** 5-hydroxymethylfurfural, **FA:** furfural, 2-furaldehyde.

For the reduction of the furan keto-group into an alcohol, alcohol dehydrogenases (ketoreductases, ADHs, EC 1.1.1.1) have been investigated. The ADH from *Lactobacillus kefir* (Lk-ADH) is a well-studied enzyme with a broad substrate range [123]. This biocatalyst has been successfully implemented in a three-step enzymatic redox cascade to produce lactones from α, β-unsaturated alcohols in vitro [124]. Redox enzymes, such as ADHs, often require cofactors (NADPH/NADP+) that are continuously consumed during product formation. However, by implementing a regeneration system, the cofactor can be reused, and addition of expensive cofactors can thus be kept at a minimum, ensuring higher yield at reduced costs. For instance, adding glucose and glucose dehydrogenase (GDH) for the regeneration of NADPH by reduction of NADP+ represents a strategy frequently employed in the pharmaceutical and chemical industries [125].

HMF and FA conversion to BHMF and FOL, respectively, have been both reported to be catalyzed by whole cells, however ADH has been identified as the main enzyme implied in these biotransformation reactions. Many fungal and bacterial strains are widely known for detoxification of furan compounds with aryl alcohol oxidases and dehydrogenases [126], including *Pseudomonas putida*, *Saccharomyces cerevisiae*, *Amorphotheca resinae* that have been shown metabolize FA by conjugated reduction and oxidation reactions [127]. Moreover, the reduction of furans to corresponding furan alcohols has been demonstrated for *Pichia stipites*, and *Bacillus coagulans* under anaerobic conditions, providing the basis for the production of FOL through biotransformation [94,128]. The first study of the properties of an enzyme that catalyzes the conversion of FA was reported by Gutiérrez et al. [129]; in that study, a FA reductase (FFR) was purified from an *E. coli* strain and was used to reduce FA to FOL. In 2011, Li et al. reported a furfural reductase (FurX) from *Cupriavidus necator,* belonging in the Zn-dependent alcohol dehydrogenase family, that efficiently produced FOL from FA [130]. The sequence of this reductase was used as a model gene to seek for homologues, coming up with three ADHs, one from yeast (YADH1), one from *E. coli* (EcADH) and one from *Pseudomonas aeruginosa* (PaADH); all enzymes showed ADH activity on FA to produce FOL [131]. ADHs from *Cupriavidus basilensis* and *P. putida* have been identified for the NADH-dependent reduction of FA and HMF to the corresponding alcohols [94,132,133], while a variety of other bacterial enzymes act in the presence of either ethanol or NADH as electron donors.

Recently, a novel HMF-tolerant yeast strain *Meyerozyma guilliermondii* was isolated and was used for bio-catalytic whole-cell reduction of HMF into BHMF [134], with a high BHMF yield of 86% and excellent selectivity of >99%. He et al. reported a chemo-enzymatic conversion of biomass-derived xylose to FA by sequential acid catalyzed dehydration combined with reduction by recombinant *E. coli* cells harboring a NADH-dependent reductase, which was able to catalyze the bioreduction of FA to FOL [135]. The two processes were successfully combined to convert the xylose-rich hydrolysate to FA, and then to FOL with 44% yield based on the starting material xylose (100% FOL yield for the bioreduction step). Finally, recycling experiments for the carrageenan immobilized whole-cell and solid acid catalyst in one-pot FOL production are conducted; both catalysts showed excellent recyclability and no obvious decrease in activity was detected after five cycles of reaction.

ADHs from *M. guilliermondii,* heterologously produced in yeast have been reported for the reduction of HMF to produce BHMF with a high yield of 94% conversion and 99% selectivity within 24 h. For the first time, corncob hydrolysate was proposed as a cheap alternative to glucose as co-substrate for biocatalytic synthesis of BHMF, thus resulting in a significantly reduced production cost [131]. Various ADHs from yeast are capable of catalyzing FA reduction at the expense of ethanol [136]. To increase the catalytic activity and thermostability, protein engineering of ADHs has been investigated. Novel enzymes have been generated and were proved to be more effective towards the reduction of desired carbonyl substrates, with increased regioselectivity for furan substrates [137,138].

3.4. Other Enzymatic Activities and Future Perspectives

Different furan-based monomers described above can serve as starting materials for the synthesis of different compounds with unique chemical properties and different applications. These synthetic routes involve multiple steps and combine chemical catalysis with enzymatic reactions. Biocatalysts hold a key role in these transformations, offering a great selectivity in mild reaction conditions. As an example, a wide range of polyamides and polyesters can be produced by lipases and proteases, thus paving the way for furan valorization via enzymes [139]. Epoxides, another class of valuable versatile intermediate compounds, can undergo various reactions such as nucleophilic substitution and hydrolysis reactions, thus producing a repertoire of products [140,141]. Several scientific groups have reported the chemistry of epoxidation reactions and highlighted the role of the epoxide ring-opening process in synthetic reactions. The furanic compounds functionalized with epoxide groups can be used for bonding polycarbonate by cationic photo-curing and other compounds having a phenyl ring to prepare adhesives due to their rigid structure and hydrophobic property. Furanic diglycidyl esters from FDCA represent a viable bio-based alternative to their petrochemical aromatic counterpart [142]. Although the addition of epoxide groups has been reported by employing chemical steps, the use of enzymes for these transformations remains an undiscovered field of study. Apart from epoxy-functionalization, furan ring epoxidation can lead to rearrangement and ring opening, thus producing activated monomers, providing novel perspectives for synthesis. It has been reported that furan is oxidized by cytochrome P450 enzymes to the activated monomer metabolite cis-2-butene-1,4-diol [143]. P450 epoxidases contain a heme coenzyme that allows the addition of oxygen atom from O_2 to the alkene double bond [144], catalyzing epoxidations generating reactive electrophiles [145]. Oxidation of furan by P450 has been proposed to lead either to an epoxide-1 or a cis-enedione as epoxide intermediates [146]. Another class of enzymes that has recently attracted much attention and catalyze oxy-functionalization reactions are peroxygenases, as they combine the versatility of P450 monooxygenases with the simplicity of cofactor-independent enzymes, albeit with lower selectivity [147]. Regarding furan transformations, it is possible that upon addition of enzymes such as peroxygenases or P450 cytochrome monooxygenases, furan derivatives can undergo epoxidation and epoxide ring opening, thus forming reactive epoxide metabolites. After rearrangement, HMF–epoxide derivatives are transformed to tricarbonyl metabolites and this can be a possible route towards the synthesis of maleic acid or other compounds from furans. Synthesis of maleic acid has been reported to occur through a FA epoxide monomer [148]. For the biotransformation and enrichment of targeted bio-bricks, as well as their functionalization into activated monomers, selective oxidative biocatalysts with high specific activities are required. By expanding the use of enzymes, combining discovery of novel enzyme activities or design catalysts with desired properties, innovative processes can be developed towards the production of bio-based monomers and reactive compounds.

4. Conclusions and Future Prospects

Lignocellulosic biomass has been in the center of attention throughout the years as a feedstock for the production of second-generation biofuels. Recently, there has been a trend to extend the range of products that can be derived from biomass on behalf of the biorefinery concept, focusing on chemicals that can be utilized as starting materials for the production of polymers. By shifting the interest towards furans, which are otherwise viewed as inhibitors to the valorization process, it is possible to combine a renewable source together with biocatalysis in order to create an environmentally friendly process to boost the production of bio-based polymers. The main aspect for the development of such processes lies in the conjunction of physicochemical processes with biocatalysis. An efficient fractionation method will allow for obtaining pure biomass fractions that can be easily processed downstream. The investigation and development of novel, robust biocatalysts with high activity and stability is the main challenge for the subsequent biotransformation of furans. Biocatalysts require high selectivity, in order to eliminate the formation of byproducts and boost the reaction yield, while the overall process has to be economically viable. The latter can achieved by employing a cost-effective strategy for the production of enzymes, together with immobilization and recycling.

Author Contributions: Conceptualization, G.D. and A.K.; methodology, G.D.; investigation, G.D.; writing—original draft preparation, G.D. and A.K.; writing—review and editing, A.K. and E.T.; supervision, E.T. All authors have read and agreed to the published version of the manuscript.

Abbreviations:

HMF: 5-hydroxymethylfurfural, **DFF**: furan-2,5-dicarbaldehyde, **HMFCA**: 5-hydroxymethyl-2-furancarboxylic acid, **FFCA**: 2,5-formylfurancarboxylic acid, **FDCA**: 2,5-furandicarboxylic acid, **HMFA**: 5-(hydroxymethyl) furfurylamine, **BHMF**: 2,5-bis(hydroxymethyl)furan, **FA**: furfural, 2-furaldehyde, **FCA**: 2-furancarboxylic acid or furoic acid, **FFA**: 2-furfurylamine, **FOL**: 2-(hydroxymethyl)furan.

References

1. Zhang, J.; Li, J.; Tang, Y.; Lin, L.; Long, M. Advances in catalytic production of bio-based polyester monomer 2,5-furandicarboxylic acid derived from lignocellulosic biomass. *Carbohydr. Polym.* **2015**, *130*, 420–428. [CrossRef] [PubMed]

2. de Bhowmick, G.; Sarmah, A.K.; Sen, R. Lignocellulosic biorefinery as a model for sustainable development of biofuels and value added products. *Bioresour. Technol.* **2018**, *247*, 1144–1154. [CrossRef] [PubMed]

3. Kudakasseril Kurian, J.; Raveendran Nair, G.; Hussain, A.; Vijaya Raghavan, G.S. Feedstocks, logistics and pre-treatment processes for sustainable lignocellulosic biorefineries: A comprehensive review. *Renew. Sustain. Energy Rev.* **2013**, *25*, 205–219. [CrossRef]

4. Sousa, A.F.; Vilela, C.; Fonseca, A.C.; Gruter, G.-J.M.; Coelho, J.F.J.; Silvestre, A.J.; Matos, M.; Freire, C.S. Biobased polyesters and other polymers from 2,5-furandicarboxylic acid: A tribute to furan excellency. *Polym. Chem.* **2015**, *6*, 5961–5983. [CrossRef]

5. Kumar, P.; Barrett, D.M.; Delwiche, M.J.; Stroeve, P. Methods for pretreatment of lignocellulosic biomass for efficient hydrolysis and biofuel production. *Ind. Eng. Chem. Res.* **2009**, *48*, 3713–3729. [CrossRef]

6. Alvira, P.; Tomás-Pejó, E.; Ballesteros, M.; Negro, M.J. Pretreatment technologies for an efficient bioethanol production process based on enzymatic hydrolysis: A review. *Bioresour. Technol.* **2010**, *101*, 4851–4861. [CrossRef] [PubMed]

7. Xu, N.; Liu, S.; Xin, F.; Zhou, J.; Jia, H.; Xu, J.; Jiang, M.; Dong, W. Biomethane production from lignocellulose: Biomass recalcitrance and its impacts on anaerobic digestion. *Front. Bioeng. Biotechnol.* **2019**, *7*, 191. [CrossRef]

8. Abdel-Rahman, M.A.; Tashiro, Y.; Sonomoto, K. Lactic acid production from lignocellulose-derived sugars using lactic acid bacteria: Overview and limits. *J. Biotechnol.* **2011**, *156*, 286–301. [CrossRef]

9. Liu, R.; Liang, L.; Li, F.; Wu, M.; Chen, K.; Ma, J.; Jiang, M.; Wei, P.; Ouyang, P. Efficient succinic acid production from lignocellulosic biomass by simultaneous utilization of glucose and xylose in engineered *Escherichia coli*. *Bioresour. Technol.* **2013**, *149*, 84–91. [CrossRef]

10. Karnaouri, A.; Chalima, A.; Kalogiannis, K.G.; Varamogianni-Mamatsi, D.; Lappas, A.; Topakas, E. Utilization of lignocellulosic biomass towards the production of omega-3 fatty acids by the heterotrophic marine microalga *Crypthecodinium cohnii*. *Bioresour. Technol.* **2020**, *303*, 122899. [CrossRef]

11. Putro, J.N.; Soetaredjo, F.E.; Lin, S.-Y.; Ju, Y.-H.; Ismadji, S. Pretreatment and conversion of lignocellulose biomass into valuable chemicals. *RSC Adv.* **2016**, *6*, 46834–46852. [CrossRef]

12. Huang, R.; Qi, W.; Su, R.; He, Z. Integrating enzymatic and acid catalysis to convert glucose into 5-hydroxymethylfurfural. *Chem. Commun. (Camb.)* **2010**, *46*, 1115–1117. [CrossRef] [PubMed]

13. Nguyen, C.; Van Lewis, D.; Chen, W.H.; Huang, H.W.; ALOthman, Z.A.; Yamauchi, Y.; Wu, K.C.W. Combined treatments for producing 5-hydroxymethylfurfural (HMF) from lignocellulosic biomass. *Catal. Today* **2016**, *278*, 344–349. [CrossRef]

14. Werpy, T.; Petersen, G. Top Value-Added Chemicals from Biomass: Volume I—Results of Screening for Potential Candidates from Sugars and Synthesis Gas. 2004. Available online: http://www.nrel.gov/docs/fy04osti/35523.pdf (accessed on 3 July 2020).

15. Moreau, C.; Belgacem, M.N.; Gandini, A. Recent catalytic advances in the chemistry of substituted furans from carbohydrates and in the ensuing polymers. *Top. Catal.* **2004**, *27*, 11–30. [CrossRef]

16. Delidovich, I.; Hausoul, P.J.C.; Deng, L.; Pfützenreuter, R.; Rose, M.; Palkovits, R. Alternative monomers based on lignocellulose and their use for polymer production. *Chem. Rev.* **2016**, *116*, 1540–1599. [CrossRef] [PubMed]

17. Jadhav, H.; Taarning, E.; Pedersen, C.M.; Bols, M. Conversion of d-glucose into 5-hydroxymethylfurfural (HMF) using zeolite in [Bmim]Cl or tetrabutylammonium chloride (TBAC)/CrCl 2. *Tetrahedron Lett.* **2012**, *53*, 983–985. [CrossRef]

18. Ståhlberg, T.; Sørensen, M.G.; Riisager, A. Direct conversion of glucose to 5-(hydroxymethyl)furfural in ionic liquids with lanthanide catalysts. *Green Chem.* **2010**, *12*, 321–325. [CrossRef]

19. Zhao, H.; Holladay, J.E.; Brown, H.; Zhang, Z.C. Metal chlorides in ionic liquid solvents convert sugars to 5-hydroxymethylfurfural. *Science* **2007**, *316*, 1597–1600. [CrossRef]

20. Yuan, H.; Liu, H.; Du, J.; Liu, K.; Wang, T.; Liu, L. Biocatalytic production of 2,5-furandicarboxylic acid: Recent advances and future perspectives. *Appl. Microbiol. Biotechnol.* **2020**, *104*, 527–543. [CrossRef]

21. Cherubini, F. The biorefinery concept: Using biomass instead of oil for producing energy and chemicals. *Energy Convers. Manag.* **2010**, *51*, 1412–1421. [CrossRef]

22. Axelsson, L.; Franzén, M.; Ostwald, M.; Berndes, G.; Lakshmi, G.; Ravindranath, N.H. Perspective: Jatropha cultivation in southern India: Assessing farmers' experiences. *Biofuel Bioprod. Biore-fining* **2012**, *6*, 246–256. [CrossRef]

23. Ks, L.; Raghavendran, V.; Yakimenko, O.; Persson, G.; Olsson, E.; Rova, U.; Olsson, L.; Christakopoulos, P. Lignin-first biomass fractionation using a hybrid organosolv—Steam explosion pretreatment technology improves the saccharification and fermentability of spruce biomass. *Bioresour. Technol.* **2019**, *273*, 521–528. [CrossRef]

24. Brandt, A.; Gräsvik, J.; Hallett, J.P.; Welton, T. Deconstruction of lignocellulosic biomass with ionic liquids. *Green Chem.* **2013**, *15*, 550–583. [CrossRef]

25. Taherzadeh, M.J.; Karimi, K. Pretreatment of lignocellulosic wastes to improve ethanol and biogas production: A review. *Int. J. Mol. Sci.* **2008**, *9*, 1621–1651. [CrossRef] [PubMed]

26. Nhuchhen, D.; Basu, P.; Acharya, B.A. Comprehensive review on biomass torrefaction. *Int. J. Renew. Energ. Biofuels* **2014**, 1–56. [CrossRef]

27. Menon, V.; Rao, M. Trends in bioconversion of lignocellulose: Biofuels, platform chemicals & biorefinery concept. *Prog. Energy Combust.* **2012**, *38*, 522–550. [CrossRef]

28. Arsène, M.A.; Bilba, K.; Junior, H.S.; Ghavami, K. Treatments of non-wood plant fibres used as reinforcement in composite materials. *Mater. Res.* **2013**, *16*, 903–923. [CrossRef]

29. Smit, A.; Huijgen, W. Effective fractionation of lignocellulose in herbaceous biomass and hardwood using a mild acetone organosolv process. *Green Chem.* **2017**, *19*, 5505–5514. [CrossRef]

30. O'Sullivan, A.C. Cellulose: The structure slowly unravels. *Cellulose* **1997**, *4*, 173–207. [CrossRef]

31. Karnaouri, A.; Matsakas, L.; Krikigianni, E.; Christakopoulos, P.; Rova, U. Valorization of waste forest biomass towards the production of cello-oligosaccharides with prebiotic potential by utilizing customized enzyme cocktails. *Biotechnol. Biofuels* **2020**, *12*, 285. [CrossRef]

32. Menegazzo, F.; Ghedini, E.; Signoretto, M. 5-Hydroxymethylfurfural (HMF) production from real biomasses. *Molecules* **2018**, *23*, 2201. [CrossRef] [PubMed]

33. Timell, T.E. Recent progress in the chemistry of wood hemicelluloses. *Wood Sci. Technol.* **1967**, *1*, 45–70. [CrossRef]

34. Katsimpouras, C.; Dedes, G.; Bistis, P.; Kekos, D.; Kalogiannis, K.G.; Topakas, E. Acetone/water oxidation of corn stover for the production of bioethanol and prebiotic oligosaccharides. *Bioresour. Technol.* **2018**, *270*, 208–215. [CrossRef] [PubMed]

35. Yang, B.; Wyman, C.E. Effect of xylan and lignin removal by batch and flowthrough pretreatment on the enzymatic digestibility of corn stover cellulose. *Biotechnol. Bioeng.* **2004**, *86*, 88–95. [CrossRef] [PubMed]

36. Luo, Y.; Li, Z.; Li, X.; Liu, X.; Fan, J.; Clark, J.H.; Hu, C. The production of furfural directly from hemicellulose in lignocellulosic biomass: A review. *Catal. Today* **2019**, *319*, 14–24. [CrossRef]

37. Calvo-Flores, F.G.; Dobado, J.A. Lignin as renewable raw material. *ChemSusChem* **2010**, *3*, 1227–1235. [CrossRef]

38. Sticklen, M.B. Plant genetic engineering for biofuel production: Towards affordable cellulosic ethanol. *Nat. Rev. Genet.* **2008**, *9*, 433. [CrossRef]

39. Palonen, H.; Thomsen, A.B.; Tenkanen, M.; Schmidt, A.S.; Viikari, L. Evaluation of wet oxidation pretreatment for enzymatic hydrolysis of softwood. *Appl. Biochem. Biotechnol.* **2004**, *117*, 1–17. [CrossRef]

40. Sewalt, V.J.H.; Ni, W.; Jung, H.G.; Dixon, R.A. Lignin impact on fiber degradation: Increased enzymatic digestibility of genetically engineered tobacco (*Nicotiana tabacum*) stems reduced in lignin content. *J. Agric. Food Chem.* **1997**, *45*, 1977–1983. [CrossRef]

41. Yang, B.; Wyman, C.E. BSA treatment to enhance enzymatic hydrolysis of cellulose in lignin containing substrates. *Biotechnol. Bioeng.* **2006**, *94*, 611–617. [CrossRef]

42. Zhao, X.; Peng, F.; Cheng, K.; Liu, D. Enhancement of the enzymatic digestibility of sugarcane bagasse by alkali-peracetic acid pretreatment. *Enzyme Microb. Technol.* **2009**, *44*, 17–23. [CrossRef]

43. Kienberger, M. Potential Applications of Lignin. In *Economics of Bioresources*; Krozer, Y., Narodoslawsky, M., Eds.; Springer: Cham, Switzerland, 2019; pp. 183–193.

44. Klinke, H.B.; Ahring, B.K.; Schmidt, A.S.; Thomsen, A.B. Characterization of degradation products from alkaline wet oxidation of wheat straw. *Bioresour. Technol.* **2002**, *82*, 15–26. [CrossRef]

45. Palmqvist, E.; Hahn-Hägerdal, B. Fermentation of lignocellulosic hydrolysates. II: Inhibitors and mechanisms of inhibition. *Bioresour. Technol.* **2000**, *74*, 25–33. [CrossRef]

46. Arora, A.; Martin, E.M.; Pelkki, W.H.; Carrier, D.J. Effect of formic acid and furfural on the enzymatic hydrolysis of cellulose powder and dilute acid-pretreated poplar hydrolysates. *ACS Sustain. Chem. Eng.* **2013**, *1*, 23–28. [CrossRef]

47. Katsimpouras, C.; Kalogiannis, K.G.; Kalogianni, A.; Lappas, A.A. Production of high concentrated cellulosic ethanol by acetone/water oxidized pretreated beech wood. *Biotechnol. Biofuels* **2017**, *10*, 1–16. [CrossRef]

48. Karunanithy, C.; Muthukumarappan, K.; Julson, J.L. Influence of high shear bioreactor parameters on carbohydrate release from different biomasses. In Proceedings of the American Society of Agricultural and Biological Engineers Annual International Meeting, ASABE, Providence, RI, USA, 29 June–2 July 2008; Volume 6, pp. 3562–3577.

49. Carvalheiro, F.; Duarte, L.C.; Girio, F.M. Hemicellulose biorefineries: A review on biomass pretreatments. *J. Sci. Ind. Res. India* **2008**, *67*, 849–864.

50. Taherzadeh, M.J.; Karimi, K. Enzyme-based hydrolysis processes for ethanol from lignocellulosic materials: A review. *BioResources* **2007**, *2*, 707–738.

51. Mosier, N.; Wyman, C.; Dale, B.; Elander, R.; Lee, Y.Y.; Holtzapple, M.; Ladisch, M. Features of promising technologies for pretreatment of lignocellulosic biomass. *Bioresour. Technol.* **2005**, *96*, 673–686. [CrossRef]

52. Saha, B.C.; Iten, L.B.; Cotta, M.A.; Wu, Y.V. Dilute acid pretreatment, enzymatic saccharification and fermentation of wheat straw to ethanol. *Process Biochem.* **2005**, *40*, 3693–3700. [CrossRef]

53. Mosier, N.; Hendrickson, R.; Ho, N.; Sedlak, M.; Ladisch, M.R. Optimization of pH controlled liquid hot water pretreatment of corn stover. *Bioresour. Technol.* **2005**, *96*, 1986–1993. [CrossRef]

54. Kootstra, A.M.J.; Beeftink, H.H.; Scott, E.L.; Sanders, J.P.M. Comparison of dilute mineral and organic acid pretreatment for enzymatic hydrolysis of wheat straw. *Biochem. Eng. J.* **2009**, *46*, 126–131. [CrossRef]

55. Zhao, X.; Cheng, K.; Liu, D. Organosolv pretreatment of lignocellulosic biomass for enzymatic hydrolysis. *Appl. Microbiol. Biotechnol.* **2009**, *82*, 815–827. [CrossRef] [PubMed]

56. Sun, Y.; Cheng, J. Hydrolysis of lignocellulosic materials for ethanol production: A review. *Bioresour. Technol.* **2002**, *83*, 1–11. [CrossRef]

57. Pielhop, T.; Amgarten, J.; von Rohr, P.R.; Studer, M.H. Steam explosion pretreatment of softwood: The effect of the explosive decompression on enzymatic digestibility. *Biotechnol. Biofuels* **2016**, *9*, 152. [CrossRef]

58. Wyman, C.E.; Dale, B.E.; Elander, R.T.; Holtzapple, M.; Ladisch, M.R.; Lee, Y.Y. Coordinated development of leading biomass pretreatment technologies. *Bioresour. Technol.* **2005**, *96*, 1959–1966. [CrossRef]

59. Teymouri, F.; Laureano-Perez, L.; Alizadeh, H.; Dale, B.E. *Optimization of the Ammonia Fiber Explosion (AFEX) Treatment Parameters for Enzymatic Hydrolysis of Corn Stover*; Cambridge University Press: Cambridge, UK, 2014; pp. 1–30. [CrossRef]

60. Martín, C.; Thomsen, M.H.; Hauggaard-Nielsen, H.; BelindaThomsen, A. Wet oxidation pretreatment, enzymatic hydrolysis and simultaneous saccharification and fermentation of clover-ryegrass mixtures. *Bioresour. Technol.* **2008**, *99*, 8777–8782. [CrossRef]

61. Kumar, R.; Wyman, C.E. Effects of cellulase and xylanase enzymes on the deconstruction of solids from pretreatment of poplar by leading technologies. *Biotechnol. Prog.* **2009**, *25*, 302–314. [CrossRef]

62. Itoh, H.; Wada, M.; Honda, Y.; Kuwahara, M.; Watanabe, T. Bioorganosolve pretreatments for simultaneous saccharification and fermentation of beech wood by ethanolysis and white rot fungi. *J. Biotechnol.* **2003**, *103*, 273–280. [CrossRef]

63. Eriksson, T.; Karlsson, J.; Tjerneld, F. A model explaining declining rate in hydrolysis of lignocellulose substrates with cellobiohydrolase I (Cel7A) and endoglucanase I (Cel7B) of *Trichoderma reesei*. *Appl. Biochem. Biotechnol.* **2002**, *101*, 41–60. [CrossRef]

64. Väljamäe, P.; Kipper, K.; Pettersson, G.; Johansson, G. Synergistic cellulose hydrolysis can be described in terms of fractal-like kinetics. *Biotechnol. Bioeng.* **2003**, *84*, 254–257. [CrossRef]

65. Karnaouri, A.; Muraleedharan, M.N.; Dimarogona, M.; Topakas, E.; Rova, U.; Sandgren, M.; Christakopoulos, P. Recombinant expression of thermostable processive MtEG5 endoglucanase and its synergism with MtLPMO from *Myceliophthora thermophila* during the hydrolysis of lignocellulosic substrates. *Biotechnol. Biofuels* **2017**, *10*, 126. [CrossRef] [PubMed]

66. Rodrigues, A.C.; Haven, M.Ø.; Lindedam, J.; Felby, C.; Gama, M. Celluclast and Cellic® CTec2: Saccharification/fermentation of wheat straw, solid-liquid partition and potential of enzyme recycling by alkaline washing. *Enzyme Microb. Technol.* **2015**, *79–80*, 70–77. [CrossRef] [PubMed]

67. Karnaouri, A.C.; Topakas, E.; Christakopoulos, P. Cloning, expression, and characterization of a thermostable GH7 endoglucanase from *Myceliophthora thermophila* capable of high-consistency enzymatic liquefaction. *Appl. Microbiol. Biotechnol.* **2014**, *98*, 231–242. [CrossRef] [PubMed]

68. Karnaouri, A.; Topakas, E.; Matsakas, L.; Rova, U.; Christakopoulos, P. Fine-tuned enzymatic hydrolysis of organosolv pretreated forest materials for the efficient production of cellobiose. *Front. Chem.* **2018**, *6*, 128. [CrossRef]

69. Karnaouri, A.; Topakas, E.; Paschos, T.; Taouki, I.; Christakopoulos, P. Cloning, expression and characterization of an ethanol tolerant GH3 β-glucosidase from *Myceliophthora thermophila*. *PeerJ.* **2013**, *1*, e46. [CrossRef] [PubMed]

70. Katsimpouras, C.; Dimarogona, M.; Petropoulos, P.; Christakopoulos, P.; Topakas, E. A thermostable GH26 endo-β-mannanase from *Myceliophthora thermophila* capable of enhancing lignocellulose degradation. *Appl. Microbiol. Biotechnol.* **2016**, *100*, 8385–8397. [CrossRef]

71. Kumar, G.P.; Pushpa, A.; Prabha, H. A Review on Xylooligosaccharides. *IRJP* **2012**, *3*, 71–74.

72. Katsimpouras, C.; Dedes, G.; Thomaidis, N.S.; Topakas, E. A novel fungal GH30 xylanase with xylobiohydrolase auxiliary activity. *Biotechnol. Biofuels* **2019**, *12*, 120. [CrossRef]

73. Moukouli, M.; Topakas, E.; Christakopoulos, P. Cloning and optimized expression of a GH-11 xylanase from *Fusarium oxysporum* in *Pichia pastoris*. *New Biotechnol.* **2011**, *28*, 369–374. [CrossRef]

74. Várnai, A.; Mäkelä, M.R.; Djajadi, D.T.; Rahikainen, J.; Hatakka, A.; Viikari, L. Carbohydrate-binding modules of fungal cellulases: Occurrence in nature, function, and relevance in industrial biomass conversion. *Adv. Appl. Microbiol.* **2014**, *88*, 103–165. [CrossRef]

75. Decker, C.H.; Visser, J.; Schreier, P. β-Glucosidases from five black *Aspergillus* species: Study of their physico-chemical and biocatalytic properties. *J. Agric. Food Chem.* **2000**, *48*, 4929–4936. [CrossRef] [PubMed]

76. Holtzapple, M.; Cognata, M.; Hendrickson, C. Inhibition of *Trichoderma reesei* cellulase by sugars and Solvents. *Biotechnol. Bioeng.* **1990**, *36*, 275–287. [CrossRef]

77. Panagiotou, G.; Olsson, L. Effect of compounds released during pretreatment of wheat straw on microbial growth and enzymatic hydrolysis rates. *Biotechnol. Bioeng.* **2007**, *96*, 250–258. [CrossRef] [PubMed]

78. Danon, B.; Gianluca Marcotullio, G.; de Jong, W. Mechanistic and kinetic aspects of pentose dehydration towards furfural in aqueous media employing homogeneous catalysis. *Green Chem.* **2014**, *16*, 39–54. [CrossRef]

79. Román-Leshkov, Y.; Chheda, J.N.; Dumesic, J.A. Phase modifiers promote efficient production of hydroxymethylfurfural from fructose. *Science* **2006**, *312*, 1933–1937. [CrossRef] [PubMed]

80. Choudhary, V.; Pinar, A.B.; Sandler, S.I.; Vlachos, D.G.; Lobo, R.F. Xylose isomerization to xylulose and its dehydration to furfural in aqueous media. *ACS Catal.* **2011**, *1*, 1724–1728. [CrossRef]

81. Seyhan Tükel, S.; Alagöz, D. Catalytic efficiency of immobilized glucose isomerase in isomerization of glucose to fructose. *Food Chem.* **2008**, *111*, 658–662. [CrossRef]

82. Dehkordi, A.M.; Tehrany, M.S.; Safari, I. Kinetics of glucose isomerization to fructose by immobilized glucose isomerase (Sweetzyme IT). *Ind. Eng. Chem. Res.* **2009**, *48*, 3271–3278. [CrossRef]

83. Takasaki, Y. Studies on sugar-isomerizing enzymes effect of borate on glucose-fructose isomerization catalyzed by glucose isomerase. *Agric. Biol. Chem.* **1971**, *35*, 1371–1375. [CrossRef]

84. an den Berg, R.; Peters, J.A.; van Bekkum, H. The structure and (local) stability constants of borate esters of mono- and di-saccharides as studied by 11B and 13C NMR spectroscopy. *Carbohydr. Res.* **1994**, *253*, 1–12. [CrossRef]

85. Wang, W.; Mittal, A.; Pilath, H.; Chen, X.; Tucker, M.P.; Johnson, D.K. Simultaneous upgrading of biomass-derived sugars to HMF/furfural via enzymatically isomerized ketose intermediates. *Biotechnol. Biofuels* **2019**, *12*, 253. [CrossRef] [PubMed]

86. Takagaki, A.; Ohara, M.; Nishimura, S.; Ebitani, K. One-pot formation of furfural from xylose via isomerization and successive dehydration reactions over heterogeneous acid and base catalysts. *Chem. Lett.* **2010**, *39*, 838–840. [CrossRef]

87. Yang, Y.; Hu, C.W.; Abu-Omar, M.M. Synthesis of furfural from xylose, xylan, and biomass using AlCl $3 \cdot 6H_2O$ in biphasic media via xylose isomerization to xylulose. *ChemSusChem* **2012**, *5*, 405–410. [CrossRef] [PubMed]

88. Wierckx, N.; Koopman, F.; Ruijssenaars, H.J.; de Winde, J.H. Microbial degradation of furanic compounds: Biochemistry, genetics, and impact. *Appl. Microbiol. Biotechnol.* **2011**, *92*, 1095–1105. [CrossRef]

89. Krystof, M.; Pérez-Sánchez, M.; de María, P.D. Lipase-mediated selective oxidation of furfural and 5-hydroxymethylfurfural. *ChemSusChem* **2013**, *6*, 826–830. [CrossRef]

90. Carro, J.; Ferreira, P.; Rodríguez, L.; Prieto, A.; Serrano, A.; Balcells, B.; Ardá, A.; Jiménez-Barbero, J.; Gutiérrez, A.; Ullrich, R.; et al. 5-Hydroxymethylfurfural conversion by fungal aryl-alcohol oxidase and unspecific peroxygenase. *FEBS J.* **2015**, *282*, 3218–3229. [CrossRef]

91. Serrano, A.; Calviño, E.; Carro, J.; Sánchez-Ruiz, M.I.; Cañada, J.F.; Martínez, A.T. Complete oxidation of hydroxymethylfurfural to furandicarboxylic acid by aryl-alcohol oxidase. *Biotechnol. Biofuels* **2019**, *12*, 1–12. [CrossRef]

92. Dijkman, W.P.; Groothuis, D.E.; Fraaije, M.W. Enzyme-catalyzed oxidation of 5-hydroxymethylfurfural to furan-2,5-dicarboxylic acid. *Angew. Chem.* **2014**, *126*, 6633–6636. [CrossRef]

93. Koopman, F.; Wierckx, N.; De Winde, J.H.; Ruijssenaars, H.J. Identification and characterization of the furfural and 5-(hydroxymethyl)furfural degradation pathways of *Cupriavidus basilensis* HMF14. *Proc. Natl. Acad. Sci. USA* **2010**, *107*, 4919–4924. [CrossRef]

94. Viña-Gonzalez, J.; Martinez, A.T.; Guallar, V.; Alcalde, M. Sequential oxidation of 5-hydroxymethylfurfural to furan-2,5-dicarboxylic acid by an evolved aryl-alcohol oxidase. *Biochim. Biophys. Acta* **2020**, *1868*, 140293. [CrossRef]

95. McKenna, S.M.; Leimkühler, S.; Herter, S.; Turner, N.J.; Carnell, A.J. Enzyme cascade reactions: Synthesis of furandicarboxylic acid (FDCA) and carboxylic acids using oxidases in tandem. *Green Chem.* **2015**, *17*, 3271–3275. [CrossRef]

96. McKenna, S.M.; Mines, P.; Law, P.; Kovacs-Schreiner, K.; Birmingham, W.R.; Turner, N.J.; Leimkühler, S.; Carnell, A.J. The continuous oxidation of HMF to FDCA and the immobilisation and stabilisation of periplasmic aldehyde oxidase (PaoABC). *Green Chem.* **2017**, *19*, 4660–4665. [CrossRef]

97. Qin, Y.Z.; Li, Y.M.; Zong, M.H.; Wu, H.; Li, N. Enzyme-catalyzed selective oxidation of 5-hydroxymethylfurfural (HMF) and separation of HMF and 2,5-diformylfuran using deep eutectic solvents. *Green Chem.* **2015**, *17*, 3718–3722. [CrossRef]

98. Karich, A.; Kleeberg, S.B.; Ullrich, R.; Hofrichter, M. Enzymatic preparation of 2,5-furandicarboxylic acid (FDCA)—A substitute of terephthalic acid—By the joined action of three fungal enzymes. *Microorganisms* **2018**, *6*, 5. [CrossRef]

99. Daou, M.; Yassine, B.; Wikee, S.; Record, E.; Duprat, F.; Bertrand, E.; Faulds, C.B. *Pycnoporus cinnabarinus* glyoxal oxidases display differential catalytic efficiencies on 5-hydroxymethylfurfural and its oxidized derivatives. *Fungal Biol. Biotechnol.* **2019**, *6*, 1–15. [CrossRef]

100. Mathieu, Y.; Offen, W.A.; Forget, S.M.; Ciano, L.; Viborg, A.H.; Bloagova, E.; Henrissat, B.; Walton, P.H.; Davies, G.J.; Brumer, H. Discovery of a fungal copper radical oxidase with high catalytic efficiency toward 5-hydroxymethylfurfural and benzyl alcohols for bioprocessing. *ACS Catal.* **2020**, *10*, 3042–3058. [CrossRef]

101. Dong, J.; Fernández-Fueyo, E.; Hollmann, F.; Paul, C.E.; Pesic, M.; Schmidt, S.; Wang, Y.; Younes, S.; Zhang, W. Biocatalytic oxidation reactions: A chemist's perspective. *Angew. Chem.* **2018**, *57*, 9238–9261. [CrossRef] [PubMed]

102. Pérez, H.I.; Manjarrez, N.; Solís, A.; Luna, H.; Ramírez, M.A.; Cassani, J. Microbial biocatalytic preparation of 2-furoic acid by oxidation of 2-furfuryl alcohol and 2-furanaldehyde with *Nocardia corallina*. *Afr. J. Biotechnol.* **2009**, *8*, 2279–2282.

103. Ran, H.; Zhang, J.; Gao, Q.; Lin, Z.; Bao, J. Analysis of biodegradation performance of furfural and 5-hydroxymethylfurfural by *Amorphotheca resinae* ZN1. *Biotechnol. Biofuels* **2014**, *7*, 1–12. [CrossRef] [PubMed]

104. Kumar, H.; Fraaije, M.W. Conversion of furans by Baeyer-Villiger monooxygenases. *Catalysts* **2017**, *7*, 179. [CrossRef]

105. Lankenaua, A.W.; Kanan, M.W. Polyamide monomers via carbonate-promoted C–H carboxylation of furfurylamine. *Chem. Sci.* **2020**, *11*, 248–252. [CrossRef]

106. Wang, H.; Wang, J.; He, X.; Feng, T.; Ramdani, N.; Luan, M.; Liu, W.; Xu, X. Synthesis of novel furan-containing tetrafunctional fluorene-based benzoxazine monomer and its high performance thermoset. *RSC Adv.* **2014**, *4*, 64798–64801. [CrossRef]

107. Haas, T.; Pfeffer, J.C.; Faber, K.; Fuchs, M. (Evonik Degussa Gmbh) Enzymatic Amination. U.S. Patent WO2012171666, 20 December 2012.

108. Schaub, T.; Buschhaus, B.; Brinks, M.K.; Schelwies, M.; Paciello, R.; Melder, J.-P.; Merger, M. (BASF SE) Process for the Preparation of Primary Amines by Homogenously Catalysed Alcohol Amination. U.S. Patent 8785693, 22 July 2014.

109. Meng, J.; Zeng, Y.; Zhu, G.; Zhang, J.; Chen, P.; Cheng, Y.; Fang, Z.; Guo, K. Sustainable bio-based furan epoxy resin with flame retardancy. *Polym. Chem.* **2019**, *10*, 2370–2375. [CrossRef]

110. Wei, H.; Yao, K.; Chu, H.; Li, Z.C.; Zhu, J.; Zhao, Z.X.; Feng, Y.L. Click synthesis of the thermo- and pH-sensitive hydrogels containing β-cyclodextrins. *J. Mater. Sci.* **2012**, *47*, 332–340. [CrossRef]

111. Roylance, J.J.; Choi, K.S. Electrochemical reductive amination of furfural-based biomass intermediates. *Green Chem.* **2016**, *18*, 5412–5417. [CrossRef]

112. Dunbabin, A.; Subrizi, F.; Ward, J.M.; Sheppard, T.D.; Hailes, H.C. Furfurylamines from biomass: Transaminase catalysed upgrading of furfurals. *Green Chem.* **2017**, *19*, 397–404. [CrossRef]

113. Bornscheuer, U.T.; Huisman, G.W.; Kazlauskas, R.J.; Lutz, S.; Moore, J.C.; Robins, K. Engineering the third wave of biocatalysis. *Nature* **2012**, *485*, 185–194. [CrossRef] [PubMed]

114. Kelly, S.A.; Moody, T.S.; Gilmore, B.F. Biocatalysis in seawater: Investigating a halotolerant ω-transaminase capable of converting furfural in a seawater reaction medium. *Eng. Life Sci.* **2019**, *19*, 721–725. [CrossRef]

115. Deska, J.; Blume, F.; Albeiruty, M. Alkylative amination of biogenic furans through imine-to-azaallyl anion umpolung. *Synthesis* **2015**, *47*, 2093–2099. [CrossRef]

116. Neto, W.; Schürmann, M.; Panella, L.; Vogel, A.; Woodley, J.M. Immobilisation of ω-transaminase for industrial application: Screening and characterisation of commercial ready to use enzyme carriers. *J. Mol. Catal. B Enzym.* **2015**, *117*, 54–61. [CrossRef]

117. Petri, A.; Masia, G.; Piccolo, O. Biocatalytic conversion of 5- hydroxymethylfurfural: Synthesis of 2,5-bis(hydroxymethyl)furan and 5-(hydroxymethyl)furfurylamine. *Catal. Commun.* **2018**, *114*, 15–18. [CrossRef]

118. Meng, Q.; Capra, N.; Palacio, C.M.; Lanfranchi, E.; Otzen, M.; van Schie, L.Z.; Rozeboom, H.J.; Thunnissen, A.-M.W.H.; Wijma, H.J.; Janssen, D.B. Robust ω-transaminases by computational stabilization of the subunit interface. *ACS Catalysis* **2020**, *10*, 2915–2928. [CrossRef]

119. Höhne, M.; Bornscheuer, U.T. Biocatalytic routes to optically active amines. *ChemCatChem* **2009**, *1*, 42–51. [CrossRef]

120. Knaus, T.; Böhmer, W.; Mutti, F.G. Amine dehydrogenases: Efficient biocatalysts for the reductive amination of carbonyl compounds. *Green Chem.* **2017**, *9*, 453–463. [CrossRef]

121. Jiang, Y.; Woortman, A.J.J.; van Alberda Ekenstein, G.O.R.; Petrović, D.M.; Loos, K. Enzymatic synthesis of biobased polyesters using 2,5-bis(hydroxymethyl)furan as the building block. *Biomacromolecules* **2014**, *15*, 2482–2493. [CrossRef]

122. Baraldi, S.; Fantin, G.; Di Carmine, G.; Ragno, D.; Brandolese, A.; Massi, A.; Bortolini, O.; Marchetti, N.; Giovannini, P.P. Enzymatic synthesis of biobased aliphatic–aromatic oligoesters using 5,5′-bis(hydroxymethyl)furoin as a building block. *RSC Adv.* **2019**, *9*, 29044–29050. [CrossRef]

123. Bradshaw, C.W.; Hummel, W.; Wong, C.H. *Lactobacillus* kefir alcohol dehydrogenase: A useful catalyst for synthesis. *Org. Chem.* **1992**, *57*, 1532–1536. [CrossRef]

124. Oberleitner, N.; Peters, C.; Rudroff, F.; Bornscheuer, U.T.; Mihovilovic, M.D. In vitro characterization of an enzymatic redox cascade composed of an alcohol dehydrogenase, an enoate reductases and a Baeyer-Villiger monooxygenase. *J. Biotechnol.* **2014**, *192*, 393–399. [CrossRef]

125. Kaswurm, V.; Van Hecke, W.; Kulbe, K.D.; Ludwig, R. Engineering of a bi-enzymatic reaction for efficient production of the ascorbic acid precursor 2-keto-L-gulonic acid. *Adv. Synth. Catal.* **2013**, *355*, 1709–1714. [CrossRef]

126. Feldman, D.; Kowbel, D.J.; Glass, N.L.; Yarden, O.; Hadar, Y. Detoxification of 5-hydroxymethylfurfural by the *Pleurotus ostreatus* lignolytic enzymes aryl alcohol oxidase and dehydrogenase. *Biotechnol. Biofuels* **2015**, *8*, 63. [CrossRef]

127. Dominguez, P.D.M.; Guajardo, N.V. Biocatalytic valorization of furans: Opportunities for inherently unstable substrates. *Chemsuschem* **2017**, *10*, 4123–4134. [CrossRef] [PubMed]

128. Yan, Y.; Bu, C.; He, Q.; Zheng, Z.; Ouyang, J. Efficient bioconversion of furfural to furfuryl alcohol by *Bacillus coagulans* NL01. *RSC Adv.* **2018**, *8*, 26720–26727. [CrossRef]

129. Gutiérrez, T.; Ingram, L.O.; Preston, J.F. Purification and characterization of a furfural reductase (FFR) from *Escherichia coli* strain LYO1—An enzyme important in the detoxification of furfural during ethanol production. *J. Biotechnol.* **2006**, *121*, 154–164. [CrossRef] [PubMed]

130. Li, Q.; Metthew Lam, L.K.; Xun, L. *Cupriavidus necator* JMP134 rapidly reduces furfural with a Zn-dependent alcohol dehydrogenase. *Biodegradation* **2011**, *22*, 1215–1225. [CrossRef] [PubMed]

131. Li, Q.; Metthew Lam, L.K.; Xun, L. Biochemical characterization of ethanol-dependent reduction of furfural by alcohol dehydrogenases. *Biodegradation* **2011**, *22*, 1227. [CrossRef] [PubMed]

132. Jiang, T.; Qiao, H.; Zheng, Z.; Chu, Q.; Li, X.; Qiang, Y.; Jia, O. Lactic acid production from pretreated hydrolysates of corn stover by a newly developed *Bacillus coagulans* strain. *PLoS ONE* **2016**, *11*, e0149101. [CrossRef]

133. Laadan, B.; Almeida, J.R.; Radstrom, P.; Hahn-Hagerdal, B.; Gorwa-Grauslund, M. Identification of an NADH-dependent 5-hydroxymethylfurfural-reducing alcohol dehydrogenase in *Saccharomyces cerevisiae*. *Yeast* **2008**, *25*, 191–198. [CrossRef] [PubMed]

134. Li, Y.M.; Zhang, X.Y.; Li, N.; Xu, P.; Lou, W.Y.; Zong, M.H. Biocatalytic reduction of HMF to 2,5-bis(hydroxymethyl)furan by HMF-tolerant whole cells. *ChemSusChem* **2017**, *10*, 372–378. [CrossRef]

135. He, Y.; Ding, Y.; Ma, C.; Di, J.; Jiang, C.; Li, A. One-pot conversion of biomass-derived xylose to furfuralcohol by a chemo-enzymatic sequential acid-catalyzed dehydration and bioreduction. *Green Chem.* **2017**, *19*, 3844–3850. [CrossRef]

136. Xia, Z.-H.; Zong, M.-H.; Li, N. Catalytic synthesis of 2,5-bis(hydroxymethyl)furan from 5-hydroxymethylfurfual by recombinant *Saccharomyces cerevisiae*. *Enzyme Microb. Technol.* **2019**, *134*, 109491. [CrossRef]

137. Gong, X.M.; Qin, Z.; Li, F.L.; Zeng, B.B.; Zheng, G.W.; Xu, J.H. Development of an engineered ketoreductase with simultaneously improved thermostability and activity for making a bulky atorvastatin precursor. *ACS Catal.* **2019**, *9*, 147–153. [CrossRef]

138. Zheng, G.-W.; Liu, Y.-Y.; Chen, Q.; Huang, L.; Yu, H.-L.; Lou, W.-Y.; Li, C.-X.; Bai, Y.-P.; Li, A.; Xu, J.-H. Preparation of structurally diverse chiral alcohols by engineering ketoreductase CgKR1. *ACS Catal.* **2017**, *7*, 7174–7181. [CrossRef]

139. Jiang, Y.; Loos, K. Enzymatic synthesis of biobased polyesters and polyamides. *Polymers* **2016**, *8*, 243. [CrossRef] [PubMed]

140. Lambert, S.; Wagner, M. Environmental performance of bio-based and biodegradable plastics: The road ahead. *Chem. Soc. Rev.* **2017**, *46*, 6855–6871. [CrossRef]

141. Gandini, A. The irruption of polymers from renewable resources on the scene of macromolecular science and technology. *Green Chem.* **2011**, *13*, 1061–1083. [CrossRef]

142. Marotta, A.; Ambrogi, V.; Cerruti, P.; Mija, A. Green approaches in the synthesis of furan-based diepoxy monomers. *RSC Adv.* **2018**, *8*, 16330. [CrossRef]

143. Peterson, L.A.; Cummings, M.E.; Vu, C.C.; Matter, B.A. Glutathione trapping to measure microsomal oxidation of furan to *cis*-2-butene-1,4-dial. *Drug Metab Dispos.* **2005**, *33*, 1453–1458. [CrossRef]

144. Thibodeaux, C.J.; Chang, W.C.; Liu, H.W. Enzymatic chemistry of cyclopropane, epoxide, and aziridine biosynthesis. *Chem Rev.* **2012**, *112*, 1681–1709. [CrossRef]

145. Guengerich, F.P. Cytochrome P450 oxidations in the generation of reactive electrophiles: Epoxidation and related reactions. *Arch. Biochem. Biophys.* **2003**, *409*, 59–71. [CrossRef]

146. Peterson, L.A. Reactive metabolites in the biotransformation of molecules containing a furan ring. *Chem Res Toxicol.* **2013**, *26*, 6–25. [CrossRef]

147. Wang, Y.; Lan, D.; Durrani, R.; Hollmann, F. Peroxygenases *en route* to becoming dream catalysts. What are the opportunities and challenges? *Curr. Opin. Chem. Biol.* **2017**, *37*, 1–9. [CrossRef] [PubMed]

148. Alonso-Fagúndez, N.; Agirrezabal-Telleria, I.; Aria, P.L.; Fierro, J.L.G.; Mariscal, R.; López Granados, M. Aqueous-phase catalytic oxidation of furfural with H2O2: High yield of maleic acid by using titanium silicalite-1. *RSC Adv.* **2014**, *4*, 54960–54972. [CrossRef]

6

Characterization of a Carbonyl Reductase from *Rhodococcus erythropolis* WZ010 and its Variant Y54F for Asymmetric Synthesis of (*S*)-*N*-Boc-3-Hydroxypiperidine

Xiangxian Ying [1],[*], Jie Zhang [1], Can Wang [1], Meijuan Huang [1], Yuting Ji [1], Feng Cheng [1], Meilan Yu [2], Zhao Wang [1] and Meirong Ying [3],[*]

[1] Key Laboratory of Bioorganic Synthesis of Zhejiang Province, College of Biotechnology and Bioengineering, Zhejiang University of Technology, Hangzhou 310014, China; m15958047548@163.com (J.Z.); m17816035735@163.com (C.W.); meyroline.huang@gmail.com (M.H.); LJ15957189939@163.com (Y.J.); fengcheng@zjut.edu.cn (F.C.); hzwangzhao@163.com (Z.W.)

[2] College of Life Sciences, Zhejiang Sci-Tech Univeristy, Hangzhou 310018, China; meilanyu@zstu.edu.cn

[3] Grain and Oil Products Quality Inspection Center of Zhejiang Province, Hangzhou 310012, China

[*] Correspondence: yingxx@zjut.edu.cn (X.Y.); hz85672100@163.com (M.Y.).

Academic Editor: Stefano Serra

Abstract: The recombinant carbonyl reductase from *Rhodococcus erythropolis* WZ010 (ReCR) demonstrated strict (*S*)-stereoselectivity and catalyzed the irreversible reduction of *N*-Boc-3-piperidone (NBPO) to (*S*)-*N*-Boc-3-hydroxypiperidine [(*S*)-NBHP], a key chiral intermediate in the synthesis of ibrutinib. The NAD(H)-specific enzyme was active within broad ranges of pH and temperature and had remarkable activity in the presence of higher concentration of organic solvents. The amino acid residue at position 54 was critical for the activity and the substitution of Tyr54 to Phe significantly enhanced the catalytic efficiency of ReCR. The k_{cat}/K_m values of ReCR Y54F for NBPO, (*R/S*)-2-octanol, and 2-propanol were 49.17 s^{-1} mM^{-1}, 56.56 s^{-1} mM^{-1}, and 20.69 s^{-1} mM^{-1}, respectively. In addition, the (*S*)-NBHP yield was as high as 95.92% when whole cells of *E. coli* overexpressing ReCR variant Y54F catalyzed the asymmetric reduction of 1.5 M NBPO for 12 h in the aqueous/(*R/S*)-2-octanol biphasic system, demonstrating the great potential of ReCR variant Y54F for practical applications.

Keywords: (*S*)-*N*-Boc-3-hydroxypiperidine; carbonyl reductase; asymmetric reduction; rational design; *Rhodococcus erythropolis*

1. Introduction

Many natural products and active pharmaceutical ingredients share a common piperidine core, and the introduction of a chiral hydroxyl group on the C3-position of the piperidine ring may alter the bioactivity of the molecule [1–3]. (*S*)-*N*-Boc-3-hydroxypiperidine ((*S*)-NBHP) is a key chiral intermediate in the synthesis of ibrutinib as the inhibitor of Bruton's tyrosine kinase [4]. In the chemical synthesis of (*S*)-NBHP, employed strategies include the synthesis of racemic 3-hydroxypiperidine followed by chiral resolution and the enantiospecific synthesis of (*S*)-NBHP from chiral precursors. The former only achieves a maximum yield of 50%, making the process economically unviable, while the latter appears to be limited because of the lengthy procedure, rather poor yields of the products, and the use of potentially hazardous reagents [1,5,6]. Alternatively, the carbonyl-reductase-catalyzed asymmetric reduction of *N*-Boc-3-piperidone (NBPO) has gained increasing focus due to its mild reaction conditions, high yield, and remarkable enantioselectivity [4,7–9].

Coenzymes are required in carbonyl reductase-catalyzed reactions, and well-established approaches for coenzyme regeneration include the use of a second enzyme and a second substrate (i.e., glucose dehydrogenase and glucose), and the use of the second substrate catalyzed by the same enzyme (i.e., 2-propanol) [10]. Recently, an NADPH-dependent carbonyl reductase from *Saccharomyces cerevisiae* (YDR541C) was employed for the efficient synthesis of (S)-NBHP from NBPO by adopting a biphasic system to alleviate product inhibition and using glucose/glucose dehydrogenase to achieve coenzyme regeneration [8]. The glucose/glucose dehydrogenase system yields to the continuous production of gluconic acid; thus, pH adjustment is needed during the reaction, eventually making the process more complex and forming a large quantity of solid waste salt. Alternatively, the 2-propanol oxidation catalyzed by the same carbonyl reductase was widely used for coenzyme regeneration in order to simplify the operating process and increase the solubility of the substrates [11]. An efficient process catalyzed by the commercially-available ketoreductase KR-110 has been demonstrated to reduce 0.5 M NBPO to render the (S)-NBHP yield of 97.6% after a 24-h reaction [4]. The enzyme KR-110 was heat-sensitive and the substrate inhibition was obviously observed at a substrate concentration of 0.5 M. In addition, the 2-propanol concentration is usually required in excess to increase the product yield. Thus, high concentrations of the co-substrate together with the substrate further aggravate the inhibition of the enzyme activity in the 2-propanol-coupled strategy [4,11].

To overcome the inhibition from the high load of substrate/co-substrate, protein engineering is one of the promising approaches expanding the upper limit of the substrate/co-substrate concentration on a larger preparative scale [12,13]. Variants of the phenylacetaldehyde reductase from *Rhodococcus* sp. ST-10 (PAR) have been constructed through directed evolution, fully converting 200 g/L ethyl 4-chloro-3-oxobutanoate into ethyl (S)-4-chloro-3-hydroxybutyrate in the presence of 15% (v/v) 2-propanol [14,15]. Furthermore, attempts with biphasic catalysis in the presence of water-immiscible organic solvents have demonstrated an intriguing potential for overcoming the inhibition from substrate/co-substrate, increasing the solubility of substrates, easy product removal, decreasing the spontaneous hydrolysis of substrate/product, and avoiding unfavorable equilibria [16–19]. In an aqueous/octanol biphasic system, the biosynthesis process of ethyl (R)-4-chloro-3-hydroxybutyrate using a stereoselective carbonyl reductase from *Burkholderia gladioli* was established, in which 1.2 M ethyl 4-chloro-3-oxobutanoate was completely converted to afford ethyl (R)-4-chloro-3-hydroxybutyrate through the substrate fed-batch strategy [20]. In addition, the integration of protein engineering and medium engineering can further improve the effectiveness of asymmetric reduction at a high substrate load [20–22].

Although several processes for the efficient biosynthesis of (S)-NBHP have been developed, the pivot carbonyl reductases as biocatalysts still lack an in-depth characterization. Our previous genome mining enabled the discovery of chiral ketoreductases from *Rhodococcus erythropolis* WZ010 and the exploration of its application in the synthesis of chiral alcohols [23,24]. Here, a strictly (S)-enantioselective carbonyl reductase from *R. erythropolis* WZ010 (ReCR) and its variant Y54F were characterized for the efficient bioreduction of NBPO to (S)-NBHP, providing a basis for process development with an efficient coenzyme regeneration employing (R/S)-2-octanol or 2-propanol as the co-substrate (Scheme 1).

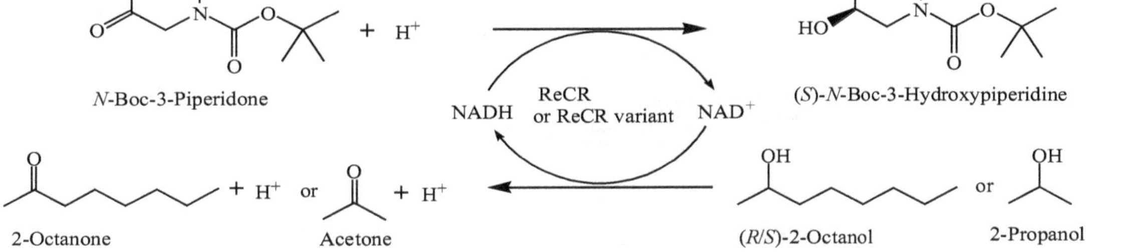

Scheme 1. Asymmetric bioreduction of N-Boc-3-piperidone (NBPO) using (R/S)-2-octanol or 2-propanol as co-substrate for NADH regeneration.

2. Results and Discussion

2.1. Characterization of Recombinant ReCR

The 1044-bp-long gene encoding ReCR was PCR-amplified from the genomic DNA of *R. erythropolis* WZ010 and over-expressed in *E. coli* BL21(DE3) in the form of the recombinant plasmid pEASY-E2-*recr*. The recombinant ReCR with C-terminal His-tag was subsequently purified by Ni-NTA chromatography. The gene *recr* encoded 348 amino acids with a deduced mass of 36.17 kDa, and the purified recombinant ReCR was verified with a single band of around 44 kDa by SDS-PAGE (Figure 1). The encoded amino acid sequence of ReCR displayed a 98% identity to that of PAR or alcohol dehydrogenase from *R. erythropolis* DSM 43297 (ReADH) [25–28], with five amino acids Arg67, Ser94, Lys110, Ser233, and Arg336 in ReCR different from Lys67, Asn94, Gln110, Lys233, and Gly336 in PAR or ReADH (Figure 2). The structure-related sequence alignment revealed that the enzyme belonged to the superfamily of zinc-containing alcohol dehydrogenases and had all conserved residues for the binding of catalytic and structural zinc ions [29]. It should be noted that the activity of the enzyme was severely inhibited by the exogenous zinc ion (Table S1), similar to what was observed with other zinc-containing alcohol dehydrogenases [24,30].

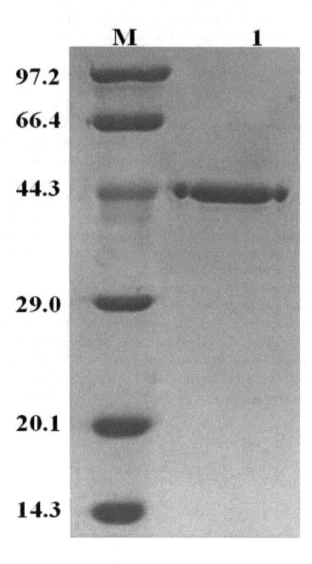

Figure 1. SDS-PAGE (12.5%) analysis of the purified recombinant ReCR. Lane 1, 2 μg purified ReCR with C-terminal His-tag; lane M, molecular weight marker. Coomassie Brilliant Blue R-250 was used to visualize the protein bands in the SDS-PAGE gel.

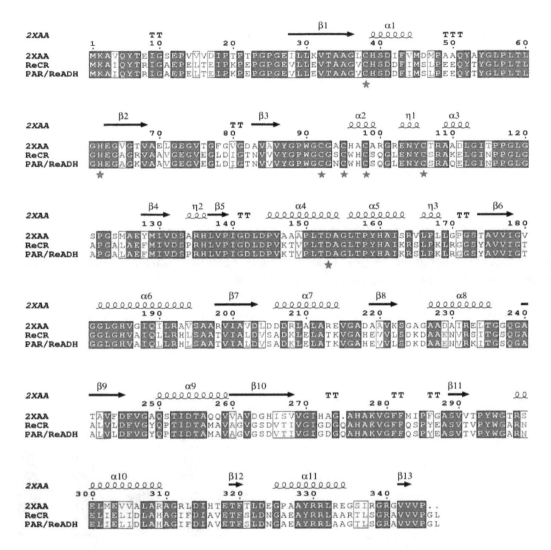

Figure 2. Structure-related sequence alignment between ReCR and its homologous proteins. 2XAA, PDB code of alcohol dehydrogenase from *Rhodococcus ruber* DSM 44541; PAR, alcohol dehydrogenase from *Rhodococcus* sp. ST-10 (GenBank accession No.: AB020760.3); ReADH, alcohol dehydrogenase from *R. erythropolis* DSM 43297 (GenBank accession No.: AY161280.1). The amino acid sequences of both PAR and ReADH are identical. Shown above the alignments are elements of the secondary structure of 2XAA. The numbering shown is from 2XAA. Red stars, putative catalytic residues; blue stars, residues for the coordination of structural zinc. Strictly conserved residues are highlighted with red boxes.

The recombinant ReCR was strictly NAD$^+$-dependent, since the enzyme activity was not detectable when NADP(H) was used as a coenzyme. The effect of pH on the activity was investigated within the pH range of 5.5–10.5. The maximum activities for NBPO reduction and (R/S)-2-octanol oxidation were observed at pH 6.0 and 10.0, respectively (**Figure 3A**), indicating that ReCR-catalyzed oxidation/reduction was pH-dependent [24]. The optimal temperature was 60 °C for NBPO reduction and 50 °C for (R/S)-2-octanol oxidation (Figure 3B). The enzyme activity in NBPO reduction was stable at 35 °C, whereas the remaining activity decreased to 50% of the initial activity after heat treatment at 60 °C for 1.5 h or 55 °C for 6.5 h (Figure 4A), demonstrating that its thermostability was superior to the heat-sensitive enzyme KR-110 [4]. Among the tested organic solvents, 20% (v/v) 2-propanol drastically decreased the activity of ReCR, similar to the performance of PAR in the presence of >10% (v/v) 2-propanol [18]. In contrast to 20% (v/v) 2-propanol, the enzyme displayed higher stability after 3.5 h incubation with 40% (v/v) (R/S)-2-octanol (**Figure 4B**).

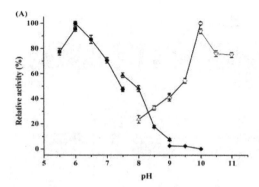
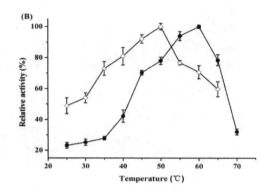

Figure 3. Effect of pH (**A**) and temperature (**B**) on the activity of recombinant ReCR. The relative activities of 100% represent 85.8 U/mg for NBPO reduction (solid symbols) and 88.3 U/mg for (R/S)-2-octanol oxidation (open symbols). The buffers 2-(N-morpholino)ethanesulfonic acid (MES, ■), piperazine-1,4-bisethanesulfonic acid(PIPES, ●), Tris-HCl (▲), and 3-(cyclohexylamino)-2-hydroxy-1-propanesulfonic acid (CAPSO, ◆) were used for the reduction reaction, while the buffers Tris-HCl (△), CAPSO (◇), and 3-(cyclohexylamino)-1-propanesulfonic acid (CAPS, ▽) were used for the oxidation reaction.

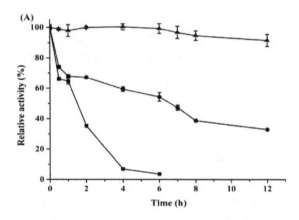
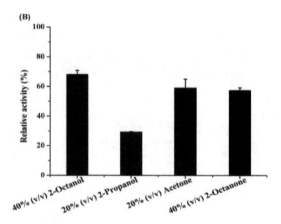

Figure 4. The stability of ReCR against heat (**A**) and organic solvents (**B**). Symbols: (■) for 60 °C, (●) for 55 °C, (▲) for 35 °C. The relative activity of 100% represents 85.8 U/mg for NBPO reduction. The enzyme was incubated with organic solvent (40% (v/v) (R/S)-2-octanol, 40% (v/v) 2-octanone, 20% (v/v) 2-propanol, or 20% (v/v) acetone) at 35 °C for 3.5 h prior to the stability test against organic solvent.

The substrate specificity of ReCR was tested using a set of alcohols and ketones (Table 1). Among the tested substrates, the enzyme exhibited the highest activities with 2,3-butanedione in the ketone reduction and (R/S)-2-octanol in the alcohol oxidation. The purified ReCR presented an activity of 85.8 U/mg towards NBPO reduction at pH 6.0 and 60 °C. Distinct from PAR and its variants [14], the activity of ReCR toward N-Boc-3-pyrrolidone reduction was relatively low. Particularly, the activity towards the oxidation of either (S)- or (R)-NBHP was not detectable at various temperatures (25–75 °C) and pHs (6.0–10.0), suggesting that the ReCR-catalyzed NBPO reduction was irreversible. A similar case was the secondary alcohol dehydrogenase SdcA from *R. erythropolis* DSM 44534 catalyzing the irreversible (S)-2-octanol oxidation [31]. The K_m and k_{cat}/K_m values for NBPO were 1.74 mM and 35.98 s^{-1} mM^{-1}, respectively (Table 2). The k_{cat}/K_m value for (R/S)-2-octanol and 2-propanol was 13.04 s^{-1} mM^{-1} and 9.74 s^{-1} mM^{-1}, respectively, implying that the use of (R/S)-2-octanol or 2-propanol as a co-substrate could be feasible to regenerate NADH in the NBPO reduction.

Table 1. Substrate spectrum of recombinant ReCR against ketones and alcohols [a].

Substrate	Relative Activity (%)	Substrate	Relative Activity (%)
N-Boc-3-Piperidone	100.0 [b] $\pm$ 2.6	(R/S)-2-Octanol	100.0 [c] $\pm$ 1.6
2,3-Butanedione	189.0 $\pm$ 3.4	(R/S)-2-Pentanol	61.8 $\pm$ 2.3
2-Octanone	169.2 $\pm$ 2.9	2-Propanol	47.4 $\pm$ 0.5
p-Bromoacetophenone	143.9 $\pm$ 4.3	(R/S)-2-Butanol	43.8 $\pm$ 1.1
Acetoin	47.2 $\pm$ 0.7	DL-1-Phenylethanol	31.5 $\pm$ 2.1
β-Ionone	34.8 $\pm$ 1.2	Cyclohexanol	8.0 $\pm$ 1.0
4-Hydroxy-2-butanone	31.8 $\pm$ 1.1	2-Buten-1-ol	6.8 $\pm$ 0.2
3-Octen-2-one	25.7 $\pm$ 0.7	(S)-N-Boc-3-Pyrrolidinol	2.7 $\pm$ 0.4
Acetophenone	25.3 $\pm$ 1.0	(S)-N-Boc-3-Hydroxypiperidine	0
Hydroxyacetone	23.6 $\pm$ 0.6	(R)-N-Boc-3-Hydroxypiperidine	0
N-Boc-3-Pyrrolidone	9.2 $\pm$ 0.5		
Acetone	4.8 $\pm$ 0.3		
2-Bromoacetophenone	1.8 $\pm$ 0.1		

[a] Data present mean values $\pm$ SD from two independent experiments. [b] Relative activity of 100% represents 85.8 U/mg for NBPO reduction at pH 6.0 and 60 °C; [c] Relative activity of 100% represents 88.3 U/mg for (R/S)-2-octanol oxidation at pH 10.0 and 50 °C.

Table 2. Kinetic parameters of recombinant ReCR [a].

Substrate	Coenzyme (mM)	V_{max} (U mg^{-1})	K_m (mM)	k_{cat} (s^{-1})	k_{cat}/K_m (s^{-1} mM^{-1})
NBPO	NADH (0.4)	103.57 $\pm$ 2.46	1.74 $\pm$ 0.08	62.61 $\pm$ 1.49	35.98 $\pm$ 0.86
(S)-NBHP	NAD$^+$ (0.4)	ND [b]	ND [b]	ND [b]	ND [b]
Acetone	NADH (0.4)	66.30 $\pm$ 3.27	46.06 $\pm$ 2.62	40.08 $\pm$ 1.98	0.87 $\pm$ 0.04
2-Propanol	NAD$^+$ (0.4)	23.54 $\pm$ 0.27	1.46 $\pm$ 0.06	14.22 $\pm$ 0.16	9.74 $\pm$ 0.11
2-Octanone	NADH (0.4)	235.54 $\pm$ 5.95	3.29 $\pm$ 0.05	142.38 $\pm$ 3.11	43.28 $\pm$ 0.95
(R/S)-2-Octanol	NAD$^+$ (0.4)	106.57 $\pm$ 2.74	4.94 $\pm$ 0.45	64.42 $\pm$ 1.66	13.04 $\pm$ 0.34

[a] Data present mean values $\pm$ SD from three independent experiments. [b] ND, not detectable.

2.2. Rational Design and Characterization of ReCR Variant Y54F

For the in-depth characterization, attempts of rational design of ReCR were conducted to improve its activity. The ReCR homology model was built based on the X-ray crystal structure of ADH-A from *Rhodococcus ruber* (PDB: 2XAA). Sequence identity of ReCR towards ADH-A was 60%. The QMEAN and Z-score values were used for the quality evaluation of the models. The QMEAN and Z-score values of the ReCR homology model were 0.822 and 0.533, respectively, which indicated satisfactory quality. In Ramachandran Plot analysis, 91.5% of residues were located in a favorable region, and only 0.4% were found in the sterically disallowed region. This ReCR homology model was selected for subsequent docking studies.

Furthermore, substrate docking was employed to predict potentially beneficial amino acid positions on ReCR. Figure 5A shows that NBPO was ideally accommodated in the ligand binding pocket of ReCR composed by zinc ion, NADH, and Tyr54 (in the vicinity of the entrance to the active site). Similar to the binding mode of ADH-A with the substrate [29], the carbonyl oxygen atom of NBPO in ReCR was bound to the Zn^{2+} ion with a distance of 4.1 Å, and the carbonyl carbon atom was in close proximity to the C4-atom of NADH. Thus, the hydride was transferred onto the *re*-face of the carbonyl group, consistent with the strict (S)-enantioselectivity of ReCR. On the other hand, the bulky Boc group of NBPO was close to the hydroxyl group of Tyr54 (distance of 4.3 Å between the hydroxyl oxygen of Tyr and the tertiary carbon of the Boc group), which might cause a steric hindrance during the substrate binding (Figure 5). Therefore, Tyr54 was selected to be mutated to Phe.

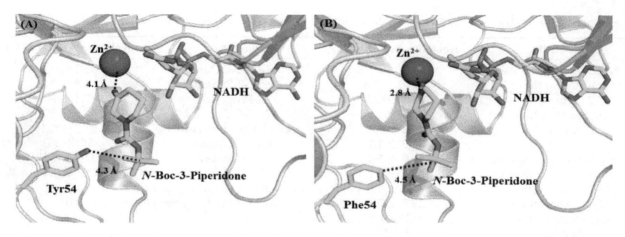

Figure 5. Protein-ligand structures of ReCR with NBPO (**A**) and ReCR Y54F with NBPO (**B**). ReCR and ReCR Y54F are represented in cartoon format. Tyr54, Phe54, NADH, and NBPO are highlighted in sticks. The zinc ion is shown as a magenta sphere.

As anticipated, the substitution of Tyr54 to Phe significantly improved the catalytic performance of ReCR, implying that the amino acid residue at position 54 could be critical for the enzyme activity. In the ketone reduction, the k_{cat}/K_m values of ReCR Y54F for NBPO (49.17 s^{-1} mM^{-1}), acetone (1.47 s^{-1} mM^{-1}), and 2-octanone (53.21 s^{-1} mM^{-1}) were 1.37, 1.69, and 1.23 times higher than those of ReCR (35.98 s^{-1} mM^{-1}, 0.87 s^{-1} mM^{-1}, and 43.28 s^{-1} mM^{-1}), respectively (Tables 2 and 3). In the alcohol oxidation, the k_{cat}/K_m values of ReCR Y54F for (R/S)-2-octanol (56.56 s^{-1} mM^{-1}) and 2-propanol (20.69 s^{-1} mM^{-1}) were 4.34 and 2.12 times higher than those of ReCR (13.04 s^{-1} mM^{-1} and 9.74 s^{-1} mM^{-1}), respectively (Tables 2 and 3). Although the K_m value of ReCR Y54F for NBPO (1.74 mM) was similar to that of ReCR, the K_m values of ReCR Y54F for other tested substrates were lowered to a certain extent.

Table 3. Kinetic parameters of ReCR variant Y54F [a].

Substrate	Coenzyme (mM)	V_{max} (U mg^{-1})	K_m (mM)	k_{cat} (s^{-1})	k_{cat}/K_m (s^{-1} mM^{-1})
NBPO	NADH (0.4)	140.72 ± 6.52	1.73 ± 0.05	85.07 ± 3.94	49.17 ± 2.28
(S)-NBHP	NAD$^+$ (0.4)	ND [b]	ND [b]	ND [b]	ND [b]
Acetone	NADH (0.4)	90.46 ± 1.69	37.32 ± 0.56	54.68 ± 1.02	1.47 ± 0.03
2-Propanol	NAD$^+$ (0.4)	35.29 ± 0.88	1.03 ± 0.05	21.32 ± 0.53	20.69 ± 0.51
2-Octanone	NADH (0.4)	273.75 ± 7.58	3.11 ± 0.12	165.48 ± 4.58	53.21 ± 1.47
(R/S)-2-Octanol	NAD$^+$ (0.4)	128.19 ± 3.12	1.37 ± 0.08	77.49 ± 1.88	56.56 ± 1.37

[a] Data present mean values $\pm$ SD from three independent experiments. [b] ND, not detectable.

Consistently with kinetic parameters, the productivity of asymmetric bioreduction of NBPO was significantly enhanced when whole cells overexpressing ReCR Y54F instead of ReCR were used as biocatalyst (Table 4). In contrast to the free enzyme, the use of a whole-cell biocatalyst was chosen because of higher enzyme stability and simpler procedure of biocatalyst preparation [4,11,32]. Both 2-propanol and (R/S)-2-octanol were investigated as co-substrates for the NADH regeneration. In the presence of 10% (v/v) 2-propanol, the bioreduction of 0.5 M NBPO catalyzed by whole cells overexpressing ReCR Y54F gave a (S)-NBHP yield of 98.08% after 12 h, which was 1.34 times higher than that of ReCR (72.15%). The whole-cell biphasic system has been demonstrated to be effective at a higher substrate load, in which (R/S)-2-octanol instead of 2-propanol was used not only as co-substrate for coenzyme regeneration but also as the organic phase for the substrate reservoir and product sink [33,34]. In the aqueous/(R/S)-2-octanol biphasic system, the (S)-NBHP yield was increased from 77.78% to 95.92% when ReCR Y54F replaced ReCR in the whole-cell biocatalyst. The corresponding total turnover number value of 1199, the calculated space-time yield of 579.15 g L^{-1} day^{-1}, and the remarkable stereoselectivity (*e.e.*$_p$ > 99.9%) together with the substrate concentration (up to 1.5 M) demonstrated a great potential of ReCR variant Y54F in the practical synthesis of (*S*)-NBHP.

Table 4. Asymmetric reduction of *N*-Boc-3-piperidone catalyzed by whole cells overexpressing ReCR or ReCR Y54F [a].

Enzyme [b]	Substrate (M)	Co-substrate (*v/v*)	Yield (%)	*e.e.*$_p$ (%) [c]
ReCR	NBPO, 0.5	2-Propanol, 10%	72.15 ± 3.51	>99.9 (*S*)
ReCR Y54F	NBPO, 0.5	2-Propanol, 10%	98.08 ± 1.65	>99.9 (*S*)
ReCR	NBPO, 1.5	(*R/S*)-2-Octanol, 60%	77.78 ± 2.23	>99.9 (*S*)
ReCR Y54F	NBPO, 1.5	(*R/S*)-2-Octanol, 60%	95.89 ± 2.37	>99.9 (*S*)

[a] Data present mean values ± SD from two independent experiments. [b] Whole cells overexpressing ReCR or ReCR variant Y54F. [c] The *e.e.*$_p$ value (>99.9%) means that no (*R*)-NBHP peak was detected during GC analyses.

3. Materials and Methods

3.1. Strain and Growth Condition

The strain *R. erythropolis* WZ010 was deposited in the China Center for Type Culture Collection (CCTCC M 2011336) and used as the donor of the gene *recr* encoding the carbonyl reductase ReCR [35]. The host strains *E. coli* Trans1-T1 and *E. coli* BL21(DE3) were used for the purposes of cloning and over-expression, respectively. Both *R. erythropolis* WZ010 and *E. coli* strains were cultured at 30 °C and 200 rpm for 24 h in Luria-Bertani (LB) medium with a NaCl concentration of 5 g/L, unless stated otherwise.

3.2. Construction, Expression, and Purification of Recombinant Enzyme ReCR

The gene *recr* was PCR-amplified from the genomic DNA of *R. erythropolis* WZ010 using forward and reverse primers: *recr*F1 (5′-ATGAAGGCAATCCAGTACAC-3′) and *recr*R1 (5′-CTACAGACCAG GGACCACA-3′). The PCR conditions were listed as follows: denaturalization, 94 °C for 4 min; 30 cycles of 94 °C for 30 s, 53.5 °C for 30 s, and 72 °C for 1 min; and the final extension, 72 °C for 10 min. According to TA cloning strategy from the instructions of the pEASY-E2 expression kit (TransGen Biotech Co., Ltd., Beijing, China), the PCR product was subcloned into the expression vector pEASY-E2 to form the recombinant vector pEASY-E2-*recr* with the C-terminal His-tag. The recombinant plasmid was then transformed into Trans1-T1 competent cells and the recombinant cells were cultured at 37 °C and 200 rpm in LB medium with 100 μg/mL ampicillin (Amp). The recombinant cell named as *E. coli* Trans1-T1/pEASY-E2-*recr* was selected by colony PCRs and the recombinant plasmid pEASY-E2-*recr* was further extracted and verified by DNA sequencing (Sunny Biotechnology, Shanghai, China).

The recombinant plasmid pEASY-E2-*recr* was extracted and then transformed into *E. coli* BL21(DE3) competent cells. The positive recombinant cell named as *E. coli* BL21(DE3)/pEASY-E2-*recr* was cultured at 37 °C and 200 rpm in LB medium with 100 μg/mL Amp. When the OD$_{600}$ reached 0.6, isopropyl β-D-1-thiogalactopyranoside (IPTG) was added to the culture at a final concentration of 0.3 mM, and the temperature was maintained at 20 °C. After 20 h incubation, the *E. coli* cells were harvested by centrifugation and the expression level was analyzed by sodium dodecyl sulfate-polyacrylamide gel electrophoresis (SDS-PAGE). Following the same procedure in the study of 2,3-butanediol dehydrogenase from *R. erythropolis* WZ010 [24], the recombinant ReCR with C-terminal His-tag was purified to homogeneity by nickel affinity chromatography, desalted with 50 mM Tris-HCl (pH 8.0) by ultrafiltration, and stored at −20 °C for further characterization. The subunit molecular mass and purity of ReCR were verified by SDS-PAGE as described previously [36].

3.3. Enzyme Activity Assays and Characterization of Recombinant ReCR

The ReCR enzyme activity was measured by the reduction of NAD$^+$ or oxidation of NADH at 340 nm (ε_{340} = 6.3 mM^{-1} cm^{-1}). Unless otherwise specified, the standard enzyme activity assay for the ketone reduction was performed at 60 °C in duplicate using the assay mixture (2.5 mL) containing 10 mM NBPO, 0.4 mM NADH, and 50 mM PIPES buffer (pH 6.0). The standard assay mixture (2.5 mL) for the alcohol oxidation at 50 °C contained 50 mM (R/S)-2-octanol, 0.4 mM NAD$^+$, and 50 mM CAPSO buffer (pH 10.0). Unless stated otherwise, the reduction and oxidation reactions were initiated by the addition of 5 µg purified enzyme, respectively. One unit of activity was defined as the amount of enzyme that oxidized or reduced 1 µmol NADH or NAD$^+$ per minute under optimal pH and temperature. The protein concentrations of ReCR samples were determined using the Bradford reagent with bovine serum albumin as the standard protein.

The optimal temperature of ReCR activity was determined at a series of temperatures ranging from 25 to 70 °C using 50 mM PIPES buffer (pH 6.0) for NBPO reduction or 50 mM CAPSO buffer (pH 10.0) for (R/S)-2-octanol oxidation. The optimal pH of ReCR activity was determined over a range of pH from 5.5 to 11.0 at 60 °C for NBPO reduction or 50 °C for (R/S)-2-octanol oxidation. The buffers (50 mM) used were 2-(N-morpholino)ethanesulfonic acid (MES, pH 5.5–6.0), piperazine-1,4-bisethanesulfonic acid (PIPES, pH 6.1–7.5), Tris-HCl (pH 7.5–9.0), 3-(cyclohexylamino)-2-hydroxy-1-propanesulfonic acid (CAPSO, pH 9.0–10.0), and 3-(cyclohexylamino)-1-propanesulfonic acid (CAPS, pH 10.0–11.0). All the pH values of the buffers used were determined at 25 °C using a Mettler Toledo FE20 FiveEasy pH Meter (Mettler-Toledo (Schweiz) GmbH, Greifensee, Switzerland).

The thermostability of the ReCR was investigated by determining its residual activities when the enzyme samples were incubated at 35 °C, 55 °C, or 60 °C. To determine the stability in the presence of organic solvents, the enzyme was incubated with organic solvent at 35 °C for 3.5 h and then the residual activities were assayed for NBPO reduction. The determination of kinetic constants for ReCR was carried out using different substrates. The substrates were NBPO (0–20 mM), acetone (0–1 M), 2-propanol (0–70 mM), 2-octanone (0–30 mM), and (R/S)-2-octanol (0–20 mM). Apparent values of K_m and V_{max} were calculated using a non-linear regression curve fitting to the Michaelis-Menten equation with the software Origin 8.0 (OriginLab Corporation, Northampton, UK). Data of kinetic parameters present mean values ± SD from three independent experiments.

3.4. Asymmetric Reduction of NBPO Catalyzed by Whole Cells of E. coli BL21(DE3)/pEASY-E2-recr

The asymmetric reduction of NBPO was carried out using (R/S)-2-octanol or 2-propanol as a co-substrate for the coenzyme regeneration. In the case of 2-propanol as the co-substrate, the reaction mixture (5 mL) contained 0.5 M NBPO, 10% (v/v) 2-propanol, 0.4 mM NAD$^+$, and 0.4 g wet cells in 50 mM Tris-HCl buffer (pH 8.0). In the aqueous/(R/S)-2-octanol biphasic system, the reaction mixture (5 mL) contained 1.5 M NBPO, 60% (v/v) (R/S)-2-octanol, 1.2 mM NAD$^+$, and 1.2 g wet cells in 50 mM Tris-HCl buffer (pH 8.0). The reactions were carried out in a C76 Water Bath Shaker (New Brunswick, Edison, NJ, USA) at 35 °C and 300 rpm for 12 h.

The reaction mixture was extracted with 5 mL of ethyl acetate under strong vibration. The organic phase in the samples was separated by centrifugation and dehydrated with anhydrous sodium sulfate; then, 1 µL dehydrated sample was applied onto the injector (250 °C) for GC analysis. The reactants were determined with an Agilent 6890N (Santa Clara, CA, USA) gas chromatograph equipped with a chiral GC column (BGB174, 30 m × 250 µm × 0.25 µm). The temperature program for GC analysis was set as follows: 5 °C/min from 100 °C to 125 °C, hold 3 min; 2 °C/min to 140 °C, hold 8 min; 1 °C/min to 150 °C. The peak areas were quantitated using specific external standards. The standards NBPO, (S)-NBHP, and (R)-NBHP were purchased from Sigma-Aldrich Corporation (Shanghai, China). Retention times of the reactants were listed as follows: 26.997 min for NBPO, 28.452 min for (S)-NBHP, and 28.739 min for (R)-NBHP (Figure S1). Specifically, the (S)-NBHP peak was further determined by GC-MS analysis (Figure S2).

3.5. Construction, Characterization, and Docking Analysis of ReCR Variant Y54F

Site-specific mutagenesis was carried out by inverse PCR using native pEASY-E2-*recr* as a template and a pair of primers Y54F F1 (5'-TACACCTTCGGCCTTCCTCTCACGC-3') and Y54F

R1 (5'-AAGGCCGAAGGTGTACTGCTCCTCG-3') under conditions as follows: denaturation, 95 °C for 2 min; 30 cycles of 95 °C for 20 s, 68 °C for 20 s, and 72 °C for 3 min; and the final extension, 72 °C for 8 min. The PCR product was digested at 37 °C for 2 h to digest the native template with the help of *Dpn* I. The digested product was directly transformed into *E. coli* BL21(DE3) competent cells. The positive recombinant cells were cultured at 37 °C and 200 rpm in LB medium with 100 μg/mL Amp. The recombinant cell named as *E. coli* Trans1-T1/pEASY-E2-*recr-mut* was selected by colony PCRs and the recombinant plasmid pEASY-E2-*recr-mut* was further extracted and verified by DNA sequencing (Sunny Biotechnology, Shanghai, China). Following the same procedure for the recombinant ReCR, the positive recombinant cell named as *E. coli* BL21(DE3)/pEASY-E2-*recr-mut* was obtained and the ReCR variant Y54F was purified for further characterization including kinetic parameters and catalytic performance in NBPO reduction.

The homology model of ReCR was built on the X-ray crystallographic structures of ADH-A from *Rhodococcus ruber* (PDB: 2XAA, resolution of 2.8 Å) by HHpred server [37]. Water molecules, ligands, and other hetero atoms (except the NAD$^+$ coenzyme and the zinc ion) were removed from the protein molecule. The coenzyme was remodeled as NADH. The charge of the catalytic zinc ion was assigned to +2, and the ligating side chain of Cys 38 was set as deprotonated and negatively charged. For the homology model of ReCR Y54F, the substitution of Tyr54 to Phe was introduced by FoldX [38]. A structure energy minimization of the proteins was performed to remove improper torsions of the side-chain conformation and correct the covalent geometry. The ligand molecule structures (NBPO and NADH) were directly drawn in ChemBioDraw and followed by an energy minimization. Global docking was performed using AutoDock Vina under the default docking parameters [39]. Point charges were initially assigned according to the AMBER03 force field [40], and then damped to mimic the less polar Gasteiger charges. Subsequently, local docking was executed to predict the binding energy and fine-tune the ligand placement in the binding site.

3.6. Nucleotide Sequence Accession Number

The nucleotide sequence of ReCR has been submitted to the GenBank database under the accession number of KX827723.

4. Conclusions

The enzyme ReCR showed high specific activity, moderate thermostability, and strict (*S*)-stereoselectivity for asymmetric bioreduction of NBPO to (*S*)-NBHP. The NAD(H)-specific enzyme was active over broad pH and temperature ranges, and tolerated a higher concentration of organic solvents, offering greater flexibility in practical biocatalysis. Particularly, the reduction of NBPO to (*S*)-NBHP was irreversible, which was kinetically in favor of both coenzyme regeneration and formation of (*S*)-NBHP. The substitution of Tyr54 to Phe further improved the catalytic efficiency of ReCR including kinetic parameters and the productivity of (*S*)-NBHP. The k_{cat}/K_m values of ReCR Y54F for NBPO (49.17 s^{-1} mM^{-1}), (*R/S*)-2-octanol (56.56 s^{-1} mM^{-1}), and 2-propanol (20.69 s^{-1} mM^{-1}) were 1.37, 4.34, and 2.12 times higher than those of ReCR (35.98 s^{-1} mM^{-1}, 13.04 s^{-1} mM^{-1}, and 9.74 s^{-1} mM^{-1}), respectively. Furthermore, the (*S*)-NBHP yield was increased from 77.78% to 95.89% in the aqueous/(*R/S*)-2-octanol biphasic system when asymmetric reduction of 1.5 M NBPO was catalyzed for 12 h by whole cells of *E. coli* overexpressing ReCR Y54F instead of ReCR. Taken as a whole, ReCR variant Y54F has a great potential in the asymmetric synthesis of (*S*)-NBHP using (*R/S*)-2-octanol or 2-propanol as a co-substrate.

Author Contributions: Conceptualization, X.Y. and M.Y. (Meirong Ying); Data curation, X.Y. and M.Y. (Meilan Yu); Formal analysis, X.Y., F.C. and M.Y. (Meirong Ying); Funding acquisition, X.Y. and M.Y. (Meilan Yu); Investigation, J.Z., C.W., M.H., Y.J. and F.C.; Supervision, Z.W.; Writing – original draft, X.Y. and M.Y. (Meirong Ying).

References

1. Babu, M.S.; Raghunadh, A.; Ramulu, K.; Dahanukar, V.H.; Kumar, U.K.S.; Dubey, P.K. A practical and enantiospecific synthesis of (−)-(R)- and (+)-(S)-piperidin-3-ols. *Helv. Chim. Acta* **2014**, *97*, 1507–1515. [CrossRef]

2. Vitaku, E.; Smith, D.T.; Njardarson, J.T. Analysis of the structural diversity, substitution patterns, and frequency of nitrogen heterocycles among U.S. FDA approved pharmaceuticals. *J. Med. Chem.* **2014**, *57*, 10257–10274. [CrossRef] [PubMed]

3. Chen, L.-F.; Zhang, Y.-P.; Fan, H.-Y.; Wu, K.; Lin, J.-P.; Wang, H.-L.; Wei, D.-Z. Efficient bioreductive production of (R)-N-Boc-3-hydroxypiperidine by a carbonyl reductase. *Catal. Commun.* **2017**, *97*, 5–9. [CrossRef]

4. Ju, X.; Tang, Y.; Liang, X.; Hou, M.; Wan, Z.; Tao, J. Development of a biocatalytic process to prepare (S)-N-Boc-3-hydroxypiperidine. *Org. Process Res. Dev.* **2014**, *18*, 827–830. [CrossRef]

5. Amat, M.; Llor, N.; Huguet, M.; Molins, E.; Espinosa, E.; Bosch, J. Unprecedented oxidation of a phenylglycinol-derived 2-pyridone: Enantioselective synthesis of polyhydroxypiperidines. *Org. Lett.* **2001**, *3*, 3257–3260. [CrossRef] [PubMed]

6. Zhang, Y.-J.; Zhang, W.-X.; Zheng, G.-W.; Xu, J.-H. Identification of an ε-keto ester reductase for the efficient synthesis of an (R)-α-lipoic acid precursor. *Adv. Synth. Catal.* **2015**, *357*, 1697–1702. [CrossRef]

7. Lacheretz, R.; Pardo, D.G.; Cossy, J. *Daucus carota* mediated-reduction of cyclic 3-oxo-amines. *Org. Lett.* **2009**, *11*, 1245–1248. [CrossRef] [PubMed]

8. Chen, L.-F.; Fan, H.-Y.; Zhang, Y.-P.; Wu, K.; Wang, H.-L.; Lin, J.-P.; Wei, D.-Z. Development of a practical biocatalytic process for (S)-N-Boc-3-hydroxypiperidine synthesis. *Tetrahedron Lett.* **2017**, *58*, 1644–1650. [CrossRef]

9. Xu, G.-P.; Wang, H.-B.; Wu, Z.-L. Efficient bioreductive production of (S)-N-Boc-3-hydroxypiperidine using ketoreductase ChKRED03. *Proc. Biochem.* **2016**, *51*, 881–885. [CrossRef]

10. Hummel, W.; Groger, H. Strategies for regeneration of nicotinamide coenzymes emphasizing self-sufficient closed-loop recycling systems. *J. Biotechnol.* **2014**, *191*, 22–31. [CrossRef] [PubMed]

11. Stamper, W.; Kosjek, B.; Faber, K.; Kroutil, W. Biocatalytic asymmetric hydrogen transfer employing *Rhodococcus ruber* DSM 44541. *J. Org. Chem.* **2003**, *68*, 402–406. [CrossRef] [PubMed]

12. Huang, L.; Ma, H.-M.; Yu, H.-L.; Xu, J.-H. Altering the substrate specificity of reductase CgKR1 from *Candida glabrata* by protein engineering for bioreduction of aromatic α-keto esters. *Adv. Synth. Catal.* **2014**, *356*, 1943–1948. [CrossRef]

13. Turner, N.J.; O'Reilly, E. Biocatalytic retrosynthesis. *Nat. Chem. Biol.* **2013**, *9*, 285–288. [CrossRef] [PubMed]

14. Itoh, N.; Isotani, K.; Nakamura, M.; Inoue, K.; Isogai, Y.; Makino, Y. Efficient synthesis of optically pure alcohols by asymmetric hydrogen-transfer biocatalysis: Application of engineered enzymes in a 2-propanol-water medium. *Appl. Microbiol. Biotechnol.* **2012**, *93*, 1075–1085. [CrossRef] [PubMed]

15. Makino, Y.; Inoue, K.; Dairi, T.; Itoh, N. Engineering of phenylacetaldehyde reductase for efficient substrate conversion in concentrated 2-propanol. *Appl. Environ. Microbiol.* **2005**, *71*, 4713–4720. [CrossRef] [PubMed]

16. Au, S.K.; Bommarius, B.R.; Bommarius, A.S. Biphasic reaction system allows for conversion of hydrophobic substrates by amine dehydrogenases. *ACS Catal.* **2014**, *4*, 4021–4026. [CrossRef]

17. De Gonzalo, G.; Lavandera, I.; Faber, K.; Kroutil, W. Enzymatic reduction of ketones in "micro-aqueous" media catalyzed by ADH-A from *Rhodococcus ruber*. *Org. Lett.* **2007**, *9*, 2163–2166. [CrossRef] [PubMed]

18. Itoh, N.; Matsuda, M.; Mabuchi, M.; Dairi, T.; Wang, J. Chiral alcohol production by NADH-dependent phenylacetaldehyde reductase coupled with in situ regeneration of NADH. *Eur. J. Biochem.* **2002**, *269*, 2394–2402. [CrossRef] [PubMed]

19. Xu, G.-C.; Tang, M.-H.; Ni, Y. Asymmetric synthesis of lipitor chiral intermediate using a robust carbonyl reductase at high substrate to catalyst ratio. *J. Mol. Catal. B: Enzym.* **2016**, *123*, 67–72. [CrossRef]

20. Chen, X.; Liu, Z.-Q.; Lin, C.-P.; Zheng, Y.-G. Efficient biosynthesis of ethyl (R)-4-chloro-3-hydroxybutyrate using a stereoselective carbonyl reductase from *Burkholderia gladioli*. *BMC Biotechnol.* **2016**, *16*, 70. [CrossRef] [PubMed]

21. Nealon, C.M.; Musa, M.M.; Patel, J.M.; Phillips, R.S. Controlling substrate specificity and stereospecificity of alcohol dehydrogenases. *ACS Catal.* **2015**, *5*, 2100–2114. [CrossRef]

22. Stepankova, V.; Bidmanova, S.; Koudelakova, T.; Prokop, Z.; Chaloupkova, R.; Damborsky, J. Strategies for stabilization of enzymes in organic solvents. *ACS Catal.* **2013**, *3*, 2823–2836. [CrossRef]

23. Wang, Z.; Song, Q.; Yu, M.; Wang, Y.; Xiong, B.; Zhang, Y.; Zheng, J.; Ying, X. Characterization of a stereospecific acetoin(diacetyl) reductase from *Rhodococcus erythropolis* WZ010 and its application for the synthesis of (2*S*,3*S*)-2,3-butanediol. *Appl. Microbiol. Biotechnol.* **2014**, *98*, 641–650. [CrossRef] [PubMed]

24. Yu, M.; Huang, M.; Song, Q.; Shao, J.; Ying, X. Characterization of a (2*R*,3*R*)-2,3-butanediol dehydrogenase from *Rhodococcus erythropolis* WZ010. *Molecules* **2015**, *20*, 7156–7173. [CrossRef] [PubMed]

25. Abokitse, K.; Hummel, W. Cloning, sequence analysis, and heterologous expression of the gene encoding a (*S*)-specific alcohol dehydrogenase from *Rhodococcus erythropolis* DSM 43297. *Appl. Microbiol. Biotechnol.* **2003**, *62*, 380–386. [CrossRef] [PubMed]

26. Itoh, N.; Morihama, R.; Wang, J.; Okada, K.; Mizuguchi, N. Purification and characterization of phenylacetaldehyde reductase from a styrene-assimilating *Corynebacterium* strain, ST-10. *Appl. Environ. Microbiol.* **1997**, *63*, 3783–3788. [PubMed]

27. Kasprzak, J.; Bischoff, F.; Rauter, M.; Becher, K.; Baronian, K.; Bode, R.; Schauer, F.; Vorbrodt, H.-M.; Kunze, G. Synthesis of 1-(*S*)-phenylethanol and ethyl (*R*)-4-chloro-3-hydroxybutanoate using recombinant *Rhodococcus erythropolis* alcohol dehydrogenase produced by two yeast species. *Biochem. Eng. J.* **2016**, *106*, 107–117. [CrossRef]

28. Makino, Y.; Dairi, T.; Itoh, N. Engineering the phenylacetaldehyde reductase mutant for improved substrate conversion in the presence of concentrated 2-propanol. *Appl. Microbiol. Biotechnol.* **2007**, *77*, 833–843. [CrossRef] [PubMed]

29. Karabec, M.; Łyskowski, A.; Tauber, K.C.; Steinkellner, G.; Kroutil, W.; Grogan, G.; Gruber, K. Structural insights into substrate specificity and solvent tolerance in alcohol dehydrogenase ADH-A from *Rhodococcus ruber* DSM 44541. *Chem. Commun.* **2010**, *46*, 6314–6316. [CrossRef] [PubMed]

30. Ying, X.; Wang, Y.; Xiong, B.; Wu, T.; Xie, L.; Yu, M.; Wang, Z. Characterization of an allylic/benzyl alcohol dehydrogenase from *Yokenella* sp. strain WZY002, an organism potentially useful for the synthesis of α,β-unsaturated alcohols from allylic aldehydes and ketones. *Appl. Environ. Microbiol.* **2014**, *80*, 2399–2409. [CrossRef] [PubMed]

31. Martinez-Rojas, E.; Kurt, T.; Schmidt, U.; Meyer, V.; Garbe, L.-A. A bifunctional enzyme from *Rhodococcus erythropolis* exhibiting secondary alcohol dehydrogenase-catalase activities. *Appl. Microbiol. Biotechnol.* **2014**, *98*, 9249–9258. [CrossRef] [PubMed]

32. Kratzer, R.; Woodley, J.M.; Nidetzky, B. Rules for biocatalyst and reaction engineering to implement effective, NAD(P)H-dependent, whole cell bioreductions. *Biotechnol. Adv.* **2015**, *33*, 1641–1652. [CrossRef] [PubMed]

33. Glonke, S.; Sadowski, G.; Brandenbusch, C. Applied catastrophic phase inversion: A continuous non-centrifugal phase separation step in biphasic whole-cell biocatalysis. *J. Ind. Microbiol. Biotechnol.* **2016**, *43*, 1527–1535. [CrossRef] [PubMed]

34. Wei, L.; Zhang, M.; Zhang, X.; Xin, H.; Yang, H. Pickering emulsion as an efficient platform for enzymatic reactions without stirring. *ACS Sustain. Chem. Eng.* **2016**, *4*, 6838–6843. [CrossRef]

35. Yang, C.; Ying, X.; Yu, M.; Zhang, Y.; Xiong, B.; Song, Q.; Wang, Z. Towards the discovery of alcohol dehydrogenases: NAD(P)H fluorescence-based screening and characterization of the newly isolated *Rhodococcus erythropolis* WZ010 in the preparation of chiral aryl secondary alcohols. *J. Ind. Microbiol. Biotechnol.* **2012**, *39*, 1431–1443. [CrossRef] [PubMed]

36. Laemmli, U.K. Cleavage of structural proteins during the assembly of the head of bacteriophage T4. *Nature* **1970**, *227*, 680–685. [CrossRef] [PubMed]

37. Söding, J.; Biegert, A.; Lupas, A.N. The HHpred interactive server for protein homology detection and structure prediction. *Nucleic Acids Res.* **2005**, *33*, W244–W248. [CrossRef] [PubMed]

38. Schymkowitz, J.; Borg, J.; Stricher, F.; Nys, R.; Rousseau, F.; Serrano, L. The FoldX web server: An online force field. *Nucleic Acids Res.* **2005**, *33*, W382–W388. [CrossRef] [PubMed]

39. Trott, O.; Olson, A.J. AutoDock Vina: Improving the speed and accuracy of docking with a new scoring function, efficient optimization, and multithreading. *J. Comput. Chem.* **2010**, *31*, 455–461. [CrossRef] [PubMed]

40. Duan, Y.; Wu, C.; Chowdhury, S.; Lee, M.C.; Xiong, G.; Zhang, W.; Yang, R.; Cieplak, P.; Luo, R.; Lee, T.; et al. A point-charge force field for molecular mechanics simulations of proteins based on condensed-phase quantum mechanical calculations. *J. Comput. Chem.* **2003**, *24*, 1999–2012. [CrossRef] [PubMed]

7

Antifungal Activity against *Botrytis cinerea* of 2,6-Dimethoxy-4-(phenylimino)cyclohexa-2,5-dienone Derivatives

Paulo Castro [1,*], Leonora Mendoza [1], Claudio Vásquez [2], Paz Cornejo Pereira [1], Freddy Navarro [1], Karin Lizama [1], Rocío Santander [3] and Milena Cotoras [1]

[1] Laboratorio de Micología, Facultad de Química y Biología, Universidad de Santiago de Chile, Avenida Libertador Bernardo O'Higgins 3363, Santiago 518000, Chile; leonora.mendoza@usach.cl (L.M.); paz.cornejo@usach.cl (P.C.P.); freddy.navarro@usach.cl (F.N.); karin.lizama@usach.cl (K.L.); milena.cotoras@usach.cl (M.C.)
[2] Laboratorio de Microbiología Molecular, Departamento de Biología, Facultad de Química y Biología, Universidad de Santiago de Chile, Santiago 518000, Chile; claudio.vasquez@usach.cl
[3] Departamento de Ciencias del Ambiente, Facultad de Química y Biología, Universidad de Santiago de Chile, Casilla 40 Correo 33, Santiago 518000, Chile; rocio.santanderm@usach.cl
* Correspondence: paulo.castro@usach.cl.

Academic Editor: Stefano Serra

Abstract: In this work the enzyme laccase from *Trametes versicolor* was used to synthetize 2,6-dimethoxy-4-(phenylimino)cyclohexa-2,5-dienone derivatives. Ten products with different substitutions in the aromatic ring were synthetized and characterized using ^{1}H- and ^{13}C-NMR and mass spectrometry. The 3,5-dichlorinated compound showed highest antifungal activity against the phytopathogen *Botrytis cinerea*, while the *p*-methoxylated compound had the lowest activity; however, the antifungal activity of the products was higher than the activity of the substrates of the reactions. Finally, the results suggested that these compounds produced damage in the fungal cell wall.

Keywords: *Botrytis cinerea*; antifungal activity; laccase; 2,6-dimethoxy-4-(phenylimino)cyclohexa-2,5-dienone derivatives

1. Introduction

Botrytis cinerea is a phytopathogenic fungus promoted by the presence of free surface water or high relative humidity and causing significant crop losses in a wide variety of plant species [1]. Regarding the control, methods aiming to reduce humidity can be combined to help decrease this disease, in addition to chemical fungicides or biocontrol treatments [1]. Chemical control is the most common way to manage *B. cinerea*, mainly using synthetic compounds [1]. The restriction of this type of control becomes necessary to reduce the impact on the environment [2] and to avoid the acquired resistance to botrycides [3–7]. For this reason, the development of new antifungal compounds is essential. Natural products can be a good alternative to commercial fungicides [8,9]. For instance, phenolic compounds, terpenoids, nitrogen-containing compounds, and aliphatic compounds isolated from plants have shown antifungal activities [10–12]. Additionally, new antifungal compounds against *B. cinerea* derived of natural products have been synthesized, such as derivatives of natural stilbene resveratrol [13], chlorophenyl derivatives [14], or different clovanes [15].

Several phenolic metabolites found in grape pomace have shown low antifungal activity against *B. cinerea* [16], therefore, it is possible to increase the biological activity of phenolic compounds using the enzyme laccase [17]. These enzymes (benzenediol: oxygen oxidoreductase, EC 1.10.3.2) belong to the oxidase group, and they are also used for cleaner industrial application [18]. Laccases are also known as multicopper oxidases, they belong to the family of copper-containing phenol oxidases [19] and can oxidize a diversity of compounds, e.g., phenolic and nonphenolic compounds [18]. Aromatic compounds can produce reactive radical intermediates, which undergo self-coupling reactions, thus forming different dimers and trimers [20–24]. This enzyme has been previously used to improve the activity of antibiotics [25,26]. On the other hand, the synthesis of a heterodimeric compound (2,6-dimethoxy-4-(phenylimino)cyclohexa-2,5-dienone) by the laccase-mediated coupling reaction between syringic acid and aniline was reported, this compound showed an antifungal effect against *B. cinerea* with an EC_{50} value of 0.14 mM [27].

Antifungal compounds have shown several inhibition mechanisms related to the molecular structure. For instance, the resveratrol derivative (*E*)-3,5-dimethoxy-β-(2-furyl)-styrene cause cell membrane damage against *B. cinerea* [13]. Phenylpyrroles induce morphological alterations of germ tubes [28]. Fungicides such as dinocap and fuazinam have been described as uncouplers of oxidative phosphorylation [29,30] and fungicides, like dicloran, cloroneb, and etazol, affect cell wall synthesis [28].

This work aimed to determine the antifungal activity against *B. cinerea* and the effect on the cell wall integrity of ten 2,6-dimethoxy-4-(phenylimino)cyclohexa-2,5-dienone derivatives (**3a–j**) obtained by reaction of syringic acid (**1**) with substituted anilines (**2a–j**). To analyze the effect of the carboxylic group in these laccase-catalyzed reactions, syringaldehyde was used instead syringic acid and the reaction product was characterized.

2. Results and Discussion

2.1. Laccase-Mediated Synthesis of 2,6-Dimethoxy-4-(phenylimino)cyclohexa-2,5-dienone Derivatives

In this work, laccase catalyzed reactions using **1** and **2a–j** were carried out. It has been previously reported that using laccases from different fungal sources (*Trametes* sp. and *Rhizoctonia praticola*), catalyze reactions between phenolic compounds and anilines, heterodimeric compounds are formed [25,26,31,32], similarly found in this work (Scheme 1).

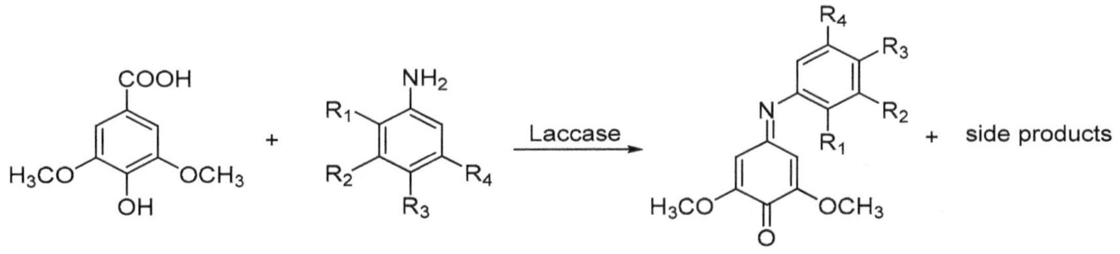

Compounds	R_1	R_2	R_3	R_4
2a/3a	H	H	Cl	H
2b/3b	H	Cl	H	H
2c/3c	H	H	OCH_3	H
2d/3d	H	OCH_3	H	H
2e/3e	H	H	NO_2	H
2f/3f	H	Cl	Cl	H
2g/3g	H	H	CF_3	H
2h/3h	H	Cl	H	Cl
2i/3i	Cl	H	H	Cl
2j/3j	Cl	H	H	H

Scheme 1. Reaction scheme for laccase-mediated synthesis between **1** and **2a–j**.

To determine the reaction yields in the formation of the products, different substrate ratios were analyzed. Excluding reactions 1, 7 and 10, most reactions reached higher yields using ratio 1:1 (syringic acid:aniline) (Table 1). Moreover, when aniline was used as substrate, the same result was reported [27], indicating that the increase of the concentration of one of them decrease the yield of the obtained compounds.

Highest yields were obtained using 3-chloroaniline and 3,5-dichloroaniline as substrates (**2b** and **2h**) (Table 1). This high yield could be explained because the oxidation by laccase (from *Trametes versicolor*) of 3-chloroaniline does not occur [33]. On the other hand, using methoxyanilines (**2c** and **2d**) low yields were obtained, due to a high amount of side products (data not shown).

On the other hand, yield did not increase when the enzyme concentration was increased (data not shown). Bollag et al. [31] showed that the prolonged incubation or higher enzyme amounts caused further polymerization reaction decreasing cross-coupling formation. Furthermore, Itoh et al. [34] concluded that reactivity of laccase mediated reaction between phenolic acids and chlorophenols is due to the substrate specificity of the laccase rather than the chemical property of the substrates, which could explain the lack of relations among electron donating and withdrawing groups and yield of the reactions.

Table 1. Percentage yields of products at different reactant ratios.

Reaction	Product	Yield (%) Substrate Ratio (Syringic Acid: Aniline)		
		1:2	1:1	2:1
1	3a	10.4	36.9	55.5
2	3b	38.1	72.0	24.1
3	3c	15.5	26.7	ND [1]
4	3d	8.2	24.2	4.1
5	3e	22.6	56.8	36.4
6	3f	41.8	50.4	44.8
7	3g	12.7	10.2	8.2
8	3h	15.8	74.0	29.0
9	3i	6.9	38.3	23.9
10	3j	39.7	13.6	44.9

[1] ND: Not determined.

The ten synthetized compounds were purified using semipreparative chromatography and were identified (Figure 1) using ^{1}H-NMR and ^{13}C-NMR spectra and mass spectrometry (Figure S1–S30, Supplementary Materials). Compound **3b** showed two aliphatic proton signals (δ 3.670 (s, 3H, H8) and δ 3.874 (s, 3H, H7)) and two aliphatic carbon signals (δ 56.212 and δ 56.321) that determined the presence of two methoxy moieties. Two olefin hydrogen signals at higher fields (δ 6.010 (d, 1H, H3 J = 1.9 Hz) and δ 6.368 (d, 1H, H5 J = 1.9 Hz)), the olefin carbon signals (δ 98.583 (C3) and δ 111.717 (C5)) and one carbon signal at δ 176.633 (C1) indicated the quinonoid character of the products. Table 2 presents the NMR data (chemical shift assignments for short and long-range heteronuclear coupling) of compound **3b**.

The only difference in spectra signals between compounds **3a** and **3b** (Figures S2, S3, S5 and S6) was in the aromatic region of the spectra. Compound **3b** showed four aromatic proton signals δ 6.737 (d, 1H, H6', J = 8.0 Hz), δ 6.888 (s, 1H, H2'), δ 7.159 (d, 1H, H4', J = 8.0 Hz), and δ 7.316 (t, 1H, H5', J = 8.0 Hz) that indicated the existence of a *m*-substituted aromatic fragment. The assignation of the entire molecule was achieved by using two-dimensional NMR analysis. Therefore, identifying this compound as 4-(3'-chlorophenylimino)-2,6-dimethoxycyclohexa-2,5-dienone (Figure 1).

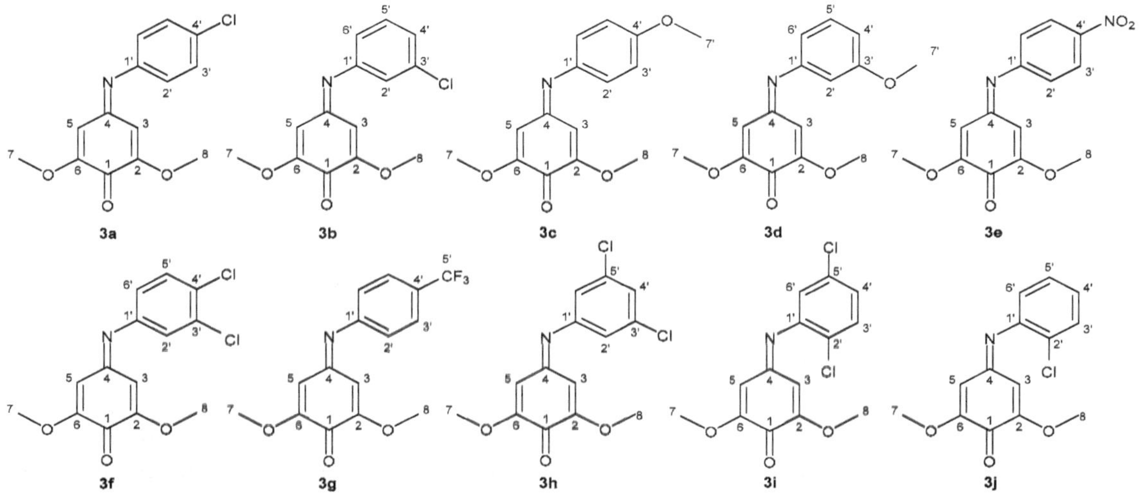

Figure 1. Structure of synthetized compounds **3a–j** 2,6-dimethoxy-4-(phenylimino)cyclohexa-2,5-dienone derivatives.

Table 2. ^{1}H- and ^{13}C- chemical shifts assignments for compound **3b**. Information from HSQC and HMBC experiments are also provided.

	^{13}C	^{1}H	HMBC	HSQC
1	176,633		5, 3	
4	157,816			
2	155.741		8	
6	154.881		7	
1′	151.483		5′	
3′	134.928		5′	
5′	130.247	7.316 (t, J 8.0 Hz, 1H)		5′
4′	125.023	7.159 (d, J 8.0 Hz, 1H)	6′, 2′	4′
2′	120.635	6.888 (s, 1H)	6′, 4′	2′
6′	118.705	6.737 (d, J 7.9 Hz, 1H)	4′, 2′	6′
5	111.717	6.368 (d, J 1.9 Hz, 1H)	3	5
3	98.583	6.010 (d, J 1.9 Hz, 1H)	5	3
7	56.321	3.874 (s, 3H)		7
8	56.212	3.670 (s, 3H)		8

The spectra of compounds **3c–j** (Figures S8, S9, S11, S12, S14, S15, S17, S18, S20, S21, S23, S24, S26, S27, S29 and S30) only showed differences in the aromatic region; the assignment of the ^{1}H and ^{13}C-NMR spectra can be found in the spectroscopic data section (Section 3.4). Figure 1 shows the structures of the ten synthetized compounds in this work. To our knowledge compounds **3a** and **3f** were previously synthesized [31], ^{1}H-NMR spectra for compounds **3a** and **3f** (spectroscopic data Section 3.4) have the same number of signals and comparable chemical shifts and coupling constants like those found by Bollag et al. [31]; furthermore, the mass spectra of **3a** and **3f** showed a base peak with m/z 277 and 311, respectively, corresponding to the molecular ions, equivalent to the previously described data [31]. Therefore, the other eight compounds (**3b**, **3c**, **3d**, **3e**, **3g**, **3h**, **3i** and **3j**) have not been previously reported.

Interesting, compound **3a** was also obtained using syringaldehyde instead of syringic acid in the reaction with 4-chloroaniline. This could be explained with an extra step when using syringaldehyde, an oxidation of the aldehyde to a carboxylic acid (syringic acid), similar oxidations has been previously described using several aromatic aldehydes with laccase, yielding carboxylic acids [35]. Hence, syringaldehyde is oxidized to syringic acid and then the same product (compound **3a**) could be found in both reactions, starting with syringaldehyde or with syringic acid. However, this synthesis had a very low yield (data not shown).

2.2. Antifungal Activity

Antifungal activity of compounds **1** and **2a–j** and compounds **3a–j** against *B. cinerea* was measured on mycelial growth in solid media and the EC_{50} were calculated using the mycelial growth (Tables 3 and 4). The most active compound was the 3,5-dichloro-substituted product (compound **3h**), while compound **3c** had the lowest activity. It has been reported that the substituent affects the antifungal activity of a molecule [36], for instance, the position of the chlorine atom in the aromatic ring is important for the antifungal activity against *B. cinerea* since para-substituted compound (**3a**) and ortho-substituted compound (**3j**) were more active than the meta-substituted compound and unsubstituted compound (EC_{50} = 0.14 ± 0.02) [27], while activity of meta-substituted compound (compound **3b**) and unsubstituted compound are similar [27]. The number of chlorine atoms in the aromatic ring is also important, both dichlorinated compounds **3f** and **3h** showed higher antifungal activity than mono chlorinated compounds **3a**, **3b**, and **3j**, however, dichlorinated compound **3i** showed an antifungal activity comparable to the monochlorinated compounds, therefore, the number and position of chlorine atoms in the aromatic ring seems to be important for the antifungal activity of these compounds.

Table 3. Effect of compounds **3a–j** on the mycelial growth of *B. cinerea* in solid medium.

Compound	EC_{50} (mM)
3a	0.065 ± 0.003
3b	0.15 ± 0.01
3c	0.54 ± 0.06
3d	0.39 ± 0.02
3e	- *
3f	0.055 ± 0.004
3g	0.101 ± 0.014
3h	0.032 ± 0.003
3i	0.065 ± 0.011
3j	0.069 ± 0.004

* No activity.

Table 4. Effect of the substrates on the mycelial growth of *B. cinerea* in solid medium.

Compound	EC_{50} (mM)
1	>1.51
2a	0.71 ± 0.08
2b	0.59 ± 0.04
2c	>3.00
2d	>3.00
2e	0.15 ± 0.01
2f	0.047 ± 0.005
2g	2.058 ± 0.434
2h	0.12 ± 0.02
2i	0.414 ± 0.031
2j	1.09 ± 0.08

Furthermore, the methoxy derivative compounds (**3c** and **3d**) were less active against the fungus than the other compounds, even the nonsubstituted compound **3** [27], the same effect was observed for aspirin derivatives, where the methoxy para-substituted derivative showed almost 30% less antifungal activity against *B. cinerea* than the chlorinated para-substituted compound [37]. Similar behavior was previously reported for oxadiazole derivatives when tested the activity of the methoxy meta-substituted oxadiazole derivative against *B. cinerea*, and its activity was less than half compared to the nonsubstituted compound [38]. Usually, the chloro-substituted compounds have higher antifungal activity in commercial fungicides, for example, chlorine compounds like boscalid, chlorothalonil, and iprodione have been used to control *B. cinerea* [6]. The antifungal activity of iprodione has been tested against this strain of *B. cinerea*, showing an EC_{50} of 0.015 ± 0.003 mM [27], this antifungal activity is in the same order of magnitude than the most active compound obtained in this work (**3h**). Additionally, *p*-nitro and *p*-trifluoromethyl compounds (**3e** and **3g**) were tested against this fungus, **3e** showed no antifungal activity, probably because of the low solubility of this molecule, for this reason **3e** was not used in further assays. Compound **3g** only showed an intermediate antifungal activity compared to the rest of the synthetized molecules in this work. Lastly, most of the substrates used in the reactions (i.e., **1** and **2a–j**) showed lower antifungal activity than the products (Table 4), only **2f** was more active than **3f**.

2.3. Effect on the Cell Wall Integrity of B. cinerea

To analyze the effect of the compounds on the cell wall integrity, the dye calcofluor white (CFW) was used. This dye binds to β-1,3 and β-1,4 polysaccharides, for example chitin, which is a primary component of the cell wall in fungi, and fluorescence of the hyphae can be detected [39]. Figure 2 shows the effect of compound **3a** on the cell wall of *B. cinerea*. Treatment with this compound showed lower fluorescence intensity than the negative control (acetone), indicating that this compound can damage the cell wall of this fungus. The same assay was performed using compounds **3b–j**. The ten synthesized compounds caused a decrease of the fluorescence intensity compared to the control; relative fluorescence intensity is observed in Figure 3. This result could be attributed to the toxicity of quinones, which could be connected to the production of reactive oxygen species (ROS) which cause oxidation of cell molecules [40]. Quinone derivative *N*-acetyl-*p*-benzoquinone imine (NAPQI) can react with nucleophiles such as thiol groups of proteins or glutathione [41,42]; this last molecule is an important antioxidant molecule in fungi [43]. On the other hand, some aromatic antifungal compounds have shown effects on cell wall synthesis [6] by inhibiting chitin and glucan synthases [44], enzymes that catalyze the synthesis of the main polymers of the cell wall in fungi.

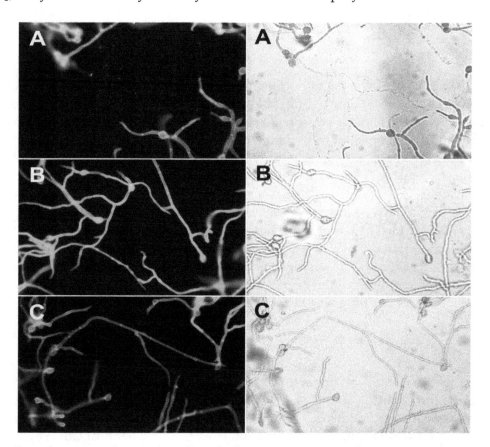

Figure 2. Effect of compound **3a** on the cell wall of *B. cinerea*. Hyphae of *B. cinerea* incubated with liquid medium along with (**A**) lysing enzymes (positive control), (**B**) acetone 5% (*v/v*) (negative control), and (**C**) compound **3a** at 0.16 mM. *B. cinerea* hyphae were treated with calcofluor white (CFW) stain. Assays were carried out in triplicate.

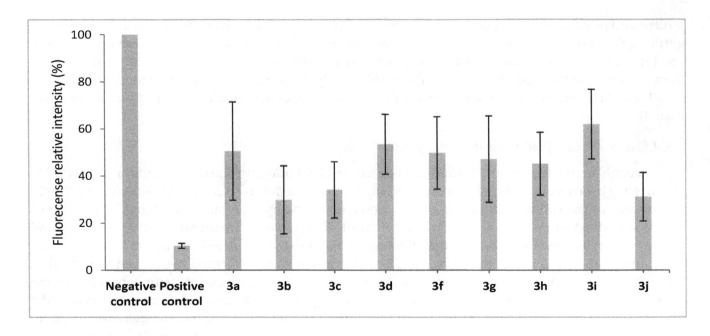

Figure 3. Effect of the compounds on the cell wall integrity of *B. cinerea*. The effect was measured as relative fluorescence intensity compared to maximum fluorescence (negative control).

3. Materials and Methods

3.1. General Experimental Procedures

The NMR spectra of **3a–j** were acquired using a Bruker Avance 400 MHz spectrometer (Bruker, Billerica, MA, USA) (400,133 MHz for ^{1}H, 100.624 MHz for ^{13}C). Measurements were done in CDCl$_3$ at 27 °C. Chemical shifts were calibrated to solvent signal: CHCl$_3$ 7.26 ppm (residual signal solvent) and 77.16 ppm for ^{1}H and ^{13}C, respectively, and informed relative to Me$_4$Si. Thin-layer chromatography was done with a Merck Kiesegel 60 F$_{254}$, 0.2 mm thick and semipreparative thin layer chromatography on Merck Kieselgel 60 F$_{254}$ 0.25 mm thick. A Thermo Scientific GC-MS system (GC: model: Trace 1300 and MS: model TSQ8000Evo) (Waltham, Massachusetts, USA) was used to analyze the sample. The separation was performed on a 60 m × 0.25 mm internal diameter fused silica capillary column coated with 0.25 μm film Rtx-5MS. The oven temperature was maintained at 40 °C for 5 min, then it was programmed from 40 to 80 at 5 °C/min for 1 min, then from 80 to 300 at 10 °C/min and finally maintained at 300 °C for 10 min. The mode used was splitless injection, helium was used as carrier gas, and flow-rate was 1.2 mL/min. Mass spectra were recorded over a range of 40 to 400 atomic mass units at 0.2 s/scan. Solvent cut time was 11 min. Ionization energy was 70 eV.

3.2. Chemical Reagents

Laccase from *Trametes versicolor* (EC 1.10.3.2), lysing enzymes from *Trichoderma harzianum*, Calcofluor white stain, 4-hydroxy-3,5-dimethoxy-benzoic acid (syringic acid), 3,5-dimethoxy-4-hydroxybenzaldehyde (syringaldehyde), 4-chloroaniline, 2,5-dichloroaniline, 3,5-dichloroaniline, and 4-nitroaniline were obtained from Sigma Chemical Co. (St. Louis, MO, USA). 3-chloroaniline, 4-methoxyaniline, 3-methoxyaniline, 3,4-dichloroaniline, 2-chloroaniline, organic solvents, and salts were obtained from Merck (Hohenbrunn, Germany). 4-(trifluoromethyl)aniline was obtained from Santa Cruz Biotechnology (Finnell St, Dallas Tx).Agar was obtained from Difco Laboratories (Detroit, MI, USA).

3.3. Laccase-Mediated Synthesis of 2,6-Dimethoxy-4-(phenylimino)cyclohexa-2,5-dienone Derivatives (Compounds 3a–j)

Syringic acid with an aniline derivative at different ratios (1:1, 1:2, and 2:1) (e.g., for ratio 1:1 means 0.1 mmol for both syringic acid and the aniline derivative were used) and different enzyme quantities (2.25, 4.5, and 9 U) were tested to increase the yield of the synthesized compounds.

For the first reaction, syringic acid (1) and 4-chloroaniline (2a) were dissolved in 1 mL ethyl acetate and laccase was dissolved in 1 mL sodium acetate buffer (20 mM, pH 4.5). Both solutions were mixed and stirred at 100 rpm for 180 min at 22 °C. Afterwards, the solvent was evaporated at 40 °C using a rotary evaporator. The synthetized compounds were purified by using semipreparative thin layer chromatography with hexane: ethyl acetate (1:1) as an eluent system. Same procedure was carried out using a different substituted aniline (2b–j).

Alternatively, compound 3a was also found when using syringaldehyde and 4-chloroaniline under the same conditions described above.

3.4. Spectroscopic Data

Compound **3a** *(4-(4′-chlorophenylimino)-2,6-dimethoxycyclohexa-2,5-dienone)* Yield 55.5%. ^{1}H-NMR (CDCl$_3$, 400 MHz) δ 3.668 (s, 3H, H8), 3.870 (s, 3H, H7), 6.040 (d, 1H, J = 1.9 Hz, H3), 6.377 (d, 1H, J = 1.9 Hz, H5), 6.816 (d, 2H, J = 8.5 Hz, H2′), 7.361(d, 2H, J = 8.5 Hz, H3′); ^{13}C-NMR (CDCl$_3$, 100 MHz) δ 56.188 (C8), 56.298 (C7), 98.529 (C3), 111.846 (C5), 122.040 (C2′), 129.333 (C3′), 130.749 (C4′), 148.777 (C1′), 154.813 (C6), 155.745 (C2), 157.642 (C4), 176.647 (C1). mp 208.0–209.1 °C. GC-MS RI$_{(Rtx-5ms)}$ = 2366, C$_{14}$H$_{12}$O$_3$NCl EI-MS m/z: 111 (15); 150 (16); 178 (15); 182 (17); 197 (35); 212 (15); 224 (22); [M]$^+$ = 277 (100); [M + 1]$^+$ = 278 (17); [M + 2]$^+$ = 279 (36).

Compound **3b** *(4-(3′-chlorophenylimino)-2,6-dimethoxycyclohexa-2,5-dienone)* Yield 72.0%. ^{1}H-NMR (CDCl$_3$, 400 MHz) δ 3.670 (s, 3H, H8), 3.874 (s, 3H, H7), 6.010 (d, 1H, J = 1.9 Hz, H3), 6.368 (d, 1H, J = 1.9 Hz, H5), 6.737 (d, 1H, J = 7.9 Hz, H6′), 6.888 (s, 1H, J = 8.0 Hz, H2′), 7.159 (d, 1H, J = 8,0 Hz, H4′), 7.316 (t, 1H, J = 7.9 Hz, H5′); ^{13}C-NMR (CDCl$_3$, 100 MHz) δ 56.212 (C8), 56.321 (C7), 98.583 (C3), 111.717 (C5), 118.705 (C6′), 120.635 (C2′), 125.023 (C4′), 130.247 (C5′), 134.928 (C3′), 151.483 (C1′), 154.881 (C6), 155.741 (C2), 157.816 (C4), 176.633 (C1). mp 155.7–156.0 °C. GC-MS RI$_{(Rtx-5ms)}$ = 2340, C$_{14}$H$_{12}$O$_3$NCl EI-MS m/z: 69 (22); 75 (52); 111 (58); 113 (21); 140 (20); 178 (26); 182 (24); 197 (43) [M]$^+$ = 277 (100); [M + 1]$^+$ = 278 (17); [M + 2]$^+$ =279 (36).

Compound **3c** *(4-(4′-methoxyphenylimino)-2,6-dimethoxycyclohexa-2,5-dienone)* Yield 23.7%. ^{1}H-NMR (CDCl$_3$, 400 MHz) δ 3.691 (s, 3H, H8), 3.846 (s, 3H, H7′), 3.870 (s, 3H, H7), 6.242 (d, 1H, J = 1.9 Hz, H3), 6.467 (d, 1H, J = 1.9 Hz, H5), 6.903 (d, 2H, J = 8.9 Hz, H2′), 6.956 (d, 2H, J = 8.9 Hz, H3′); ^{13}C-NMR (CDCl$_3$, 100 MHz) δ 55.656 (C7′), 56.104 (C8), 56.277 (C7), 98.966 (C3), 112.113 (C5), 114.594 (C3′), 123.047 (C2′), 143.196 (C1′), 154.661 (C6), 155.739 (C2), 156.809 (C4), 158.018 (C4′), 176.669 (C1). mp 111.7–112.3 °C. GC-MS RI$_{(Rtx-5ms)}$ = 2451, C$_{15}$H$_{15}$O$_4$N EI-MS m/z: 134 (10); 172 (20); 198 (12); 200 (9); 212 (9); 230 (30); 240 (12); 258 (45); [M]$^+$ = 273 (100); [M + 1]$^+$ = 274 (18).

Compound **3d** *(2,6-dimethoxy-4-(3′-methoxyphenylimino)cyclohexa-2,5-dienone)* Yield 24.24%. ^{1}H-NMR (CDCl$_3$, 400 MHz) δ 3.656 (s, 3H, H8), 3.819 (s, 3H, H7′), 3.868 (s, 3H, H7), 6.125 (d, 1H, J = 2.0 Hz, H3), 6.392 (d, 1H, J = 2.0 Hz, H5), 6.429 (d, 1H, J = 7.7 Hz, H6′), 6.452 (s, 1H, H2′), 6.736 (d, 1H, J = 8.1 Hz, H4′), 7.281 (t, 1H, J = 8.0 Hz, H5′); ^{13}C-NMR (CDCl$_3$, 100 MHz) δ 55.476 (C7′), 56.124 (C8), 56.248 (C7), 99.051 (C3), 106.387 (C2′), 110.942 (C4′), 112.023 (C5), 112.868 (C6′), 126.946 (C5′), 151.664 (C1′), 154.783 (C6), 155.569 (C2), 157.264 (C4), 160.402 (C3′), 176.785 (C1). mp 133.5–134.6 °C. GC-MS RI$_{(Rtx-5ms)}$ = 2403, C$_{15}$H$_{15}$O$_4$N EI-MS m/z: 159 (13); 187 (16); 199 (19); 200 (13); 212 (12); 215 (22); 230 (21); 242 (24) [M]$^+$ = 273 (100); [M + 1]$^+$ = 274 (18).

Compound **3e** *(2,6-dimethoxy-4-(4'-nitrophenylimino)cyclohexa-2,5-dienone)* Yield 56.8%. ^{1}H-NMR (CDCl$_3$, 400 MHz) δ 3.660 (s, 3H, H8), 3.895 (s, 3H, H7), 5.845 (d, 1H, J = 1.8 Hz, H3), 6.363 (d, 1H, J = 1.8 Hz, H5), 6.959 (d, 2H, J = 8.8 Hz, H2'), 8.279(d, 2H, J = 8.8 Hz, H3'); ^{13}C-NMR (CDCl$_3$, 100 MHz) δ 56.341 (C8), 56.439 (C7), 98.339 (C3), 111.172 (C5), 120.620 (C2'), 125.205 (C3'), 144.864 (C4'), 155.257 (C1'), 156.102 (C6), 156.170 (C2), 157.940 (C4), 176.321 (C1). mp 208.0–209.1 °C. GC-MS RI$_{(Rtx-5ms)}$ = 2690, C$_{14}$H$_{12}$O$_5$N$_2$ EI-MS m/z: 16 (33); 128 (21); 143 (29); 156 (25); 168 (18); 169 (20); 197 (38); 211 (26); [M]$^+$ = 288 (100); [M + 1]$^+$ = 289 (16).

Compound **3f** *(4-(3',4'-dichlorophenylimino)-2,6-dimethoxycyclohexa-2,5-dienone)* Yield 50.4%. ^{1}H-NMR (CDCl$_3$, 400 MHz) δ 3.690 (s, 3H, H8), 3.874 (s, 3H, H7), 5.982 (s, 1H, H3), 6.347 (s, 1H, H5), 6.715 (d, 1H, J = 8.4 Hz, H6'), 6.994 (s, 1H, H2'), 7.449 (d, 1H, J = 8.4 Hz, H5'); ^{13}C-NMR (CDCl$_3$, 100 MHz) δ 56.319 (C8), 56.355 (C7), 98.322 (C3), 111.572 (C5), 120.097 (C6'), 122.367 (C2'), 128.704 (C3'), 130.913 (C5'), 133.211 (C4'), 149.770 (C1'), 155.033 (C6), 155.973 (C2), 158.242 (C4), 176.486 (C1). mp 160.1–161.5 °C. GC-MS RI$_{(Rtx-5ms)}$ = 2556, C$_{14}$H$_{11}$O$_3$NCl$_2$ EI-MS m/z: 109 (22); 145 (18); 184 (21); 212 (20); 216 (26); 231 (41); 233 (28); 258 (24); [M]$^+$ = 311 (100); [M + 2]$^+$ = 313 (65); [M + 4]$^+$ = 315 (13).

Compound **3g** *(4-(4'-(trifluoromethyl)phenylimino)-2,6-dimethoxycyclohexa-2,5-dienone)* Yield 12.7%. ^{1}H-NMR (CDCl$_3$, 400 MHz) δ 3.643 (s, 3H, H8), 3.868 (s, 3H, H7), 5.928 (s, 1H, H3), 6.367 (s, 1H, H5), 6.933 (d, 2H, J = 8.1 Hz, H2'), 7.633 (d, 2H, J = 8.1 Hz, H3'); ^{13}C-NMR (CDCl$_3$, 100 MHz) δ 56.211 (C8), 56.302 (C7), 98.439 (C3), 111.559 (C5), 120.453 (C2'), 124.319 (q, J = 271.6 Hz, C5'), 126.425 (q, J = 3.7 Hz, C3'), 126.936 (q, J = 32.7 Hz, C4'), 153.272 (C1'), 154.987 (C6), 155.898 (C2), 157.775 (C4), 176.491 (C1). mp 130–133 °C. GC-MS RI$_{(Rtx-5ms)}$ = 2084, C$_{15}$H$_{12}$O$_3$NF$_3$ EI-MS m/z: 53 (22); 69 (43); 95 (37); 125 (30); 145 (96); 184 (29); 197 (57); 212 (52); 221 (22); [M]$^+$ = 311 (100).

Compound **3h** *(4-(3',5'-dichlorophenylimino)-2,6-dimethoxycyclohexa-2,5-dienone)* Yield 74.0%. ^{1}H-NMR (CDCl$_3$, 400 MHz) δ 3.699 (s, 3H, H8), 3.874 (s, 3H, H7), 5.935 (d, 1H, J = 2.1 Hz, H3), 6.327 (d, 1H, J = 2.1 Hz, H5), 6.758 (d, 2H, J = 1.8 Hz, H2'), 7.167 (d, 1H, J = 1.8 Hz, H4'); ^{13}C-NMR (CDCl$_3$, 100 MHz) δ 56.375 (C8 and C7), 98.324 (C3), 111.375 (C5), 118.863 (C2'), 124.712 (C4'), 135.532 (C3' and C5'), 152.147 (C1'), 155.044 (C6), 155.938 (C2), 158.388 (C4), 176.473 (C1). mp. 170–173 °C. GC-MS RI$_{(Rtx-5ms)}$ = 2472, C$_{14}$H$_{11}$O$_3$NCl$_2$ EI-MS m/z: 109 (18); 145 (19); 212 (18); 216 (19); 231 (29); 233 (22); 246 (14); [M]$^+$ = 311 (100); [M + 2]$^+$ = 313 (66); [M + 4]$^+$ = 315 (13).

Compound **3i** *(4-(2',5'-dichlorophenylimino)-2,6-dimethoxycyclohexa-2,5-dienone)* Yield 38.3%. ^{1}H-NMR (CDCl$_3$, 400 MHz) δ 3.678 (s, 3H, H8), 3.893 (s, 3H, H7), 5.794 (d, 1H, J = 1.8 Hz, H3), 6.416 (d, 1H, J = 1.8 Hz, H5), 6.838 (d, 1H, J = 2.2 Hz, H3'), 7.094 (dd, 1H, J = 8.6, 2.2 Hz, H5'), 7.382 (d, 1H, J = 8.6 Hz, H6'); ^{13}C-NMR (CDCl$_3$, 100 MHz) δ 56.418 (C8 and C7), 98.759 (C3), 111.205 (C5), 121.133 (C3'), 123.120 (C1'), 125.644 (C5'), 131.130 (C6'), 133.044 (C4'), 148.282 (C2'), 155.162 (C6), 155.840 (C2), 159.208 (C4), 176.453 (C1). Mp. 198–202 °C. GC-MS RI$_{(Rtx-5ms)}$ = 2449, C$_{14}$H$_{11}$O$_3$NCl$_2$ EI-MS m/z: 190 (24); 212 (25); 231 (29); 233 (79); 261 (25); 276 (92); 278 (29); [M]$^+$ = 311 (100); [M + 2]$^+$ = 313 (73); [M + 4]$^+$ = 315 (15).

Compound **3j** *(4-(2'-chlorophenylimino)-2,6-dimethoxycyclohexa-2,5-dienone)* Yield 44.9%. ^{1}H-NMR (CDCl$_3$, 400 MHz) δ 3.637 (s, 3H, H8), 3.887 (s, 3H, H7), 5.862 (s, 1H, H3), 6.460 (s, 1H, H5), 6.797 (d, 1H, J = 7.7 Hz, H6'), 7.119 (t, 1H, J = 7.6 Hz, H5'), 7.274 (t, 1H, J = 7.6 Hz, H4'), 7.458 (d, 1H, J = 8.0 Hz, H3'); ^{13}C-NMR (CDCl$_3$, 100 MHz) δ 56.158 (C8), 56.332 (C7), 99.002 (C3), 111.548 (C5), 121.317 (C6'), 124.984 (C1'), 125.949 (C5'), 127.326 (C4'), 130.277 (C3'), 147.388 (C2'), 154.971 (C6), 155.618 (C2), 158.576 (C4), 176.616 (C1). mp 143–146 °C. GC-MS RI$_{(Rtx-5ms)}$ = 2449, C$_{14}$H$_{12}$O$_3$NCl EI-MS m/z: 150 (17); 170 (16); 178 (21); 197 (25); 199 (54); 214 (14); 242 (54); [M]$^+$ = 277 (100); [M + 1]$^+$ = 278 (16); [M + 2]$^+$ = 279 (36).

3.5. Fungal Strain and Culture Conditions

The strain G29 of *B. cinerea* used in this work was isolated from infected grapes (*Vitis vinifera*) and genetically characterized by the INIA, La Platina, Chile [45]. It was kept on malt yeast extract agar slants with 0.2% (*w/v*) yeast extract, 2% (*w/v*) malt extract, and 1.5% (*w/v*) agar) at 4 °C. For the cell wall integrity assay, liquid minimal medium of pH 6.5 was used, containing KH$_2$PO$_4$ (1 g/L), MgSO$_4$·7H$_2$O

(0.5 g/L), KCl (0.5 g/L), K_2HPO_4 (0.5 g/L), $FeSO_4 \cdot 7H_2O$ (0.01 g/L),1% (*w/v*) glucose as a carbon source, and 4.6 g/L ammonium tartrate as a nitrogen source.

3.6. Antifungal Assay

Effect on Mycelial Growth

The antifungal activity of the compounds was evaluated in vitro as described by Caruso et al. [13]. Compounds were dissolved in acetone and then added to Petri dishes along with malt yeast agar medium. Inhibition percentages were calculated after 72 h of incubation. Antifungal activity was expressed as the concentration that reduced mycelial growth by 50% (EC_{50}), calculated by regressing the antifungal activity percentage against compound concentration. These experiments were done at least in triplicate.

3.7. Effect on the Cell Wall Integrity of B. cinerea

The effect of compounds **3a–j** on cell wall integrity was evaluated using the method described by Mendoza et al. [27]. Compounds **3a–j** were tested at 0.16 mM. To measure the effect of these compounds on the cell wall, fluorescence intensity was quantified using ImageJ (v1.80), an outline was drawn around each hypha, and mean fluorescence was measured, along with several adjacent background readings. Mean fluorescence was compared to the negative control (maximum fluorescence).

4. Conclusions

Ten compounds were synthetized; two of them (compounds **3a** and **3f**) have been previously described. All the products showed higher antifungal activity than the substrates. Chloro-substituted compounds showed the highest antifungal effect against *B. cinerea* being the 3,5-dichlorinated product **3h** the most active. Synthesis using syringic acid or syringaldehyde with *p*-chloroaniline yield the same main product (compound **3a**). Finally, regarding the inhibition mechanism of these compounds, the results suggest that these compounds damage the cell wall.

Author Contributions: Methodology, P.C., P.C.P., K.L., and R.S.; Investigation, P.C.; Formal Analysis, L.M., M.C., F.N. and P.C.; Writing-Original Draft P.C.; Writing-Review and Editing, L.M., M.C., F.N., P.C.P., and P.C.; Project Administration C.V.

Acknowledgments: CONICYT FONDEQUIP GC-MS/MS EQM 150084.

References

1. Fillinger, S.; Elad, Y. *Botrytis—The Fungus, the Pathogen and Its Management in Agricultural Systems*; Springer: Cham, Switzerland, 2016; pp. 189–216.
2. Fenner, K.; Canonica, S.; Wackett, L.P.; Elsner, M. Evaluating pesticide degradation in the environment: Blind spots and emerging opportunities. *Science* **2013**, *341*, 752–758. [CrossRef] [PubMed]
3. Brent, K.J.; Hollomon, D.W. *Fungicide Resistance: The Assessment of the Risk*; Fungicide Resistance Action Committee: Brussels, Belgium, 2007; pp. 1–53.
4. Pappas, A.C. Evolution of fungicide resistance in *Botrytis cinerea* in protected crops in Greece. *Crop Prot.* **1997**, *16*, 257–263. [CrossRef]
5. Wilson, C.L.; Solar, J.M.; Ghaouth, A.E.; Wisniewski, M.E. Rapid evaluation of plant extracts and essential oils for antifungal activity against *Botrytis cinerea*. *Plant Dis.* **1997**, *81*, 204–210. [CrossRef]
6. Elad, Y.; Williamson, B.; Tudzynski, P.; Delen, N. *Botrytis: Biology, Pathology and Control*; Springer: Dordrecht, The Netherlands, 2007; pp. 195–217.
7. Daoubi, M.; Durán-Patrón, R.; Hmamouchi, M.; Hernández-Galán, R.; Benharref, A.; Collado, I.G. Screening study for potential lead compounds for natural product-based fungicides: I. Synthesis and in vitro evaluation of coumarins against *Botrytis cinerea*. *Pest Manag. Sci.* **2004**, *60*, 927–932. [CrossRef] [PubMed]

8.   Imperato, F. *Phytochemistry: Advances in Research*; Research Signpost: Trivandrum, India, 2006; pp. 23–67.

9.   Grayer, R.J.; Kokubun, T. Plant–fungal interactions: The search for phytoalexins and other antifungal compounds from higher plants. *Phytochemistry* **2001**, *56*, 253–263. [CrossRef]

10.   Grayer, R.J.; Harborne, J.B. A survey of antifungal compounds from higher plants, 1982–1993. *Phytochemistry* **1994**, *37*, 19–42. [CrossRef]

11.   Osbourn, A.E. Preformed antimicrobial compounds and plant defense against fungal attack. *Plant Cell* **1996**, *8*, 1821–1831. [CrossRef]

12.   Harborne, J.B. The comparative biochemistry of phytoalexin induction in plants. *Biochem. Syst. Ecol.* **1999**, *27*, 335–367. [CrossRef]

13.   Caruso, F.; Mendoza, L.; Castro, P.; Cotoras, M.; Aguirre, M.; Matsuhiro, B.; Isaacs, M.; Rossi, M.; Viglianti, A.; Antonioletti, R. Antifungal activity of resveratrol against *Botrytis cinerea* is improved using 2-furyl derivatives. *PLoS ONE* **2011**, *6*, e25421. [CrossRef]

14.   Pinedo-Rivilla, C.; Bustillo, A.J.; Hernández-Galán, R.; Aleu, J.; Collado, I.G. Asymmetric preparation of antifungal 1-(4′-chlorophenyl)-1-cyclopropyl methanol and 1-(4′-chlorophenyl)-2-phenylethanol. Study of the detoxification mechanism by *Botrytis cinerea*. *J. Mol. Catal. B Enzym.* **2011**, *70*, 61–66. [CrossRef]

15.   Saiz-Urra, L.; Racero, J.C.; Macías-Sáchez, A.J.; Hernández-Galán, R.; Hanson, J.R.; Perez-Gonzalez, M.; Collado, I.G. Synthesis and quantitative structure-antifungal activity relationships of clovane derivatives against *Botrytis cinerea*. *J. Agric. Food Chem.* **2009**, *57*, 2420–2428. [CrossRef] [PubMed]

16.   Mendoza, L.; Yañez, K.; Vivanco, M.; Melo, R.; Cotoras, M. Characterization of extracts from winery by-products with antifungal activity against *Botrytis cinerea*. *Ind. Crops Prod.* **2013**, *43*, 360–364. [CrossRef]

17.   Mikolasch, A.; Schauer, F. Fungal laccases as tools for the synthesis of new hybrid molecules and biomaterials. *Appl. Microbiol. Biotechnol.* **2009**, *82*, 605–624. [CrossRef] [PubMed]

18.   Senthivelan, T.; Kanagaraj, J.; Panda, R.C. Recent trends in fungal laccase for various industrial applications: An eco-friendly approach—A review. *Biotechnol. Bioprocess Eng.* **2016**, *21*, 19–38. [CrossRef]

19.   Mogharabi, M.; Faramarzi, M.A. Laccase and laccase-mediated systems in the synthesis of organic compounds. *Adv. Synth. Catal.* **2014**, *356*, 897–927. [CrossRef]

20.   Constantin, M.-A.; Conrad, J.; Merişor, E.; Koschorreck, K.; Urlacher, V.B.; Beifuss, U. Oxidative dimerization of (E)- and (Z)-2-propenylsesamol with O_2 in the presence and absence of laccases and other catalysts: Selective formation of carpanones and benzopyrans under different reaction conditions. *J. Org. Chem.* **2012**, *77*, 4528–4543. [CrossRef] [PubMed]

21.   Koschorreck, K.; Richter, S.M.; Ene, A.B.; Roduner, E.; Schmid, R.D.; Urlacher, V.B. Cloning and characterization of a new laccase from *Bacillus licheniformis* catalyzing dimerization of phenolic acids. *Appl. Microbiol. Biotechnol.* **2008**, *79*, 217–224. [CrossRef]

22.   Adelakun, O.E.; Kudanga, T.; Green, I.R.; le Roes-Hill, M.; Burton, S.G. Enzymatic modification of 2,6-dimethoxyphenol for the synthesis of dimers with high antioxidant capacity. *Process Biochem.* **2012**, *47*, 1926–1932. [CrossRef]

23.   Adelakun, O.E.; Kudanga, T.; Parker, A.; Green, I.R.; le Roes-Hill, M.; Burton, S.G. Laccase-catalyzed dimerization of ferulic acid amplifies antioxidant activity. *J. Mol. Catal. B Enzym.* **2012**, *74*, 29–35. [CrossRef]

24.   Navarra, C.; Gavezzotti, P.; Monti, D.; Panzeri, W.; Riva, S. Biocatalyzed synthesis of enantiomerically enriched β-5-like dimer of 4-vinylphenol. *J. Mol. Catal. B Enzym.* **2012**, *84*, 115–120. [CrossRef]

25.   Mikolasch, A.; Niedermeyer, T.H.J.; Lalk, M.; Witt, S.; Seefeldt, S.; Hammer, E.; Schauer, F.; Gesell Salazar, M.; Hessel, S.; Jülich, W.-D.; et al. Novel cephalosporins synthesized by amination of 2,5-dihydroxybenzoic acid derivatives using fungal laccases II. *Chem. Pharm. Bull.* **2007**, *55*, 412–416. [CrossRef] [PubMed]

26.   Mikolasch, A.; Hessel, S.; Salazar, M.G.; Neumann, H.; Manda, K.; Gordes, D.; Schmidt, E.; Thurow, K.; Hammer, E.; Lindequist, U.; et al. Synthesis of new N-analogous corollosporine derivatives with antibacterial activity by laccase-catalyzed amination. *Chem. Pharm. Bull.* **2008**, *56*, 781–786. [CrossRef] [PubMed]

27.   Mendoza, L.; Castro, P.; Melo, R.; Campos, A.M.; Zuñiga, G.; Guerrero, J.; Cotoras, M. Improvement of the antifungal activity against *Botrytis cinerea* of syringic acid, a phenolic acid from grape pomace. *J. Chil. Chem. Soc.* **2016**, *61*, 3039–3042. [CrossRef]

28.   Leroux, P. Recent developments in the mode of action of fungicides. *Pestic. Sci.* **1996**, *47*, 191–197. [CrossRef]

29.   Guo, Z.; Miyoshi, H.; Komyoji, T.; Haga, T.; Fujita, T. Uncoupling activity of a newly developed fungicide, fluazinam [3-chloro-N-(3-chloro-2,6-dinitro-4-trifluoromethylphenyl)-5-trifluoromethyl-2-pyridinamine]. *BBA Bioenerg.* **1991** *1056*, 89–92. [CrossRef]

30. Leroux, P.; Gredt, M.; Leroch, M.; Walker, A.S. Exploring mechanisms of resistance to respiratory inhibitors in field strains of *Botrytis cinerea*, the causal agent of gray mold. *Appl. Environ. Microbiol.* **2010**, *76*, 6615–6630. [CrossRef] [PubMed]

31. Bollag, J.M.; Minard, R.D.; Liu, S.Y. Cross-linkage between anilines and phenolic humus constituents. *Environ. Sci. Technol.* **1983**, *17*, 72–80. [CrossRef] [PubMed]

32. Tatsumi, K.; Freyer, A.; Minard, R.D. Enzymatic coupling of chloroanilines, syringic acid, vanillic acid and protocatechuic acid. *Soil Biol. Biochem.* **1994**, *26*, 135–142. [CrossRef]

33. Hoff, T.; Liu, S.Y.; Bollag, J.M. Transformation of halogen-, alkyl-, and alkoxy-substituted anilines by a laccase of *Trametes versicolor*. *Appl. Environ. Microbiol.* **1985**, *49*, 1040–1045.

34. Itoh, K.; Fujita, M.; Kumano, K.; Suyama, K.; Yamamoto, H. Phenolic acids affect transformations of chlorophenols by a *Coriolus versicolor* laccase. *Soil Biol. Biochem.* **2000**, *32*, 85–91. [CrossRef]

35. Galletti, P.; Pori, M.; Funiciello, F.; Soldati, R.; Ballardini, A.; Giacomini, D. Laccase-mediator system for alcohol oxidation to carbonyls or carboxylic acids: Toward a sustainable synthesis of profens. *ChemSusChem* **2014**, *7*, 2684–2689. [CrossRef] [PubMed]

36. Saiz-Urra, L.; Bustillo-Pérez, A.J.; Cruz-Monteagudo, M.; Pinedo-Rivilla, C.; Aleu, J.; Hernández-Galán, R.; Collado, I.G. Global antifungal profile optimization of chlorophenyl derivatives against *Botrytis cinerea* and *Colletotrichum gloeosporioides*. *J. Agric. Food Chem.* **2009**, *57*, 4838–4843. [CrossRef] [PubMed]

37. Gao, S.; Xu, Z.; Wang, X.; Feng, H.; Wang, L.; Zhao, Y.; Wang, Y.; Tang, X. Synthesis and antifungal activity of aspirin derivatives. *Asian J. Chem.* **2014**, *26*, 7157–7159. [CrossRef]

38. Chen, C.-J.; Song, B.-A.; Yang, S.; Xu, G.-F.; Bhadury, P.S.; Jin, L.-H.; Hu, D.-Y.; Li, Q.-Z.; Liu, F.; Xue, W.; et al. Synthesis and antifungal activities of 5-(3,4,5-trimethoxyphenyl)-2-sulfonyl-1,3,4-thiadiazole and 5-(3,4,5-trimethoxyphenyl)-2-sulfonyl-1,3,4-oxadiazole derivatives. *Bioorg. Med. Chem.* **2007**, *15*, 3981–3989. [CrossRef] [PubMed]

39. Herth, W.; Schnepf, E. The fluorochrome, calcofluor white, binds oriented to structural polysaccharide fibrils. *Protoplasma* **1980**, *105*, 129–133. [CrossRef]

40. Bolton, J.L.; Trush, M.A.; Penning, T.M.; Dryhurst, G.; Monks, T.J. Role of quinones in toxicology. *Chem. Res. Toxicol.* **2000**, *13*, 135–160. [CrossRef] [PubMed]

41. Dahlin, D.C.; Miwa, G.T.; Lu, A.Y.H.; Nelson, S.D. N-acetyl-p-benzoquinone imine: A cytochrome P-450-mediated oxidation product of acetaminophen. *Proc. Natl. Acad. Sci. USA* **1984**, *81*, 1327–1331. [CrossRef] [PubMed]

42. Mitchell, J.R.; Jollow, D.J.; Potter, W.Z.; Davis, D.C.; Gillette, J.R.; Brodie, B.B. Acetaminophen-induced hepatic necrosis. *J. Pharmacol. Exp. Ther.* **1973**, *187*, 185–194. [PubMed]

43. Pócsi, I.; Prade, R.A.; Penninckx, M.J. Glutathione, altruistic metabolite in fungi. *Adv. Microb. Physiol.* **2004**, *49*, 1–76. [PubMed]

44. Lopez, S.N.; Castelli, M.V.; Zacchino, S.A.; Domínguez, J.N.; Lobo, G.; Charris-Charris, J.; Cortes, J.C.; Ribas, J.C.; Devia, C.; Rodriguez, A.M.; et al. In vitro antifungal evaluation and structure–activity relationships of a new series of chalcone derivatives and synthetic analogues, with inhibitory properties against polymers of the fungal cell wall. *Bioorg. Med. Chem.* **2001**, *9*, 1999–2013. [CrossRef]

45. Muñoz, G.; Hinrichsen, P.; Brygoo, Y.; Giraud, T. Genetic characterisation of *Botrytis cinerea* populations in Chile. *Mycol. Res.* **2002**, *106*, 594–601. [CrossRef]

8

Biodegradation of 7-Hydroxycoumarin in *Pseudomonas mandelii* 7HK4 via *ipso*-Hydroxylation of 3-(2,4-Dihydroxyphenyl)-propionic Acid

Arūnas Krikštaponis * and Rolandas Meškys

Department of Molecular Microbiology and Biotechnology, Institute of Biochemistry, Life Sciences Center, Vilnius University, Sauletekio al. 7, LT-10257 Vilnius, Lithuania; rolandas.meskys@bchi.vu.lt
* Correspondence: arunas.krikstaponis@bchi.vu.lt.

Abstract: A gene cluster, denoted as *hcdABC*, required for the degradation of 3-(2,4-dihydroxyphenyl)-propionic acid has been cloned from 7-hydroxycoumarin-degrading *Pseudomonas mandelii* 7HK4 (DSM 107615), and sequenced. Bioinformatic analysis shows that the operon *hcdABC* encodes a flavin-binding hydroxylase (HcdA), an extradiol dioxygenase (HcdB), and a putative hydroxymuconic semialdehyde hydrolase (HcdC). The analysis of the recombinant HcdA activity in vitro confirms that this enzyme belongs to the group of *ipso*-hydroxylases. The activity of the proteins HcdB and HcdC has been analyzed by using recombinant *Escherichia coli* cells. Identification of intermediate metabolites allowed us to confirm the predicted enzyme functions and to reconstruct the catabolic pathway of 3-(2,4-dihydroxyphenyl)-propionic acid. HcdA catalyzes the conversion of 3-(2,4-dihydroxyphenyl)-propionic acid to 3-(2,3,5-trihydroxyphenyl)-propionic acid through an *ipso*-hydroxylation followed by an internal (1,2-C,C)-shift of the alkyl moiety. Then, in the presence of HcdB, a subsequent oxidative *meta*-cleavage of the aromatic ring occurs, resulting in the corresponding linear product (2E,4E)-2,4-dihydroxy-6-oxonona-2,4-dienedioic acid. Here, we describe a *Pseudomonas mandelii* strain 7HK4 capable of degrading 7-hydroxycoumarin via 3-(2,4-dihydroxyphenyl)-propionic acid pathway.

Keywords: 7-hydroxycoumarin; 3-(2,4-dihydroxyphenyl)-propionic acid; 3-(2,3,5-trihydroxyphenyl)-propionic acid; *ipso*-hydroxylase; *Pseudomonas mandelii*

1. Introduction

The compound 7-hydroxycoumarin (**5**), also known as umbelliferone, is one of the most abundant plant-derived secondary metabolites. It is a parent compound of other, more complex natural furanocoumarins and pyranocoumarins in higher plants [1–3]. In the case of plant damage, plants produce a high diversity of natural coumarins as a defense mechanism against insect herbivores as well as fungal and microbial pathogens [4]. For example, simple hydroxycoumarins have antibacterial activity against *Ralstonia solanacearum*, *Escherichia coli*, *Klebsiella pneumoniae*, *Staphylococcus aureus*, and *Pseudomonas aeruginosa* [4–8]. Despite the toxic effects of coumarins, it has been shown that microorganisms evolve to gain the ability to metabolize such compounds.

It has been shown previously that a number of soil microorganisms, such as *Pseudomonas*, *Arthrobacter*, *Aspergillus*, *Penicillium*, and *Fusarium* spp. can grow on coumarin (**1**) as a sole source of carbon [9–16]. The key intermediate during coumarin catabolism in bacteria is 3-(2-hydroxyphenyl)-propionic acid (**4**) [10,11,16]. The bioconversion of coumarin to 3-(2-hydroxyphenyl)-propionic acid can be achieved in two different metabolic pathways, as shown in Figure 1A. Bacteria belonging to the *Arthrobacter* genus enzymatically hydrolyze the lactone moiety to give 3-(2-hydroxyphenyl)-2-propenoic acid (**2**), and then reduce a double bond by using a NADH-dependent enzyme to yield 3-(2-hydroxyphenyl)-propionic acid [11]. In the case of *Pseudomonas* sp. 30-1 and *Aspergillus niger* ATCC 11394 cells, coumarin is initially reduced to dihydrocoumarin (**3**) by a NADH-dependent oxidoreductase and only then enzymatically hydrolyzed [9,10,14]. *Arthrobacter* and *Pseudomonas* species are capable of oxidizing 3-(2-hydroxyphenyl)-propionic acid to 3-(2,3-dihydroxyphenyl)-propionic acid by using specific flavin-binding aromatic hydroxylases [16]. However, no data are available on further conversions of 3-(2,3-dihydroxyphenyl)-propionic acid in these bacteria. Also, no microorganisms or enzymes implicated in the metabolism of any hydroxycoumarin have been identified to date.

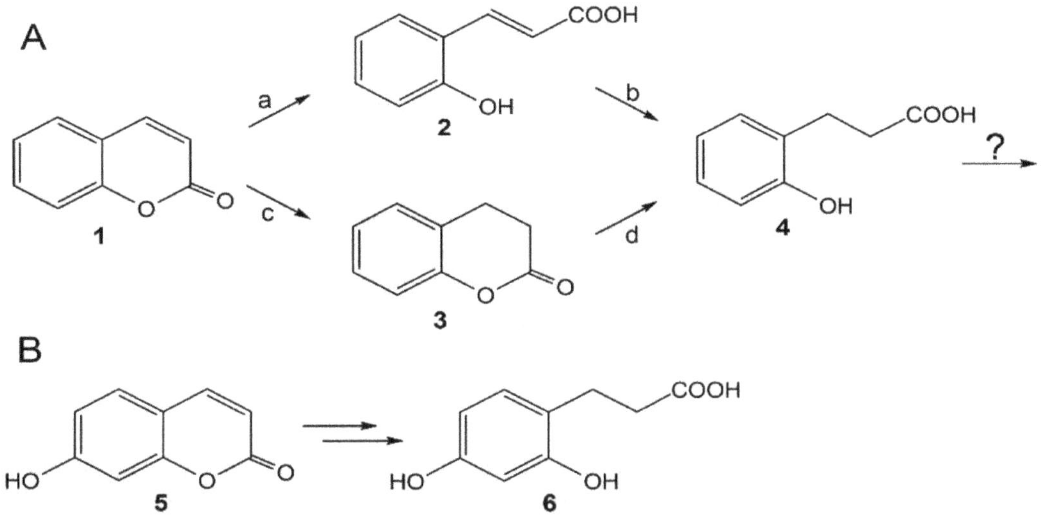

Figure 1. (**A**) Coumarin metabolic routes in *Arthrobacter* spp. (a, b), as well as *Pseudomonas* and *Aspergillus* spp. (c, d). (**B**) Proposed metabolic pathway of 7-hydroxycoumarin in *Pseudomonas* sp. 7HK4 bacteria. **1**—coumarin; **2**—3-(2-hydroxyphenyl)-2-propenoic acid; **3**—dihydrocoumarin; **4**—3-(2-hydroxyphenyl)-propionic acid; **5**—7-hydroxycoumarin; **6**—3-(2,4-dihydroxyphenyl)-propionic acid; **a**—putative coumarin hydrolase; **b**—NADH:o-coumarate oxidoreductase; **c**—dihydrocoumarin:NAD[NADP] oxidoreductase; **d**—putative dihydrocoumarin hydrolase.

Here we describe a *Pseudomonas* sp. 7HK4 strain capable of utilizing 7-hydroxycoumarin as the sole carbon and energy source, and a catabolic pathway of 7-hydroxycoumarin in these bacteria. Analysis of 7-hydroxycoumarin-inducible proteins led to the identification of the genome locus encoding 7-hydroxycoumarin catabolic proteins. The corresponding genes were cloned and heterologously expressed in *E. coli* system. The functions of the recombinant proteins were determined and enzyme activities towards various substrates were evaluated. We show that *Pseudomonas* sp. 7HK4 encodes a distinct 7-hydroxycoumarin metabolic route, which utilizes a flavin monooxygenase responsible for *ipso*-hydroxylation of 3-(2,4-dihydroxyphenyl)-propionic acid (**6**).

2. Results

2.1. Screening and Identification of 7-Hydroxycoumarin-Degrading Microorganisms

By the means of enrichment culture using various coumarin derivatives, an aerobic strain 7HK4 degrading 7-hydroxycoumarin was isolated from the garden soil in Lithuania (DSMZ accession number DSM 107615). This bacterium was tested for its ability to grow on several coumarin derivatives, such as coumarin, 3-hydroxycoumarin, 4-hydroxycoumarin, 6-hydroxycoumarin, 6-methylcoumarin, 6,7-dihydroxycoumarin, and 7-methylcoumarin, as the sole carbon and energy source in a minimal salt medium. However, of all the aforementioned compounds, the strain 7HK4 was able to utilize

7-hydroxycoumarin only. The strain utilized glucose, which was used as a control substrate in whole-cell reactions. Data on biochemical analysis of this strain are given in the Supplementary Material. The nucleotide sequence of 16S rRNA gene was determined by sequencing of the cloned DNA fragment, which was obtained by PCR amplification. The strain 7HK4 showed the highest 16S rDNA sequence similarity to that from *Pseudomonas* genus and was similar to 16S rDNA from *Pseudomonas mandelii* species according to the phylogenetic analysis, as shown in Figure S1 in the Supplementary Material.

2.2. Bioconversion Experiments by Using Whole Cells of Pseudomonas sp. 7HK4

Time course experiments using whole cells of *Pseudomonas* sp. 7HK4 pre-grown in the presence of 7-hydroxycoumarin showed that the cells converted coumarin, 6-hydroxycoumarin, and 6,7-dihydroxycoumarin in addition to 7-hydroxycoumarin, according to the changes in the UV-VIS spectra, as shown in Figure S2 in the Supplementary Material. UV absorption maxima were dropping down over time, although the biotransformation rates of 6-hydroxycoumarin, 6,7-dihydroxycoumarin, and coumarin were approximately 5-fold to 10-fold lower, respectively. After the completion of bioconversions, there were no visible spectra observed for the residual aromatic compound in the reaction mixtures with 7-hydroxycoumarin, except for the reactions with 6-hydroxycoumarin, 6,7-dihydroxycoumarin, and coumarin, which had non-disappearing UV absorption maxima at 260–270 nm wavelengths.

These spectra are similar to that of 3-phenylpropionic acid (data not shown), suggesting that 7-hydroxycoumarin-induced 7HK4 strain can only catalyze the hydrolysis and reduction of lactone moiety of coumarin, 6-hydroxycoumarin, and 6,7-dihydroxycoumarin producing 3-(2-hydroxyphenyl)-propionic, 3-(2,5-dihydroxyphenyl)-propionic, and 3-(2,4,5-trihydroxyphenyl)-propionic acids, respectively, comparable to similar biotransformations in other microorganisms [9–16]. In addition, uninduced *Pseudomonas* sp. 7HK4 cells grown in the presence of glucose showed delayed and slower conversion of 7-hydroxycoumarin. The bioconversion process only started 0.5 h after the addition of substrate, suggesting that 7-hydroxycoumarin induced its own metabolism. In addition, uninduced cells did not catalyze any conversion or showed delayed and much slower biotransformations for coumarin, 6,7-dihydroxycoumarin, and 6-hydroxycoumarin, respectively. This demonstrates that in *Pseudomonas* sp. 7HK4 bacteria, the metabolism of 7-hydroxycoumarin is an inducible process.

It has been shown previously that 3-(2-hydroxyphenyl)-2-propenoic and 3-(2-hydroxyphenyl)-propionic acids, as shown in Figure 1B, are the intermediates in a known coumarin metabolic pathway in several microorganisms [9–11,14,16]. By analogy, it was suggested that 3-(2,4-dihydroxyphenyl)-propionic acid may be an intermediate metabolite in 7-hydroxycoumarin catabolism, as shown in Figure 1B. The cells of 7HK4 strain pre-cultivated in the presence of 7-hydroxycoumarin catalyzed no conversion of 3-(2-hydroxyphenyl)-2-propenoic or 3-(2-hydroxyphenyl)-propionic acid, however 3-(2,4-dihydroxyphenyl)-propionic acid was straightforwardly consumed. The HPLC-MS analysis of the bioconversion mixtures showed that *Pseudomonas* sp. 7HK4 cells cultivated in the presence of 7-hydroxycoumarin produce 3-(2,4-dihydroxyphenyl)-propionic acid as an intermediate metabolite, as shown in Figure 2. In 7HK4 bacteria grown on glucose, the aforementioned compound was not observed suggesting that 3-(2,4-dihydroxyphenyl)-propionic acid is an intermediate metabolite during 7-hydroxycoumarin degradation.

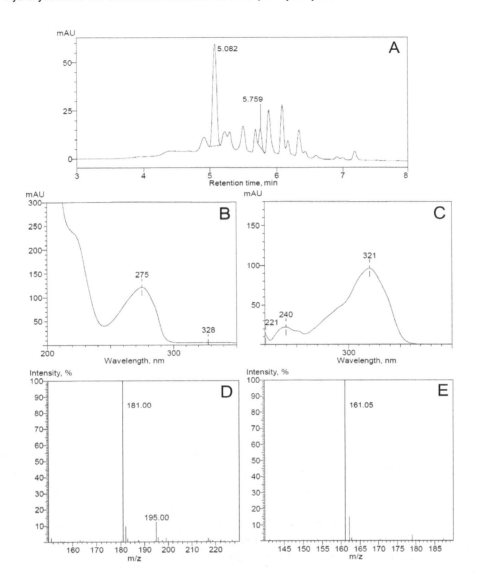

Figure 2. *Pseudomonas* sp. 7HK4 bacteria were grown in the presence of 7-hydroxycoumarin, and produced metabolites were analyzed by HPLC-MS. UV 254 nm trace of metabolites (**A**). UV and MS spectra of peaks with retention times 5.082 min (**B,D**) and 5.759 min (**C,E**). The negative ions $[M-H]^-$ generated are at m/z 181.00 (3-(2,4-dihydroxyphenyl)-propionic acid) and 161.05 (7-hydroxycoumarin).

2.3. Identification of 7-Hydroxycoumarin-Inducible Proteins

To elucidate which enzymes are involved in 7-hydroxycoumarin metabolism, *Pseudomonas* sp. 7HK4 cells were cultivated in a minimal medium supplemented with 7-hydroxycoumarin (0.3 mM) or glucose (0.3 mM) as the sole carbon and energy source. Several 7-hydroxycoumarin-inducible proteins of different molecular mass were observed by using SDS-PAGE analysis of cell-free extracts from *Pseudomonas* sp. 7HK4, as shown in Figure 3A.

Three bands corresponding to the inducible proteins of 23, 32, and 50 kDa, as shown in Figure 3A, were excised from the SDS-PAGE gel and analyzed by MS-MS sequencing. The genome sequence of strain 7HK4 was obtained via Illumina sequencing as described in Materials and Methods. For the identification of corresponding genes of inducible proteins, the sequences of the identified peptides were searched against the partial 7HK4 genome sequence. Thus, the genome fragment was discovered, encoding a 31.2 kDa protein containing 16 aa-long sequence, as shown in the bolded sequence in Figure 3A, identical to that found in 7-hydroxycoumarin-inducible ~32 kDa protein.

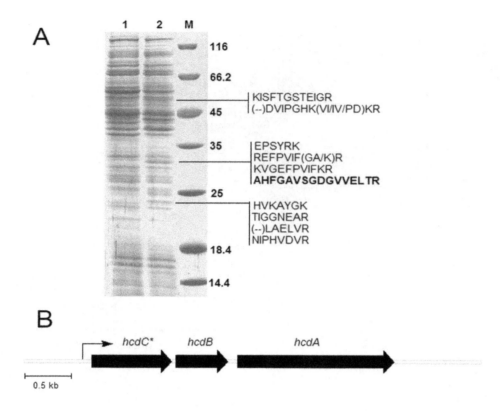

Figure 3. (**A**) SDS-PAGE analysis of cell-free extracts of *Pseudomonas* sp. 7HK4. Bacteria were cultivated in the presence of glucose (lane 1) or 7-hydroxycoumarin (lane 2). M—molecular mass ladder (kDa). The arrows indicate 7-hydroxycoumarin-inducible 23, 32, and 50 kDa proteins. The peptide sequences determined by MS-MS are given on the right. The sequence in bold was identified after the peptide sequences obtained by MS-MS were compared with the partially sequenced genome of *Pseudomonas* sp. 7HK4. (**B**) Organization of *hcd* genes in *Pseudomonas* sp. 7HK4 bacteria. The arrows indicate open reading frames (ORFs) encoding HcdA, HcdB, and HcdC. The gene for the 7-hydroxycoumarin-inducible protein of 31.2 kDa is marked by an asterisk.

2.4. Analysis of the Genome Locus Encoding the 7-Hydroxycoumarin-Inducible Protein

Analysis of *Pseudomonas* sp. 7HK4 genome sequences (35 contigs) showed that the inducible 31.2 kDa protein belongs to the fumarylacetoacetate (FAA) hydrolase family, which includes such enzymes as 2-keto-4-pentenoate hydratase, 2-oxohepta-3-ene-1,7-dioic acid hydratase, 2-hydroxy-6-oxo-6-phenylhexa-2,4-dienoate hydrolase, or bifunctional isomerases/decarboxylases (catechol pathway), as shown in Figure S3 in the Supplementary Material. The FAA family proteins are usually involved in the last stages of bacterial metabolism of aromatic compounds [17–19], suggesting that 31.2 kDa protein from *Pseudomonas* sp. 7HK4 participates in the final steps of 7-hydroxycoumarin metabolism, after oxidative cleavage of the aromatic ring.

Adjacent to the 31.2 kDa protein-encoding gene, two open reading frames (ORFs) were identified. All three genes are arranged on the same DNA strand, and are separated by short intergenic regions, suggesting that these genes are organized into an operon, as shown in Figure 3B. The putative operon was designated *hcdABC* (**h**ydroxy**c**oumarin **d**egrading operon), where *hcdC* encodes the 31.2 kDa protein. A BLAST analysis of *hcdA* and *hcdB* sequences revealed that these genes encode the putative FAD-binding hydroxylase and ring-cleavage dioxygenase, respectively. HcdA protein was not assigned to any family, but it showed similarity to a putative 2-polyprenyl-6-methoxyphenol hydroxylase, as shown in Figure S3 in the Supplementary Material. This type of enzyme belongs to class A of FAD-binding monooxygenases, which are involved in bacterial degradation of aromatic compounds [20–23]. The product of the hcdB gene belongs to the cl14632 superfamily that combines a variety of structurally related metalloproteins, including the type I extradiol dioxygenases, as shown in Figure S3 in the Supplementary Material. The type I extradiol dioxygenases catalyze the incorporation of both atoms of molecular oxygen into aromatic substrates that results in the cleavage of the aromatic rings [24,25].

2.5. Expression and Substrate Specificity of HcdA Hydroxylase

For further characterization of HcdA hydroxylase, the hcdA gene was amplified by PCR and cloned into the pET21b expression vector. The sequence was confirmed by Sanger sequencing. The recombinant C-terminally His$_6$-tagged protein was produced in Escherichia coli BL21, and purified by affinity chromatography. The purified enzyme migrated as a ~62 kDa band on SDS-PAGE, as shown in Figure S4A in the Supplementary Material, and had a bright yellow color with absorbance maxima at 380 and 450 nm wavelengths, as shown in Figure S4B in the Supplementary Material, suggesting that the protein contains a tightly bound flavin [26–28]. The gel-filtration showed that the purified HcdA protein is a monomer, as shown in Figure S4C in the Supplementary Material. The specificity for both flavin and nicotinamide cofactors was investigated. The HcdA hydroxylase was able to utilize either NADH or NADPH, although the oxidation rates of NADPH were almost 2-fold lower. Kinetic characterization of HcdA protein is presented in the Supplementary Material.

The activity of the HcdA enzyme was assayed in the presence of NADH cofactor against various substrates. The highest rate of oxidation of NADH was recorded in the presence of 3-(2,4-dihydroxyphenyl)-propionic acid. A 40-fold lower rate was observed when trans-2,4-dihydroxycinnamic acid was used as the substrate. The HcdA was not active towards trans-cinnamic, cis-2,4-dihydroxycinnamic, 3-(2-hydroxyphenyl)-propionic, 3-(2-hydroxyphenyl)-2-propenoic, 3-(4-hydroxyphenyl)-2-propenoic, 3-(3-hydroxyphenyl)-2-propenoic, 3-(2-bromophenyl)-propionic, 3-(2-nitrophenyl)-propionic, and 3-phenylpropionic acids, cinnamyl alcohol, pyrocatechol, 3-methylcatechol, 4-methylcatechol, 2-propylphenol, 2-propenylphenol, 2-ethylphenol, o-cresol, o-tyrosine, resorcinol, 2,3-dihydroxypyridine, 2-hydroxy-4-aminopyridine, N-methyl-2-pyridone, N-ethyl-2-pyridone, N-propyl-2-pyridone, N-butyl-2-pyridone, indoline, and indole. These data demonstrate that the HcdA is a highly specific monooxygenase, which shows a strong preference to 3-(2,4-dihydroxyphenyl)-propionic acid. The addition of His$_6$-tag did not affect the enzymatic activity of the HcdA protein.

The product of the reaction catalyzed by the HcdA hydroxylase was analyzed by UV-VIS absorption spectroscopy and HPLC-MS. A new UV absorption maximum was observed at 340 and 490 nm upon addition of 3-(2,4-dihydroxyphenyl)-propionic acid to the reaction mixture. The consequent red coloring was observed, which indicated the presence of para- or ortho-quinone [29], a presumed product of the autoxidation of the corresponding hydroquinone. The same coloration was also observed in vivo when Pseudomonas sp. 7HK4 cells were grown in an excess of 7-hydroxycoumarin and when Escherichia coli BL21 cells harboring the p4pmPmo plasmid were cultured in the presence of 3-(2,4-dihydroxyphenyl)-propionic acid. HPLC-MS analysis of the reaction mixtures of both in vitro and in vivo bioconversions confirmed the formation of 3-(trihydroxyphenyl)-propionic acid and quinone, found $[M-H]^-$ masses were 197.00 (traces seen only in vivo) and 195.00, respectively, as shown in Figure 4. However, the structure of the product was not confirmed at this stage by chemical analysis since it was not possible to chromatographically separate the product from the reaction mixture. The substrate and product have similar structures and chemical properties; therefore, both were detected under the same HPLC-MS chromatogram peak with retention time of ~5.5 min.

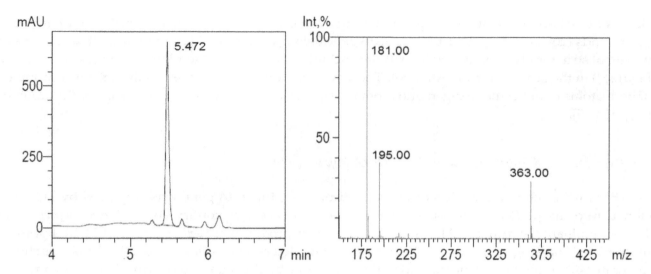

Figure 4. HPLC-MS analysis of 3-(2,4-dihydroxyphenyl)-propionic acid bioconversion mixture in vitro. UV 254 nm trace of 3-(2,4-dihydroxyphenyl)-propionic acid and its hydroxylated product under the same peak with retention time 5.472 min (on the left), and MS spectrum of the dominant peak (on the right). The negative ions $[M{-}H]^{-}$ generated are at m/z 181 (3-(2,4-dihydroxyphenyl)-propionic acid), 195 (product of 3-(trihydroxyphenyl)-propionic acid autooxidation), and 363 (dimer of 3-(2,4-dihydroxyphenyl)-propionic acid).

2.6. Expression and Substrate Specificity of HcdB Dioxygenase

HcdB dioxygenase was expressed from the plasmid pTHPPDO in *E. coli* BL21. All attempts to purify HcdB aerobically resulted in the loss of the enzymatic activity, even in the presence of organic solvents, such as glycerol, ethanol, and acetone that were known as stabilizers for similar enzymes [30–32]. The addition of dithiothreitol and ferrous sulfate to the aerobically purified protein did not restore its activity, although it had been shown to activate and/or stabilize other extradiol dioxygenases [30–34].

Therefore, due to the highly unstable nature of the HcdB enzyme, all the activity measurements were carried out in vivo using the whole cells of *E. coli* BL21 transformed with the pTHPPDO plasmid. The bioconversion of pyrocatechol, 3-methylcatechol, 3-methoxycatechol, 4-methylcatechol, and caffeic acids in each case produced yellow products with the absorbance maxima expected for the proximal *meta*-cleavage products of these catechols, as shown by the solid line in Figure 5 [35–37]. A weak yellow color was visible in bioconversion mixture containing 3-(2,3-dihydroxyphenyl)-propionic acid, and no changes in color took place, but the absorbance maxima shifts were observed during reactions with pyrogallol, gallacetophenone, 2′,3′-dihydroxy-4′-methoxyaceto-phenone hydrate, 3,4-dihydroxybenzoic acid, 2,3,4-trihydroxybenzoic acid, and 2,3,4-trihydroxy-benzophenone, as shown in Figure S11 in the Supplementary Material. No activity was observed with 1,2,4-benzenetriol and 6,7-dihydroxycoumarin. The *E. coli* cells without the *hcdB* gene showed no activity towards the aforementioned compounds. The expected shift of peaks of the UV-VIS spectra of the reaction products to a shorter wavelength were observed after acidification of the reaction mixtures containing pyrocatechol, 3-methylcatechol, 3-methoxycatechol, 4-methylcatechol, and caffeic acid, as shown by the dashed line in Figure 5 [38,39]. These findings revealed that *hcdB* encodes an extradiol dioxygenase, which can utilize a number of differently substituted catechols.

Furthermore, the HcdB dioxygenase was co-expressed with the HcdA hydroxylase in *E. coli* BL21 cells. The activity of those cells towards 3-(2,4-dihydroxyphenyl)-propionic acid was analyzed. No coloration occurred during the incubation of over 72 h, compared to the formation of a reddish bioconversion product by *E. coli* cells containing the *hcdA* gene only. Products of 3-(2,4-dihydroxyphenyl)-propionic acid conversion by HcdA and HcdB were analyzed by HPLC-MS. Ions with masses 181.00, 195.00, 197.00 ($[M{-}H]^{-}$) were not detected showing a complete conversion of substrate and its hydroxylated forms. Also, none of the expected ions ([M−H]⁻ 229.00 or [M+H]⁺ 231.00) were observed for a presumed product of the oxidative cleavage of 3-(trihydroxyphenyl)-propionic acid.

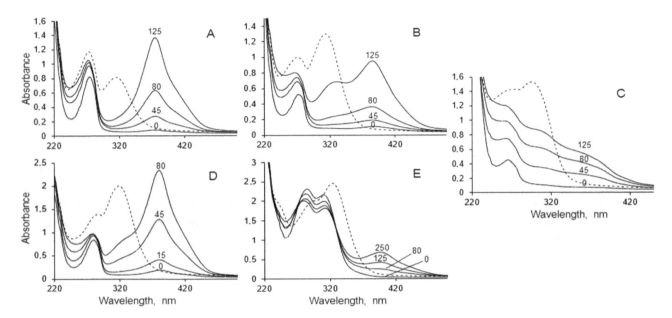

Figure 5. Biotransformations of pyrocatechol (**A**), 3-methylcatechol (**B**), 3-methoxycatechol (**C**), 4-methylcatechol (**D**), and caffeic acid (**E**) by whole cells of *E. coli* BL21 containing *hcdB* gene. Biotransformations were carried out in 50 mM potassium phosphate buffer pH 7.5 (solid line) at 30 °C with 0.5–1 mM of substrate. Incubation time is shown in min. Dashed lines indicate peak shifts to a shorter wavelength after acidification of reaction mixtures.

2.7. Isolation and Identification of 3-(2,4-Dihydroxyphenyl)-propionic Acid Bioconversion Product

Due to difficulties in detecting colorless *meta*-cleavage product of catechol derivative, and since no reasonable mass spectra could be registered, we decided to transform a cleavage product into the derivative of picolinic acid by incubation with NH_4Cl as described in Materials and Methods. The formation of a picolinic acid derivative was proven by HPLC-MS analysis, which showed the formation of [M−H]⁻ ion 210.00 mass, as shown in Figure 6, corresponding to the addition of NH_3 to the *meta*-cleavage product and the loss of two H_2O molecules [40].

The ¹H NMR spectrum of this derivative [δ 7.73 (s, 1H), 6.89 (s, 1H), 2.65 (t, *J* = 7.4 Hz, 2H), 2.38 (t, *J* = 7.4 Hz, 2H)] showed a set of two aryl protons that, from the coupling pattern (singlet + singlet), were in *meta*- or *para*-positions to each other on the aromatic ring [41], as shown in Figure S12 in the Supplementary Material. The appearance in the spectrum of two triplets with chemical shifts of 2.38 and 2.65 ppm indicated the presence of four methylene protons [41]. The ¹³C NMR spectrum [δ 181.51, 179.99, 171.01, 143.14, 136.74, 130.11, 115.22, 35.50, 24.00] showed two sp³ carbons with chemical shifts of 24.00 and 35.50 ppm, and three sp² carbons of carbonyl groups with chemical shifts of 171.01, 179.99, and 181.51 ppm, respectively, as shown in Figure S13 in the Supplementary Material. The carbonyl carbon atoms were the most strongly deshielded and their resonances formed a separate region at the highest frequency. Another four sp² carbon signals were in the aromatic carbon region [41,42]. The presence of the third carbonyl group indicated the formation of *oxo*-pyridine, for which six possible theoretical structures of *oxo*-picolinic acid derivative were presumed, as shown in Figure 7. Since the ¹H NMR spectrum showed a set of two singlet aryl protons in *meta*- or *para*-positions to each other, only structures **7** and **9**, as shown in Figure 7, were further analyzed. Besides, pyridine aromatic carbons are usually differentiated into two resonances at higher field (C-3/5, *meta* position) and three at lower field (C-2/6, *ortho* position; C-4, *para* position), where the electron-withdrawing effect of nitrogen is effective [41,42]. The chemical shift of 115.22 ppm showed that the analyzed compound had relatively strongly shielded unsubstituted aromatic carbon, which should be in *meta*-position from nitrogen, in *ortho*-position from the carbonyl group, and in *meta*- or *para*-position from the carboxyl group (C-5 atom), as shown in Figure S14 in the Supplementary Material [43,44]. This led to the conclusion that structure **9**, as shown in Figure 7, 6-(2-carboxyethyl)-4-oxo-1,4-dihydropyridine-2-carboxylic acid, was formed during incubation of *meta*-cleavage product of catechol derivative with NH₄Cl, as depicted in Scheme 1. These data allowed the reconstruction of the consecutive oxidation of 3-(2,4-dihydroxyphenyl)-propionic acid catalyzed by the HcdA and HcdB enzymes. Hence, 3-(2,3,5-trihydroxyphenyl)-propionic acid (**14**) was the product of oxidation of 3-(2,4-dihydroxyphenyl)-propionic acid by the HcdA hydroxylase. The molecular mass of 198.17 of 3-(2,3,5-trihydroxyphenyl)-propionic acid and capability to form *para*-quinone agreed with the UV-VIS and HPLC-MS data on the bioconversion of 3-(2,4-dihydroxyphenyl)-propionic by the HcdA hydroxylase. The formation of 3-(2,3,5-trihydroxyphenyl)-propionic acid from 3-(2,4-dihydroxyphenyl)-propionic acid was possible only through oxidative *ipso*-rearrangement, a unique reaction where *ipso*-hydroxylation (**13**) of the 3-(2,4-dihydroxyphenyl)-propionic acid takes place with a simultaneous shift of the propionic acid group to the vicinal position, as shown in Scheme 1 [45–48]. During the second step, 3-(2,3,5-trihydroxyphenyl)-propionic acid was cleaved by HcdB extradiol dioxygenase at the *meta*-position leading to the formation of (*2E,4E*)-2,4-dihydroxy-6-oxonona-2,4-dienedioic acid (**15**). The further imine formation and tautomerization in the presence of ammonium ions [38,49] led to 6-(2-carboxyethyl)-4-*oxo*-1,4-dihydropyridine-2-carboxylic acid (**19**), as shown in Scheme 1.

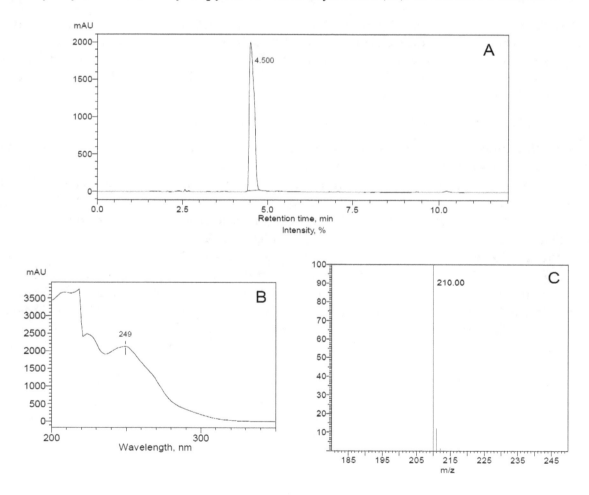

Figure 6. HPLC-MS analysis of 3-(2,4-dihydroxyphenyl)-propionic acid bioconversion mixture in vivo. UV 254 nm trace of picolinic acid derivative with retention time of 4.500 min (**A**), UV spectrum (**B**) and MS spectrum (**C**) of the dominant peak. The negative ions [M−H]⁻ generated are at *m/z* 210.00.

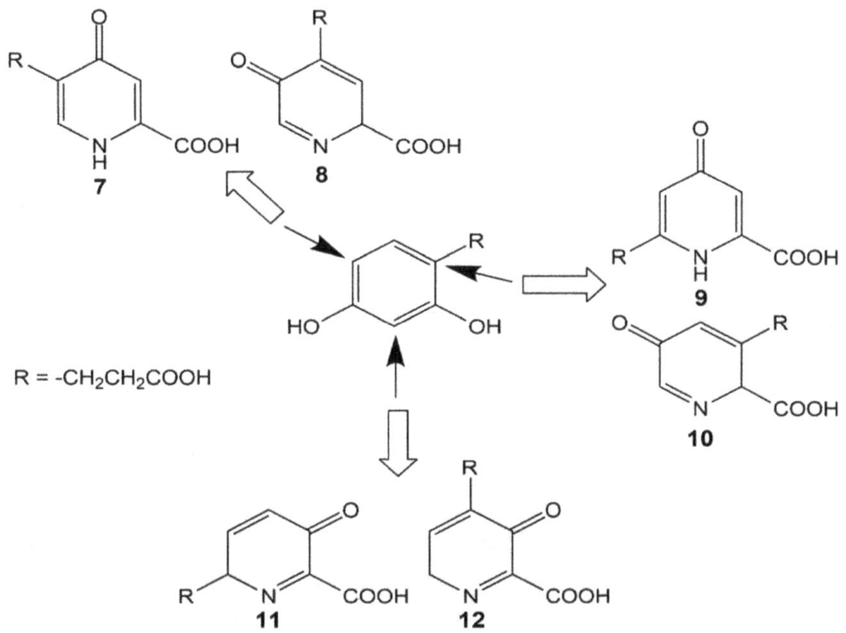

Figure 7. Suggested structures of *oxo*-picolinic acid derivative, formed during oxidative ring cleavage of 3-(2,4-dihydroxyphenyl)-propionic acid and conversion of the ring fission product. Solid lines indicate possible positions of hydroxylation by HcdA enzyme; hollow arrays indicate the probable *oxo*-picolinic acid derivatives forming after hydroxylation at each position.

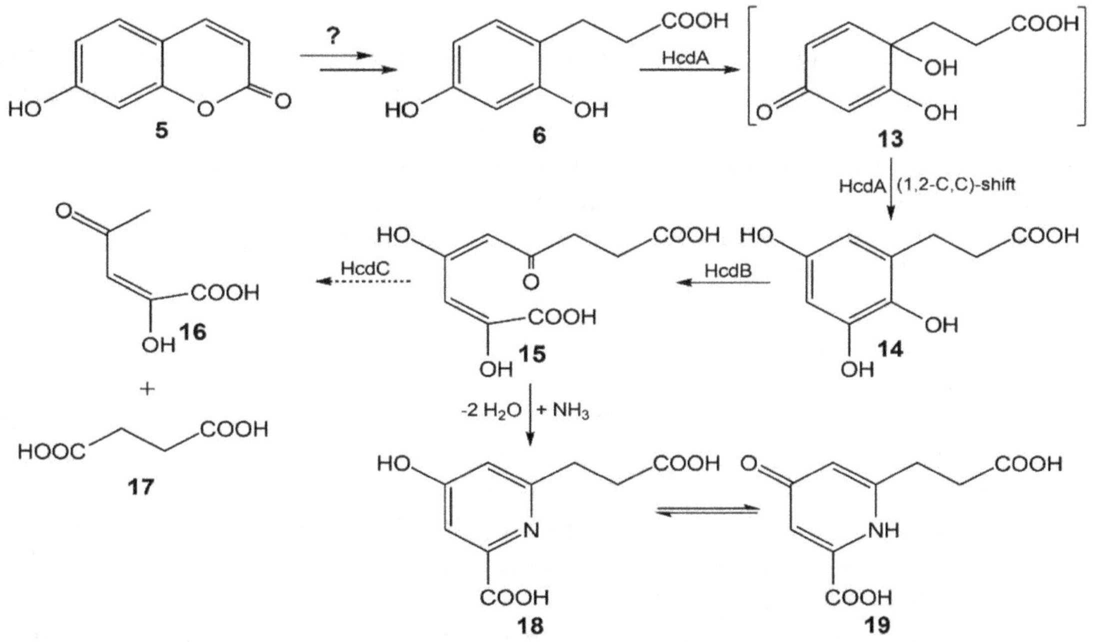

Scheme 1. The proposed metabolic pathway of 7-hydroxycoumarin in *Pseudomonas* sp. 7HK4 cells. Incubation of the compound **5** with NH₄Cl gives picolinic acid derivative. **5**—7-hydroxycoumarin; **6**—3-(2,4-dihydroxyphenyl)-propionic acid; **13**—3-(1,2-dihydroxy-4-oxocyclohexa-2,5-dienyl)-propanoic acid; **14**—3-(2,3,5-trihydroxyphenyl)-propionic acid; **15**—(2E,4E)-2,4-dihydroxy-6-oxonona-2,4-dienedioic acid; **16**—(E)-2-hydroxy-4-oxopent-2-enoic acid; **17**—succinic acid; **18**—6-(2-carboxyethyl)-4-hydroxypicolinic acid; **19**—6-(2-carboxyethyl)-4-oxo-1,4-dihydropyridine-2-carboxylic acid; HcdA—3-(2,4-dihydroxyphenyl)-propionic acid 1-monooxygenase; HcdB—3-(2,3,5-trihydroxyphenyl)-propionic acid 1,2-dioxygenase; HcdC—putative (2E,4E)-2,4-dihydroxy-6-oxonona-2,4-dienedioic acid hydrolase. The dashed arrow indicates a hypothetical reaction.

2.8. Expression of the HcdC Protein

HcdC was expressed from the plasmid p2K4PH in *E. coli* BL21. The addition of *E. coli* extracts containing the HcdC protein did not cause decolorization of the *meta*-cleavage products of 3-(2,3-dihydroxyphenyl)-propionic acid, pyrocatechol, 3-methylcatechol, or 4-methylcatechol formed in the presence of the HcdB extradiol dioxygenase. Also, the addition of NAD(P)$^+$ to these reaction mixtures did not induce decolorization of the *meta*-cleavage products [50]. To confirm the function of HcdC, *E. coli* BL21 cells were transformed with p4pmPmo and pCDF-BC plasmids. The expression of *hcdA*, *hcdB*, and *hcdC* genes in *E. coli* cells was confirmed by SDS-PAGE, the enzymes migrated as 62, 31, and 20 kDa bands, respectively, as shown in Figure S15 in the Supplementary Material.

The bioconversion of 3-(2,4-dihydroxyphenyl)-propionic acid was conducted in *E. coli* cells containing all three recombinant proteins. Later, the reaction mixture was incubated with NH$_4$Cl, and the reaction products were analyzed by HPLC-MS. The ions [M−H]$^-$ and [M+H]$^+$ with masses of 210.00 and 212.00, respectively, were not detected, compared to the bioconversion mixture with *E. coli* cells containing *hcdAB* genes only. This showed a complete conversion of *meta*-cleavage product of the catechol derivative, therefore no picolinic acid derivative could be obtained. No reaction products of (2E,4E)-2,4-dihydroxy-6-oxonona-2,4-dienedioic acid hydrolysis by the HcdC protein were identified. We suggest that the later compound was converted to succinic acid (**17**), which entered the Krebs cycle, and (*E*)-2-hydroxy-4-oxopent-2-enoic acid (**16**), which could be further converted by *E. coli* cells, thus complicating the extraction of these reaction products. Nevertheless, it may be concluded that all three enzymes encoded by the *hcdABC* operon are responsible for the catabolism of 3-(2,4-dihydroxyphenyl)-propionic acid in *Pseudomonas* sp. 7HK4 bacteria.

3. Discussion

Although coumarins are widely abundant in nature and are intensively used in biotechnology as precursory compounds [51], the metabolic pathways in microorganisms are still not known in sufficient detail. In this study, *Pseudomonas* sp. 7HK4 strain was isolated from soil and it was shown that these bacteria can utilize only 7-hydroxycoumarin as a sole source of carbon and energy. Several other coumarin derivatives were also tested, but none of those substrates support the growth of *Pseudomonas* sp. 7HK4. However, the experiments with the 7-hydroxycoumarin-induced whole cells show that *Pseudomonas* sp. 7HK4 has the enzymes that are able to transform coumarin and 6-hydroxycoumarin. The products of these biotransformations give UV spectra similar to the UV spectrum of 3-phenylpropionic acid, suggesting that *Pseudomonas* sp. 7HK4 bacteria can catalyze hydrolysis and reduction of lactone moiety of these substrates. Furthermore, 3-(2,4-dihydroxyphenyl)-propionic acid has been identified as an intermediate metabolite during biotransformations of 7-hydroxycoumarin. This finding agrees with the data published for the conversion of coumarin by *Pseudomonas* spp., *Arthrobacter* spp., and *Aspergillus* spp., which all catalyze the hydrolysis and reduction of coumarin producing 3-(2-hydroxyphenyl)-propionic acid as the main intermediate [9–11,14]. However, compared with *P. mandelii* 7HK described in this study, little is known about the bioconversion of other coumarin derivatives as well as further conversions of 3-(2-hydroxyphenyl)-propionic acid in bacteria listed above.

The analysis of the 7-hydroxycoumarin-inducible proteins lead to the identification of the genomic locus *hcdABC* encoding the enzymes required for 3-(2,4-dihydroxyphenyl)-propionic acid degradation. A BLAST search uncovered that *hcdA*, *hcdB*, and *hcdC* encode an aromatic flavin-binding hydroxylase, a ring-cleavage dioxygenase and a hydrolase, respectively. Biochemical analysis of the HcdA protein confirmed its similarity to class A FAD-binding enzymes (FMOs) [20–23,26–28]. HcdA is functionally related to previously well-described hydroxylases, such as OhpB monooxygenase from *Rhodococcus* sp. V49 [52], MhpA monooxygenase from *E. coli* K-12 [53] and *Comamonas testosteroni* TA441 [54], HppA monooxygenase from *Rhodococcus globerulus* PWD1 [31], and also *para*-hydroxybenzoate hydroxylase (PHBH) from *P. fluorescens* [55]. HcdA appears to have a high specificity for 3-(2,4-dihydroxyphenyl)-propionic acid, converting it to 3-(2,3,5-trihydroxyphenyl)-propionic acid, and a 40-fold lower activity towards *trans*-2,4-dihydroxycinnamic acid. Other structurally related substrates are not used by HcdA. Unlike HcdA, other related FMOs have a broader specificity for substrates. For example, OhpB monooxygenase is capable to oxidize 2-hydroxy-, 3-hydroxyphenylpropionic and cinnamic acids [52]. The HppA enzyme is more specific to 3-hydroxyphenylpropionic, but 4-chlorophenoxyacetic as well as 4-methyl-2-chlorophenoxyacetic acids are also oxidized [31]. On the other hand, all described FMOs including HcdA are NAD(P)H dependent, which reduces flavin for the hydroxylation of substrates [20–23]. The narrow specificity of HcdA to its natural substrate is typical for class A flavoproteins and shows the importance of HcdA enzyme in metabolism of 7-hydroxycoumarin. The kinetic analysis of HcdA yields K_m values of 50.10 ± 3.50 μM and 13.00 ± 1.20 μM for NADH and 3-(2,4-dihydroxyphenyl)-propionic acid, respectively. However, no kinetic parameters have been reported for OhpB, MhpA, and HppA hydroxylases for comparison, though PHBH has been shown to have a k_{cat} of 22.83 s^{-1} [56], which is 3-fold higher than a turnover number for the HcdA enzyme.

The hydroxylated product of the HcdA protein was analyzed by oxidizing it with the HcdB dioxygenase, followed by a chemical modification to the corresponding derivative of picolinic acid. The structure of the later compound was confirmed by ^{1}H NMR and ^{13}C NMR spectra. This allowed the reconstruction of the reaction products of both HcdA and HcdB enzymes. It was shown that a hydroxylation of 3-(2,4-dihydroxyphenyl)-propionic acid occurs at *ipso*-position of phenolic ring followed by internal rearrangements involving (1,2-C,C)-shift (NIH shift) of propionic acid moiety, hence forming 3-(2,3,5-trihydroxyphenyl)-propionic acid, as shown in Scheme 1. This would explain both the high specificity of HcdA enzyme for substrates with *para*-substituted phenol and inability of *Pseudomonas* sp. 7HK4 bacteria to utilize coumarin derivatives other than 7-hydroxycoumarin as the sole source of carbon and energy. Only a few classes of enzymes are able to catalyze *ipso*-reactions: laccases, peroxidases, dioxygenases, glutathione S-transferases (GST), cytochrome P450-dependent monooxygenases (CYP), and flavin-dependent monooxygenases. Among the known examples of *ipso*-enzymes, there are dioxygenases from *Comamonas testosteroni* T2 and *Sphingomonas* sp. strain RW1, which are involved in the desulfonation of 4-sulfobenzoate by *ipso*-substitution [57,58]. A rat liver CYP system is able to convert *p*-chloro, *p*-bromo, *p*-nitro, *p*-cyano, *p*-hydroxymethyl, *p*-formyl, and *p*-acetyl phenols to hydroquinone by *ipso*-hydroxylation [59]. GST is capable of catalyzing desulfonylation of sulfonylfuropyridine compounds by nucleophilic attack of the glutathione sulfur atom at *ipso*-position [60]. These are the examples of electrophilic or nucleophilic *ipso*-substitution reactions, however in some cases, a primary *ipso*-group is not eliminated, and instead it is shifted to *meta*-position. NIH shift restabilizes cyclohexadienone intermediate, because it leads to a rearomatization [45–48]. We showed that, similarly to flavin-dependent monooxygenases from *Sphingomonas* sp. TTNP3 and *Sphingobium xenophagum* strains, which are responsible for the degradation of alkylphenols, such as bisphenol A, octylphenol, *t*-butylphenol, *n*-octyloxyphenol, and *t*-butoxyphenol, HcdA-catalyzed reaction involves a NIH shift. Usually, NIH shift products are formed during the side reactions, and these internal rearrangements of an alkyl group upon the *ipso*-hydroxylation are spontaneous and non-enzymatic in *Sphingomonas* sp. strains [47,61]. An interesting novelty is that HcdA hydroxylase produces only one product, which has the *ipso*-group shifted to the *meta*-position. Therefore, we propose that in the case of HcdA, NIH shift occurs enzymatically, but not spontaneously or by a dienone-phenol rearrangement mechanism [62], since all bioconversions were performed under neutral or basic conditions. Although a further investigation is needed to determine the exact mechanism of the HcdA enzyme activity.

The analysis of the extradiol dioxygenase HcdB shows that this enzyme has a lower specificity for substrates. It catalyzes a conversion of the hydroxylated product of the HcdA enzyme to (2E,4E)-2,4-dihydroxy-6-oxonona-2,4-dienedioic acid, as shown in Scheme 1. Also, HcdB is capable oxidizing pyrocatechol, 3-methylcatechol, 3-methoxycatechol, 4-methylcatechol, 3-(2,3-dihydroxyphenyl)-propionic, and caffeic acids using a *meta*-cleavage mechanism forming yellow products. HcdB belongs to type I, class II extradiol dioxygenases [24,25], and is functionally related to OhpD catechol 2,3-dioxygenase from *Rhodococcus* sp. V49 [52], MhpB extradiol dioxygenase from *E. coli* K-12 [53,63], MpcI extradiol dioxygenase from *Alcaligenes eutrophus* [63], HppB extradiol dioxygenase from

Rhodococcus globerulus PWD1 [31], and DbfB 2,2′,3-trihydroxybiphenyl dioxygenase from *Sphingomonas* sp. RW1 [30].

Finally, we demonstrate that *(2E,4E)*-2,4-dihydroxy-6-oxonona-2,4-dienedioic acid, the ring cleavage product of HcdB protein, is subsequently hydrolyzed by the putative HcdC hydroxymuconic semialdehyde hydrolase. This enzyme has a low sequence homology to any of the previously characterized enzymes from *Rhodococcus* sp. V49 [52], *E. coli* K-12 [53], *Comamonas testosteroni* TA441 [55], or *Rhodococcus globerulus* PWD1 [31], therefore further investigation is needed to elucidate the exact mechanism of the hydrolysis and the specificity of the HcdC enzyme for substrates.

In summary, here we report a 7-hydroxycoumarin catabolic pathway in *Pseudomonas* sp. 7HK4 bacteria. New metabolites and genes responsible for the degradation of 3-(2,4-dihydroxyphenyl)-propionic acid have been isolated and identified. Our results show that the degradation of 7-hydroxycoumarin in *Pseudomonas* sp. 7HK4 involves a distinct metabolic pathway, compared to the previously characterized coumarin catabolic routes in *Pseudomonas*, *Arthrobacter*, and *Aspergillus* species [9–16]. It has been shown that *Pseudomonas* sp. 7HK4 bacteria employ unique flavin-binding *ipso*-hydroxylase for the oxidation of the aromatic ring of 3-(2,4-dihydroxyphenyl)-propionic acid. None of the proteins described in this paper have substantial sequence homology to the previously characterized enzymes implicated in the degradation of structurally similar substrates, such as 3-(2-hydroxyphenyl)-propionic acid in *Rhodococcus* sp. V49 [52], 3-(3-hydroxyphenyl)-propionic acid and 3-hydroxycinnamic acid in *E. coli* K-12 [53], *Comamonas testosteroni* TA441 [54], *Rhodococcus globerulus* PWD1 [31], or even 4-hydroxyphenylacetate in *Escherichia coli* W [64]. Thus, our results provide a fundamentally new insight into the degradation of hydroxycoumarins by the soil microorganisms. In addition, the discovered new bacteria and enzymes can be further employed for the development of novel biocatalytic processes useful for industry.

4. Materials and Methods

4.1. Bacterial Strains, Plasmids, and Reagents

Pseudomonas sp. 7HK4 bacterial strain capable of using 7-hydroxycoumarin as the sole source of carbon and energy was isolated from soil by enrichment in mineral medium containing 0.05% of 7-hydroxycoumarin. For cloning purposes, *E. coli* DH5α bacteria (φ80 *lacZ*ΔM15 Δ(*lacZY-argF*)U169 *deoR recA1 endA1 hsdR17*(r_K-m_K+) *supE44 thi-1 gyrA96 relA1*) (Thermo Fischer Scientific, Lithuania) were used. *E. coli* BL21 (DE3) bacteria (F- *ompT gal dcm lon hsdS*$_B$(r_B-m_B-) λ(DE3) [*lacI lacUV5-T7* gene 1, *ind1, sam7, nin5*]) (Novagen, Darmstadt, Germany) were used for gene expression studies.

All reagents used during this study are listed in Table S1 and plasmids are described in Table S2 in the Supplementary Material.

4.2. Bioconversions with Whole Cells

Pseudomonas sp. 7HK4 bacteria were grown in mineral medium supplemented with 0.05% (*w/v*) of 7-hydroxycoumarin or glucose, as the sole carbon and energy source, at 30 °C with rotary aeration (180 rpm) for 48 h. *E. coli* BL21 (DE3) bacteria containing recombinant genes were grown in 30 mL of Brain Heart Infusion (BHI) medium at 30 °C and 180 rpm overnight. High density bacterial culture was centrifuged and resuspended in 30 mL of minimal C-750501 medium, in which the synthesis of proteins was induced with 1 mM of isopropyl-β-D-1-thiogalactopyranoside (IPTG) after 1.5 h of incubation at 20 °C and 180 rpm [65]. Incubation at 20 °C was continued for another 24 h. Both *Pseudomonas* sp. 7HK4 and *E. coli* cells were sedimented by centrifugation (3220× *g*, 15 min). The collected cells were washed twice with 15 mL of 0.9% NaCl solution. For whole-cell conversion experiments, cells from 20 mL of culture were resuspended in 1 mL of 50 mM potassium phosphate buffer (pH 7.2). All small-scale bioconversions with whole cells were made in 50 mM potassium phosphate buffer, pH 7.2, which contained 1–2 mM of the substrate. The reaction mixtures were kept in a thermoblock at

30 °C and 500 rpm. Bioconversion mixtures were centrifuged for 2 min at 10,000× g, and 100 µL of the supernatant were analyzed by UV-VIS spectroscopy (range 200–600 nm). Measurements were repeated to record the changes in the absorption intensity over time. All measurements were performed with PowerWave XS microplate reader.

4.3. Preparation of cell-free extracts

Cells were sedimented by centrifugation (3220× g, 15 min). The biomass was resuspended in 3 mL of 50 mM potassium phosphate buffer (pH 7.2). The cells were disrupted by pulse-mode sonication (3 min duration and 1 s cycles) at 4 °C. Cell debris was removed by centrifugation (4 °C, 16,100× g, 15 min).

4.4. Protein Purification

Proteins were purified with Äkta purifier 900 chromatography systems (GE Healthcare, Helsinki, Finland). Cell-free extracts were loaded onto a Ni^{2+} Chelating HiTrapTM HP column (1–5 mL) (GE Healthcare, Finland) equilibrated with 50 mM potassium phosphate buffer, pH 7.2, at 1.0 mL/min. The column was washed with at least 3 volumes of the same buffer. Then the bound proteins were eluted with 0.5 M imidazole in 50 mM potassium phosphate buffer, pH 7.2. The fractions containing the purified enzyme were combined and dialyzed against the 50 mM potassium phosphate buffer, pH 7.2, at 4 °C overnight. Proteins were analyzed by SDS-PAGE according to Laemmli [66]. Protein concentration was determined by the Lowry method [67].

4.5. Preparation of Proteins from a Polyacrylamide Gel for Mass Spectrometric Analysis

Proteins were fractionated on a SDS-polyacrylamide gel. After Coomassie blue R-250 staining, protein samples were extracted from the gel as described in Reference [68] with minor changes. Protein bands were excised from the gel with a razor, and the gel was then destained twice with 200 µL of 25 mM ammonium bicarbonate and 50% acetonitrile solution for 30 min at 37 °C. Protein disulfide bonds were reduced with 40 µL of 10 mM dithiothreitol (DTT) for 45 min 60 °C, followed by incubation with 30 µL of 100 mM iodoacetamide for 1 h at room temperature in the dark to alkylate free cysteines. Gel slices were washed again twice with 100 µL 25 mM ammonium bicarbonate and 50% acetonitrile solution for 15 min at 37 °C, dehydrated by adding 50 µL 100% acetonitrile and dried using a vacuum centrifuge. Gel pieces were incubated with up to 40 µL of activated trypsin (10 ng/µL) at 37 °C overnight. The next day, the supernatant was saved and the peptides were extracted from the gel by incubating gel slices in two consecutive changes of 50 µL of 5% trifluoroacetic acid and 50% acetonitrile solution for 1 h at 37 °C. Combined supernatants were dried using a vacuum centrifuge at 30 °C. Lyophilized peptides were dissolved in 20 µL of 0.1% trifluoroacetic acid solution. Peptides purified and concentrated using Millipore C18 ZipTips.

4.6. Enzyme Assay

Activity of 3-(2,4-dihydroxyphenyl)-propionic acid hydroxylase was measured spectrophotometrically by monitoring absorption changes of the reaction mixture at 340 nm wavelength due to the oxidation of either NADH or NADPH ($\varepsilon340$ = 6220 M^{-1} cm^{-1}) after the addition of the substrate. The activity measurements were made with cell-free extracts or the purified protein. All measurements of the enzyme activity were carried out at room temperature in 1 mL of reaction mixture, containing 25–50 mM tricine or potassium phosphate buffers (pH 7.8), 100 µM NADH or NADPH, and 150 µM aromatic substrate. One unit of activity was defined as the amount of the enzyme that catalyzed the oxidation of 1 µmol of NADH or NADPH per minute.

4.7. In vivo Bioconversion of 3-(2,4-Dihydroxyphenyl)-propionic Acid

E. coli BL21 (DE3) bacteria, containing p4pmPmo and pTHPPDO plasmids, were grown in 200 mL of BHI medium at 30 °C and 180 rpm overnight. High density bacterial culture was centrifuged and resuspended in 200 mL of minimal C-750501 medium, in which synthesis of proteins was induced with 1 mM of IPTG at 20 °C and 180 rpm [65]. After 24 h of induction 3-(2,4-dihydroxyphenyl)-propionic acid was added to the final concentration of 4 mM. Bioconversion mixture was incubated for another 3 days at 30 °C with shaking. Cells were removed by centrifugation for 30 min at 3220× g and the supernatants were kept at 4 °C.

4.8. Purification of 3-(2,4-Dihydroxyphenyl)-propionic Acid Bioconversion Product

The supernatant containing the 3-(2,4-dihydroxyphenyl)-propionic acid oxidation product was incubated with 1.2 M ammonium chloride at room temperature overnight [39,50]. Reaction mixture was concentrated to ~100 mL volume and adjusted to pH 1 with concentrated HCl. The remains of substrate were extracted by five consecutive changes of 25 mL of ethyl acetate, and then aqueous fraction was purified using a reverse phase C_{18} column, equilibrated with water. Column was washed with at least 100 mL of water and then eluted with linear gradient of 0–60% methanol solution at a flow rate of 2 mL/min. Aqueous fractions were combined and evaporated (40 °C). Brownish crystals were dissolved in 0.1% formic acid solution and again loaded onto a reverse phase C_{18} column, previously equilibrated with 0.1% formic acid solution. Column was washed with at least 30 ml of 0.1% formic acid solution and then eluted with 60% methanol solution. Picolinic acid derivative was eluted with 0.1% formic acid solution. Fractions containing the product were collected, combined, and evaporated (40 °C). Picolinic acid yield from 145 mg of 3-(2,4-dihydroxyphenyl)-propionic acid fermentation was 34 mg, 24% of the theoretical yield. The product had traces of formic acid impurities, which aided the dissolution of the analyte in D_2O for NMR analysis.

4.9. High-Performance Liquid Chromatography and Mass Spectrometry

Before the analysis, the samples were mixed with an equal part of acetonitrile and centrifuged. High-performance liquid chromatography and mass spectrometry (HPLC-MS) was carried out using the system, consisting of the CBM-20 control unit, two LC-2020AD pumps, SIL-30AC auto sampler, and CTO-20AC column thermostat, using the SPD-M20A detector and LCMS-2020 mass spectrometer with ESI source (Shimadzu, Kyoto, Japan).

Chromatographic fractionation was conducted using YMC-Pack Pro C_{18} column, 150 × 3 mm (YMC, Kyoto, Japan) at 40 °C, with 0.1% formic acid solution in water and acetonitrile gradient from 5% to 95%.

Mass spectra were recorded from m/z 10 up to 500 m/z at 350 °C and ±4500 V using N_2. Mass spectrometry analysis was carried out using both the positive and negative ionization modes. The data were analyzed using LabSolutions LC/MS software (Shimadzu, Japan).

4.10. Nuclear Magnetic Resonance Spectroscopy

In total, 20 mg of the sample was dissolved in 0.5 mL D_2O. NMR spectra were recorded on an Ascend 400: ^{1}H NMR – 400 MHz, ^{13}C NMR – 100 MHz (Bruker, Billerica, MA, USA). Chemical shifts are reported in parts per million relative to the solvent resonance signal as an internal standard.

4.11. Protein MS-MS Analysis

Peptides were subjected to de novo sequencing based on matrix-assisted laser desorption ionization time of flight (MALDI-TOF/TOF) mass spectrometry (MS) and subsequent computational analysis at the Proteomics Centre of the Institute of Biochemistry, Vilnius University (Vilnius, Lithuania). The sample was purified as described previously in Materials and Methods. Tryptic digests (0.5 μL) were transferred on 384-well MALDI plate with 0.5 μL 4 mg/mL α-cyano-4-hydroxycinnamic acid (CHCA) matrix in 50% acetonitrile with 0.1% trifluoroacetic acid and analyzed with an Applied Biosystems/MDS SCIEX 4800 MALDI TOF/TOF™ mass spectrometer. Spectra were acquired in the positive reflector mode between 800 and 4000 m/z with fixed laser intensity at 3700 (Laser shots: 400; Mass accuracy: ±50 ppm). The most intense peaks of each survey scan (MS) were fragmented for sequence analysis (Collision energy: 1 keV; CID: no CID or medium air pressure CID used; Laser intensity: 4200–4400; Laser shots: 500–1000; Fragment mass accuracy: ±0.1 Da). Sequence analysis and peak lists were generated using GPS Explorer™ De Novo Explorer.

4.12. DNA Sequencing and Accession Numbers

The genome sequencing by Illumina (paired-ends) and assembling (in total, 35 contigs) was carried at Baseclear (Leiden, The Netherlands). The FASTQ sequence reads were generated using the Illumina Casava pipeline version 1.8.3. Initial quality assessment was based on data passing the Illumina Chastity filtering. Subsequently, reads containing adapters and/or PhiX control signal were removed using an in-house filtering protocol. The second quality assessment was based on the remaining reads using the FASTQC quality control tool version 0.10.0. The number of reads was 5,605,132 and an average quality score (Phred) was 37.72.

The accession number for partial 16S ribosomal RNA nucleotide sequence of *Pseudomonas* sp. 7HK4 is MH346031. Accession numbers for the sequences of *hcdA*, *hcdB*, and *hcdC* genes are MH346032, MH346033, MH346034, respectively.

Supplementary Materials: The following are available online, Table S1: Materials and reagents used in the studies, Table S2: Plasmids used in the studies, Table S3: The list of primers used in this study, Figure S1: Phylogenetic tree of *Pseudomonas* sp. 7HK4 bacteria based on partial 16S rDNA sequences, Figure S2: Biotransformation of 7-hydroxycoumarin, 6-hydroxycoumarin, coumarin and 6,7-dihydroxycoumarin by whole cells of *Pseudomonas* sp. 7HK4, Figure S3: Phylogenetic trees of HcdA, HcdB and HcdC proteins, Figure S4: SDS-PAGE, UV/Vis spectrum and analytical gel filtration chromatography of purified HcdA protein, Figure S5: Specificity of HcdA protein to flavin and nicotinamide cofactors, Figure S6: Activity of HcdA protein in different buffer systems, Figure S7: Activity of HcdA protein in different pH, Figure S8: Kinetic analysis of HcdA as determined by NADH oxidation for 3-(2,4-dihydroxyphenyl)-propionic acid, Figure S9: Kinetic analysis of HcdA as determined by NADH oxidation for NADH, Figure S10: Double reciprocal plot of NADH oxidation as a function of NADH concentration, Figure S11: Biotransformations of 3,4-dihydroxybenzoic acid, 2',3'-dihydroxy-4'-methoxyaceto-phenone hydrate, gallacetophenone, pyrogallol, 2,3,4-trihydroxybenzoic acid and 2,3,4-trihydroxy-benzophenone by whole cells of *E. coli* BL21 containing *hcdB* gene, Figure S12: ^{1}H NMR spectrum of 6-(2-carboxyethyl)-4-oxo-1,4-dihydropyridine-2-carboxylic acid, Figure S13: ^{13}C NMR spectrum of 6-(2-carboxyethyl)-4-oxo-1,4-dihydropyridine-2-carboxylic acid, Figure S14: Resonance structure of *oxo*-picolinic acid derivative showing electron densities on aromatic carbons, Figure S15: SDS-PAGE of *E. coli* BL21 cell-free extract, containing induced recombinant HcdA, HcdB and HcdC proteins, Figure S16: HPLC-MS analysis of 3-(2-hydroxyphenyl)-propionic acid bioconversion mixture.

Author Contributions: Conceptualization and Funding Acquisition, R.M.; Investigation, A.K.; Data Analysis, A.K. and R.M.; Writing, Review & Editing, A.K. and R.M.

Acknowledgments: We are grateful to Marija Ger for performing peptide sequence analysis, Daiva Tauraitė for help with chemical synthesis, and Laura Kalinienė for critical reading of the manuscript.

References

1.   Sarker, S.D.; Nahar, L. Progress in the chemistry of naturally occurring coumarins. *Prog. Chem. Org. Nat. Prod.* **2017**, *106*, 241–304. [PubMed]

2.   Murray, R.D.H.; Mendez, J.; Brown, S.A. *The Natural Coumarins—Occurrence, Chemistry and Biochemistry*; John Wiley: Chichester, UK, 1982.

3.   Bourgaud, F.; Hehn, A.; Larbat, R.; Doerper, S.; Gontier, E.; Kellner, S.; Matern, U. Biosynthesis of coumarins in plants: A major pathway still to be unravelled for cytochrome P450 enzymes. *Phytochem. Rev.* **2006**, *5*, 293–308. [CrossRef]

4.   Mazid, M.; Khan, T.A.; Mohammad, F. Role of secondary metabolites in defense mechanisms of plants. *Biol. Med.* **2011**, *3*, 232–249.

5.   Kayser, O.; Kolodziej, H. Antibacterial activity of simple coumarins: Structural requirements for biological activity. *Z. Naturforsch. C* **1999**, *54*, 169–174. [CrossRef] [PubMed]

6. De Souza, S.M.; Delle Monache, F.; Smânia, A., Jr. Antibacterial activity of coumarins. *Z. Naturforsch C* **2005**, *60*, 693–700. [CrossRef] [PubMed]

7. Yang, L.; Ding, W.; Xu, Y.; Wu, D.; Li, S.; Chen, J.; Guo, B. New insights into the antibacterial activity of hydroxycoumarins against *Ralstonia solanacearum*. *Molecules* **2016**, *21*, 468. [CrossRef] [PubMed]

8. Serghini, K.; de Lugue, A.P.; Castejon, M.M.; Garcia, T.L.; Jorrin, J.V. Sunflower (*Helianthus annuus* L.) response to broomrae (*Orobanche cernua* loefl.) parasitism: Induced synthesis and excretion of 7-hydroxylated simple coumarins. *J. Exp. Bot.* **2001**, *52*, 227–234. [CrossRef]

9. Shieh, H.S.; Blackwood, A.C. Use of coumarin by soil fungi. *Can. J. Microbiol.* **1969**, *15*, 647–648. [CrossRef] [PubMed]

10. Nakayama, Y.; Nonomura, S.; Tatsumi, C. The metabolism of coumarin by a strain of *Pseudomonas*. *Agric. Biol. Chem.* **1973**, *37*, 1423–1437. [CrossRef]

11. Levy, C.C.; Weinstein, G.D. The metabolism of coumarin by a microorganism. The reduction of *o*-coumaric acid to melilotic acid. *Biochemistry* **1964**, *3*, 1944–1947. [CrossRef] [PubMed]

12. Bellis, D.M. Metabolism of coumarin and related compounds in cultures of *Penicillium* species. *Nature* **1958**, *182*, 806–807. [CrossRef] [PubMed]

13. Sheila, M.B. The transformations of coumarin, *o*-coumaric acid and *trans*-cinnamic acid by *Aspergillus niger*. *Phytochemistry* **1967**, *6*, 127–130.

14. Aguirre-Pranzoni, C.; Orden, A.A.; Bisogno, F.R.; Ardanaz, C.E.; Tonn, C.E.; Kurina-Sanz, M. Coumarin metabolic routes in *Aspergillus* spp. *Fungal Biol.* **2011**, *115*, 245–252. [CrossRef] [PubMed]

15. Marumoto, S.; Miyazawa, M. Microbial reduction of coumarin, psoralen, and xanthyletin by *Glomerella cingulata*. *Tetrahedron* **2011**, *67*, 495–500. [CrossRef]

16. Levy, C.C.; Frost, P. The metabolism of coumarin by a microorganism. *Melilotate hydroxylase*. *J. Biol. Chem.* **1966**, *241*, 997–1003. [PubMed]

17. Roper, D.I.; Cooper, R.A. Purification, nucleotide sequence and some properties of a bifunctional isomerase/decarboxylase from the homoprotocatechuate degradative pathway of *Escherichia coli* C. *Eur. J. Biochem.* **1993**, *217*, 575–580. [CrossRef] [PubMed]

18. Díaz, E.; Timmis, K.N. Identification of functional residues in a 2-hydroxymuconic semialdehyde hydrolase. A new member of the alpha/beta hydrolase-fold family of enzymes which cleaves carbon-carbon bonds. *J. Biol. Chem.* **1995**, *270*, 6403–6411. [CrossRef] [PubMed]

19. Lim, J.C.; Lee, J.; Jang, J.D.; Lim, J.Y.; Min, K.R.; Kim, C.K.; Kim, Y. Characterization of the *pcbE* gene encoding 2-hydroxypenta-2,4-dienoate hydratase in *Pseudomonas* sp. DJ-12. *Arch. Pharm. Res.* **2000**, *23*, 187–195. [CrossRef] [PubMed]

20. Moonen, M.J.H.; Fraaije, M.W.; Rietjens, I.M.C.M.; Laane, C.; van Berkel, W.J.H. Flavoenzyme-catalyzed oxygenations and oxidations of phenolic compounds. *Adv. Synth. Catal.* **2002**, *344*, 1023–1035. [CrossRef]

21. Chaiyen, P. Flavoenzymes catalyzing oxidative aromatic ring-cleavage reactions. *Arch. Biochem. Biophys.* **2010**, *493*, 62–70. [CrossRef] [PubMed]

22. Crozier-Reabe, K.; Moran, G.R. Form Follows function: Structural and catalytic variation in the class A flavoprotein monooxygenases. *Int. J. Mol. Sci.* **2012**, *13*, 15601–15639. [CrossRef] [PubMed]

23. Romero, E.; Castellanos, J.R.G.; Gadda, G.; Fraaije, M.W.; Mattevi, A. Same substrate, many reactions: Oxygen activation in flavoenzymes. *Chem. Rev.* **2018**, *118*, 1742–1769. [CrossRef] [PubMed]

24. Abu-Omar, M.M.; Loaiza, A.; Hontzeas, N. Reaction mechanisms of mononuclear non-heme iron oxygenases. *Chem. Rev.* **2005**, *105*, 2227–2252. [CrossRef] [PubMed]

25. Cho, H.J.; Kim, K.; Sohn, S.Y.; Cho, H.Y.; Kim, K.J.; Kim, M.H.; Kim, D.; Kim, E.; Kang, B.S. Substrate binding mechanism of a type I extradiol dioxygenase. *J. Biol. Chem.* **2010**, *285*, 34643–34652. [CrossRef] [PubMed]

26. Van Berkel, W.J.; Kamerbeek, N.M.; Fraaije, M.W. Flavoprotein monooxygenases, a diverse class of oxidative biocatalysts. *J. Biotechnol.* **2006**, *124*, 670–689. [CrossRef] [PubMed]

27. Macheroux, P.; Kappes, B.; Ealick, S.E. Flavogenomics—A genomic and structural view of flavin-dependent proteins. *FEBS J.* **2011**, *278*, 2625–2634. [CrossRef] [PubMed]

28. Nakamura, S.; Nakamura, T.; Ogura, Y. Absorption spectrum of flavin mononucleotide semiquinone. *J. Biochem.* **1963**, *53*, 143–146. [CrossRef] [PubMed]

29. Koptyug, V.A. *Atlas of Spectra of Aromatic and Heterocyclic Compounds*; Science, Siberian Department of AS USSR: Novosibirsk, Russia, 1982.

30. Happe, B.; Eltis, L.D.; Poth, H.; Hedderich, R.; Timmis, K.N. Characterization of 2,2′,3-trihydroxybiphenyl dioxygenase, an extradiol dioxygenase from the dibenzofuran- and dibenzo-p-dioxin-degrading bacterium *Sphingomonas* sp. strain RW1. *J. Bacteriol.* **1993**, *175*, 7313–7320. [CrossRef] [PubMed]

31. Barnes, M.R.; Duetz, W.A.; Williams, P.A. A 3-(3-hydroxyphenyl) propionic acid catabolic pathway in *Rhodococcus globerulus* PWD1: Cloning and characterization of the *hpp* operon. *J. Bacteriol.* **1997**, *179*, 6145–6153. [CrossRef] [PubMed]

32. Vaillancourt, F.H.; Haro, M.A.; Drouin, N.M.; Karim, Z.; Maaroufi, H.; Eltis, L.D. Characterization of extradiol dioxygenases from a polychlorinated biphenyl-degrading strain that possess higher specificities for chlorinated metabolites. *J. Bacteriol.* **2003**, *185*, 1253–1260. [CrossRef] [PubMed]

33. Wolgel, S.A.; Dege, J.E.; Perkins-Olson, P.E.; Jaurez-Garcia, C.H.; Crawford, R.L.; Münck, E.; Lipscomb, J.D. Purification and characterization of protocatechuate 2,3-dioxygenase from *Bacillus macerans*: A new extradiol catecholic dioxygenase. *J. Bacteriol.* **1993**, *175*, 4414–4426. [CrossRef] [PubMed]

34. Asturias, J.A.; Eltis, L.D.; Prucha, M.; Timmisn, K.N. Analysis of three 2,3-dihydroxybiphenyl 1,2-dioxygenases found in *Rhodococcus globerulus* P6. *J. Biol. Chem.* **1994**, *269*, 7807–7815. [PubMed]

35. Bayly, R.C.; Dagley, S.; Gibson, D.T. The metabolism of cresols by species of *Pseudomonas. Biochem. J.* **1966**, *101*, 293–301. [CrossRef] [PubMed]

36. Burlingame, R.; Chapman, P.J. Catabolism of phenylpropionic acid and its 3-hydroxy derivative by *Escherichia coli. J. Bacteriol.* **1983**, *155*, 113–121. [PubMed]

37. Duggleby, C.J.; Williams, P.A. Purification and some properties of the 2-hydroxy-6-oxohepta-2,4-dienoate hydrolase (2-hydroxymuconic semialdehyde hydrolase) encoded by the Tol plasmid-pww0 from *Pseudomonas putida* mt-2. *J. Gen. Microbiol.* **1986**, *132*, 717–726. [CrossRef]

38. Riegert, U.; Heiss, G.; Fischer, P.; Stolz, A. Distal cleavage of 3-chlorocatechol by an extradiol dioxygenase to 3-chloro-2-hydroxymuconic semialdehyde. *J. Bacteriol.* **1998**, *180*, 2849–2853. [PubMed]

39. Wieser, M.; Eberspächer, J.; Vogler, B.; Lingens, F. Metabolism of 4-chlorophenol by *Azotobacter* sp. GP1: Structure of the *meta* cleavage product of 4-chlorocatechol. *FEMS Microbiol. Lett.* **1994**, *116*, 73–78. [CrossRef] [PubMed]

40. March, J. *Advanced Organic Chemistry Reactions, Mechanisms and Structure*, 3rd ed.; John Wiley & Sons Inc.: New York, NY, USA, 1985.

41. Gunther, H. *NMR Spectroscopy: Basic Principles, Concepts and Applications in Chemistry*, 3rd ed.; Wiley-VCH: Weinheim, Germany, 2013.

42. Simons, W.W. *The Sadtler Guide to Carbon-13 NMR Spectra*; Sadtler Research Laboratories: Philadelphia, PA, USA, 1983.

43. Retcofsky, H.L.; Friedel, R.A. *Carbon-13 Nuclear Magnetic Resonance Spectra of Monosubstituted Pyridines*; U.S. Dept of the Interior, Bureau of Mines: Washington, DC, USA, 1969.

44. Thomas, S.; Bruhl, I.; Heilmann, D.; Kleinpeter, E. 13C NMR chemical shift calculations for some substituted pyridines: A comparative consideration. *J. Chem. Inf. Comput. Sci.* **1997**, *37*, 726–730. [CrossRef]

45. Martin, G.; Dijols, S.; Capeillere-Blandin, C.; Artaud, I. Hydroxylation reaction catalyzed by the *Burkholderia cepacia* AC1100 bacterial strain. Involvement of the chlorophenol-4-monooxygenase. *Eur. J. Biochem.* **1999**, *261*, 533–539. [CrossRef] [PubMed]

46. Ricken, B.; Kolvenbach, B.A.; Corvini, P.F. *Ipso*-substitution—The hidden gate to xenobiotic degradation pathways. *Curr. Opin. Biotechnol.* **2015**, *33*, 220–227. [CrossRef] [PubMed]

47. Kolvenbach, B.A.; Corvini, P.F. The degradation of alkylphenols by *Sphingomonas* sp. strain TTNP3—A review on seven years of research. *N. Biotechnol.* **2012**, *30*, 88–95. [CrossRef] [PubMed]

48. Gabriel, F.L.; Heidlberger, A.; Rentsch, D.; Giger, W.; Guenther, K.; Kohler, H.P. A novel metabolic pathway for degradation of 4-nonylphenol environmental contaminants by *Sphingomonas xenophaga* Bayram: *Ipso*-hydroxylation and intramolecular rearrangement. *J. Biol. Chem.* **2005**, *280*, 15526–15533. [CrossRef] [PubMed]

49. Müller, R.; Lingens, F. Oxidative ring-cleavage of catechol in meta-position by superoxide. *Z. Naturforsch.* **1989**, *44c*, 207–211. [CrossRef]

50. Sala-Trepat, J.M.; Murray, K.; Williams, P.A. The metabolic divergence in the meta cleavage of catechols by *Pseudomonas putida* NCIB 10015. *Eur. J. Biochem.* **1972**, *28*, 347–356. [CrossRef] [PubMed]

51. Venugopala, K.N.; Rashmi, V.; Odhav, B. Review on natural coumarin lead compounds for their pharmacological activity. *Biomed. Res. Int.* **2013**, *13*, 14–19. [CrossRef] [PubMed]

52. Powell, J.A.; Archer, J.A. Molecular characterisation of a *Rhodococcus* ohp operon. *Antonie Van Leeuwenhoek* **1998**, *74*, 175–188. [CrossRef] [PubMed]

53. Díaz, E.; Ferrández, A.; Prieto, M.A.; García, J.L. Biodegradation of aromatic compounds by *Escherichia coli*. *Microbiol. Mol. Biol. Rev.* **2001**, *65*, 523–569. [CrossRef] [PubMed]

54. Arai, H.; Yamamoto, T.; Ohishi, T.; Shimizu, T.; Nakata, T.; Kudo, T. Genetic organization and characteristics of the 3-(3-hydroxyphenyl) propionic acid degradation pathway of *Comamonas testosteroni* TA441. *Microbiology* **1999**, *145*, 2813–2820. [CrossRef] [PubMed]

55. Müller, F.; Voordouw, G.; Van Berkel, W.J.H.; Steennis, P.J.; Visser, S.; Van Rooijen, P.J. A study of *p*-hydroxybenzoate hydroxylase from *Pseudomonas fluorescens*. *Eur. J. Biochem.* **1979**, *101*, 235–244. [CrossRef] [PubMed]

56. Shuman, B.; Dix, T.A. Cloning, nucleotide sequence, and expression of a *p*-hydroxybenzoate hydroxylase isozyme gene from *Pseudomonas fluorescens*. *J. Biol. Chem.* **1993**, *268*, 17057–17062. [PubMed]

57. Bunz, P.V.; Cook, A.M. Dibenzofuran 4,4a-dioxygenase from *Sphingomonas* sp. strain RW1: Angular dioxygenation by a threecomponent enzyme system. *J. Bacteriol.* **1993**, *175*, 6467–6475. [CrossRef] [PubMed]

58. Locher, H.H.; Leisinger, T.; Cook, A.M. 4-Sulphobenzoate 3,4-dioxygenase. Purification and properties of a desulphonative two-component enzyme system from *Comamonas testosteroni* T-2. *Biochem. J.* **1991**, *274*, 833–842. [CrossRef] [PubMed]

59. Vatsis, K.P.; Coon, M.J. Ipso-substitution by cytochrome P450 with conversion of *p*-hydroxybenzene derivatives to hydroquinone: Evidence for hydroperoxo-iron as the active oxygen species. *Arch. Biochem. Biophys.* **2002**, *397*, 119–129. [CrossRef] [PubMed]

60. Zhao, Z.; Koeplinger, K.A.; Peterson, T.; Conradi, R.A.; Burton, P.S.; Suarato, A.; Heinrikson, R.L.; Tomaselli, A.G. Mechanism, structure–activity studies, and potential applications of glutathione S-transferase-catalyzed cleavage of sulfonamides. *Drug Metab. Dispos.* **1999**, *27*, 992–998. [PubMed]

61. Corvini, P.F.X.; Meesters, R.J.W.; Schäffer, A.; Schröder, H.F.; Vinken, R.; Hollender, J. Degradation of a nonylphenol single isomer by *Sphingomonas* sp. strain TTNP3 leads to a hydroxylation-induced migration product. *Appl. Environ. Microbiol.* **2004**, *70*, 6897–6900. [CrossRef] [PubMed]

62. Arnold, R.T.; Buckley, J.S., Jr.; Richter, J. The dienone–phenol rearrangement. *J. Am. Chem. Soc* **1947**, *69*, 2322–2325. [CrossRef]

63. Spence, E.L.; Kawamukai, M.; Sanvoisin, J.; Braven, H.; Bugg, T.D. Catechol dioxygenases from *Escherichia coli* (MhpB) and *Alcaligenes eutrophus* (MpcI): Sequence analysis and biochemical properties of a third family of extradiol dioxygenases. *J. Bacteriol.* **1996**, *178*, 5249–5256. [CrossRef] [PubMed]

64. Prieto, M.A.; Díaz, E.; García, J.L. Molecular characterization of the 4-hydroxyphenylacetate catabolic pathway of *Escherichia coli* W: Engineering a mobile aromatic degradative cluster. *J. Bacteriol.* **1996**, *178*, 111–120. [CrossRef] [PubMed]

65. Sivashanmugam, A.; Murray, V.; Cui, C.; Zhang, Y.; Wang, J.; Li, Q. Practical protocols for production of very high yields of recombinant proteins using *Escherichia coli*. *Protein Sci.* **2009**, *18*, 936–948. [CrossRef] [PubMed]

66. Laemmli, U.K. Cleavage of structural proteins during the assembly of the head of bacteriophage T4. *Nature* **1970**, *227*, 680–685. [CrossRef] [PubMed]

67. Lowry, O.H.; Rosebrough, N.J.; Farr, A.L.; Randall, R.J. Protein measurement with the folin-phenol reagents. *J. Biol. Chem.* **1951**, *193*, 265–275. [PubMed]

68. Gundry, R.L.; White, M.Y.; Murray, C.I.; Kane, L.A.; Fu, Q.; Stanley, B.A.; Van Eyk, J.E. Preparation of proteins and peptides for mass spectrometry analysis in a bottom-up proteomics workflow. *Curr. Protoc. Mol. Biol.* **2009**, *88*, 10.25.1–10.25.23.

Effects of GGT and C-S Lyase on the Generation of Endogenous Formaldehyde in *Lentinula edodes* at Different Growth Stages

Xiaoyu Lei [1], Shuangshuang Gao [1], Xi Feng [2], Zhicheng Huang [1], Yinbing Bian [3], Wen Huang [1,*] and Ying Liu [1,*]

[1]   College of Food Science and Technology, Huazhong Agricultural University, Wuhan 430070, China; xiaoyulei1988@126.com (X.L.); 13720161459@163.com (S.G.); HuangZhiCheng1210@163.com (Z.H.)
[2]   Department of Nutrition, Food Science and Packaging, California State University, San Jose, CA 95192, USA; xi.feng@sjsu.edu
[3]   Institute of Applied Mycology, Huazhong Agricultural University, Wuhan 430070, China; bianyb.123@163.com
*   Correspondence: huangwen@mail.hzau.edu.cn (W.H.); yingliu@mail.hzau.edu.cn (Y.L.);

Academic Editor: Stefano Serra

Abstract: Endogenous formaldehyde is generated as a normal metabolite via bio-catalysis of γ-glutamyl transpeptidase (GGT) and L-cysteine sulfoxide lyase (C-S lyase) during the growth and development of *Lentinula edodes*. In this study, we investigated the mRNA and protein expression levels, the activities of GGT and C-S lyase, and the endogenous formaldehyde content in *L. edodes* at different growth stages. With the growth of *L. edodes*, a decrease was found in the mRNA and protein expression levels of GGT, while an increase was observed in the mRNA and protein expression levels of C-S lyase as well as the activities of GGT and C-S lyase. Our results revealed for the first time a positive relationship of formaldehyde content with the expression levels of *Csl* (encoding Lecsl) and Lecsl (C-S lyase protein of *Lentinula edodes*) as well as the enzyme activities of C-S lyase and GGT during the growth of *L. edodes*. This research provided a molecular basis for understanding and controlling the endogenous formaldehyde formation in *Lentinula edodes* in the process of growth.

Keywords: *Lentinula edodes*; endogenous formaldehyde; GGT; C-S lyase; expression levels

1. Introduction

Lentinula edodes (shiitake mushroom) is the second-most popular edible mushrooms in the world (the No. 1 is *Agaricus bisporus*), due to its high nutritional and medicinal values as well as the unique flavor [1–4]. Lenthionine (1,2,3,5,6-pentathiepane), the unique aroma of *L. edodes* [5,6], is derived from lentinic acid in a two-step enzymatic reaction [7,8]. In the reaction, the lentinic acid is catalyzed by γ-glutamyl transpeptidase (GGT) and L-cysteine sulfoxide lyase (C-S lyase) to generate the unique flavor compounds, including lenthionine [8–10]. Nevertheless, formaldehyde is also produced in this metabolic pathway (Figure 1).

Formaldehyde, a mutagen, can be found in the air, natural and processed foods, especially in frozen food and dry foods, and is classified as a human carcinogen by the International Agency for Research on Cancer (IARC) of World Health Organization [11]. The maximum daily dose reference for formaldehyde is defined as about 0.2 mg/kg body weight per day by the US Environmental Protection Agency. Even small doses of formaldehyde can cause various symptoms of physical discomfort [12]. However, the formaldehyde contents are 1–20 mg/kg in various fruits, vegetables, meat and fish products [13]. In recent years, the amount of formaldehyde in *L. edodes* has raised the public concerns about food safety. For instance, high levels of formaldehyde (100–300 mg/kg) have been detected in shiitake mushroom samples produced in UK and Chinese [14]. Japanese researchers have reported that formaldehyde is generated in the growth process of *L. edodes* as a normal metabolite to form its unique flavor [15].

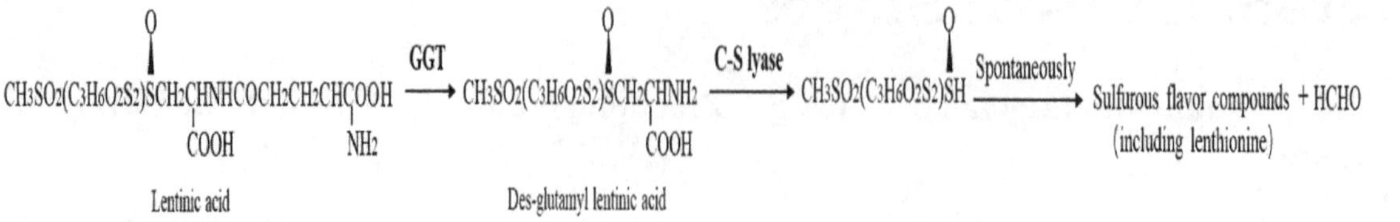

Figure 1. Proposed pathway for the generation of sulfurous flavor compounds and endogenous formaldehyde in *Lentinula edodes*.

Meanwhile, γ-glutamyl transpeptidase (GGT; EC 2.3.2.2) is an enzyme that catalyzes the transfer of the γ-glutamyl group of glutathione and related γ-glutamyl amides to water (hydrolysis) or to amino acids and peptides (transpeptidation) [16]. Cysteine sulfoxide lyase (EC 4.4.1.4) is a pyridoxal-5-phosphate (PLP) dependent enzyme and assigned to the class I family of PLP dependent enzymes [17]. In our previous work, the two enzymes were purified and characterized to play a significant biochemical role in the generation of endogenous formaldehyde in *L. edodes* [18]. Additionally, the gene of *Csl* encoding Lecsl (*L. edodes* C-S lyase) was cloned [19].

However, little is known about the relationship of GGT and C-S lyase with the production of formaldehyde in *L. edodes*. Thus, the aims of the present work were to determine the mRNA and protein expression levels and the activities of GGT and C-S lyase at different growth stages of shiitake mushrooms and to explore their correlations with endogenous formaldehyde production in *L. edodes*. This research could provide a molecular basis to understand the regulatory mechanisms of endogenous formaldehyde generation in *L. edodes* during the growth process.

2. Results and Discussion

2.1. Gene Expression of Ggtl and Csl

The expressions of *Ggtl* (encoding Leggt) and *Csl* (encoding Lecsl) during fruiting body development were analyzed by examining their transcript levels using real-time quantitative PCR. Both *Ggtl* and *Csl* were expressed at all the five stages of fruit-body development: mycelia, grey, young fruiting body, immature fruiting body and mature fruiting body, but differed in their expression patterns (Figure 2). Specifically, the transcript level of *Ggtl* decreased during the growth process, in contrast to an increase for *Csl*. Additionally, *Ggtl* showed the highest and lowest expression level in mycelia and mature fruiting body, respectively, which was just the opposite for *Csl*. There was approximately 1.5-fold difference between the highest and lowest expression levels in the two genes. Analysis of variance showed a significant difference in the *Ggtl* and *Csl* expression levels between mycelia and fruiting body stages, while *Ggtl* exhibited a significantly different expression in the four fruiting body stages ($p < 0.05$).

Based on our in-house transcriptome data, the expression pattern of *Ggtl* in different growth stages was basically in line with the quantitative RT-PCR results of our experiment. *Csl* was first reported as a gene involved in the generation of unique aroma of *L. edodes* [19]. In an early report, *Csl* displayed no obvious change in the expression level at 1, 2 and 3 h in the stage of mycelium or fruiting body during hot-air drying [20], which were basically consistent with our research results. Overall, there was a gradual decline of *Ggtl* expression level in different growth stages and an obvious increase of *Csl* expression level between mycelia and fruiting body stages.

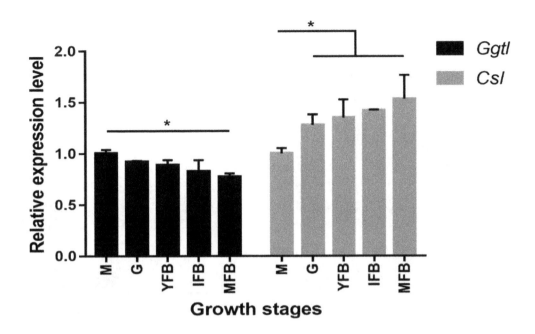

Figure 2. Expression of *Ggtl* and *Csl* during fruit-body development of *Lentinula edodes* strain W1. Relative expression during five growth stages, M, mycelia; G, grey; YFB, young fruiting body; IFB, immature fruiting body; MFB, mature fruiting body. Transcript levels of *Ggtl* (black bars) and *Csl* (gray bars) were determined by real-time quantitative RT-PCR analysis and normalized against *Actinl*. The expressions during the M stage were taken as 1. Error bars indicate standard deviation for three independent experiments. *$p < 0.05$, ANOVA tests by Duncan's indicate significant differences.

2.2. Western Blot of Leggt and Lecsl

The protein expression levels of Leggt (GGT protein of *L. edodes*) and Lecsl (C-S lyase protein of *L. edodes*) in *L. edodes* at different growth stages were determined by Western blot using β-actin as an internal reference (Figure 3A,B) [21]. Notably, Leggt exhibited three bands, which conformed to previous reports [22,23], and their gray values showed a gradual decrease in the five stages. A previous study has shown that a mature gamma-glutamyl transpeptidase consists of one polypeptide chain and can be divided into a large and a small subunit by self-catalysis at the highly conserved threonine [24]. Correspondingly, three bands were shown in our Western blot analysis of Leggt. The protein expression levels showed a decrease in Leggt while a gradual increase in Lecsl during the growth process of *L. edodes* (Figure 3C). The overall trend of protein expression levels was similar to that of mRNA expression levels. Moreover, the two proteins showed significant differences in all the five samples. The changes between the highest and lowest expression levels of Leggt and Lecsl were 2.1-fold and 1.9-fold, respectively. This is the first report on the expression levels of these two enzyme proteins in shiitake mushrooms. Our data indicated that the expression of Leggt decreased while Lecsl increased across the five growth stages of *L. edodes*.

2.3. Enzyme Activities of GGT and C-S Lyase

Figure 4A,B show the activity of GGT and C-S lyase at five different growth stages of *L. edodes*. The enzyme activities of GGT and C-S lyase were the lowest at the mycelia stage (4.7 U/g of GGT, 17.1 U/g of C-S lyase) and showed an obvious increase at the four fruiting body stages. Moreover, the GGT enzyme activity was relatively lower at the young fruiting body stage, while the C-S lyase enzyme activity showed no significant differences at the later four fruiting body stages ($p < 0.05$).

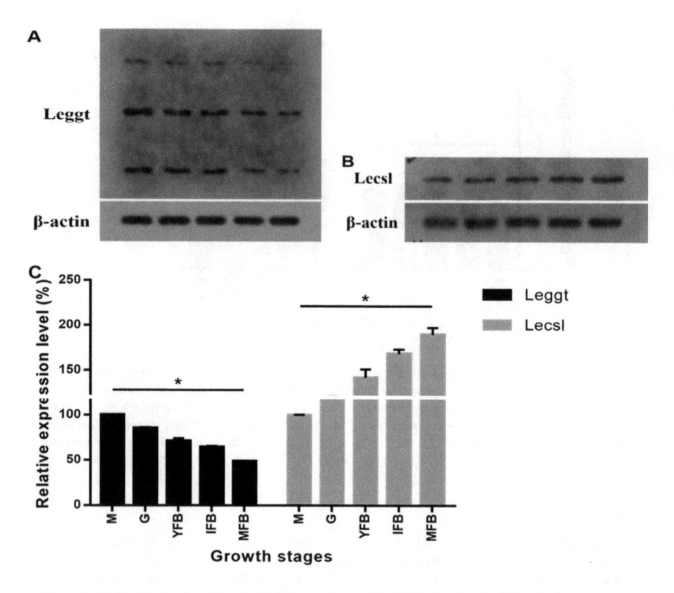

Figure 3. (**A**) Protein levels of Leggt at five stages of growth. (**B**) Protein levels of Lecsl at five stages of growth. (**C**) Relative expression of Leggt and Lecsl during five growth stages. β-actin protein was used as loading control, and the expressions during the M stage were taken as 100%. M, mycelia; G, grey; YFB, young fruiting body; IFB, immature fruiting body; MFB, mature fruiting body. Error bars indicate standard deviation for three independent experiments. *$p < 0.05$, ANOVA tests by Duncan's indicate significant differences.

Our experimental data (40.2–54.1 U/g) were consistent with the results of Huang et al., who reported that the GGT enzyme activity at the four fruiting body stages ranged from 40 to 80 U/g, with a similar difference at each stage in the fruiting body samples [25]. This is the first report about the C-S lyase enzyme activities of *L. edodes* at different growth stages. The C-S lyase enzyme activities in the present study (17.1–3575.6 U/g) were obviously lower than those determined by Xu et al. under high-temperature pre-drying (45, 55, 65 and 75 °C for 30 min) of air-dried (45 °C for 4.5 h, 60 °C for 4 h) *L. edodes* (80.17–100.54 U/mg) [26]. Liu et al. reported that the C-S lyase from *L. edodes* showed the optimum activity at 40 °C and was stable at 20–60 °C [18]. *Csl* has been demonstrated as a heat-inducible gene [27], so the activity of protein encoded by it could be improved greatly. In this study, our samples were generally collected at 25 °C, which was far below the drying treatment temperature (>45 °C), so the C-S lyase enzyme activity was lower than that treated under high temperature. Collectively, the C-S lyase enzyme activity showed no significant difference at the four fruiting body stages ($p < 0.05$).

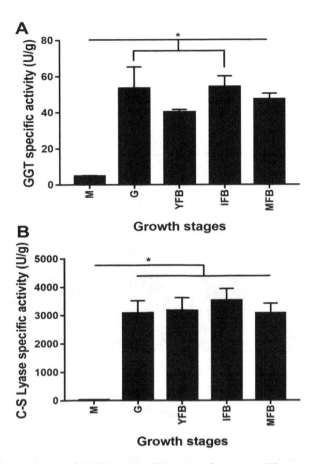

Figure 4. (A) The specific activity of GGT at the five growth stages. The reaction mixture containing the enzyme and the GPNA substrate was analyzed under standard conditions, and the residual activity was calculated. **(B)** The specific activity of C-S lyase at the five growth stages. The reaction mixture containing the enzyme and S-ethyl-L-cysteine sulfoxide substrate was analyzed under standard conditions, and the residual activity was calculated. M, mycelia; G, grey; YFB, young fruiting body; IFB, immature fruiting body; MFB, mature fruiting body. Error bars indicate standard deviation for three independent experiments. *$p < 0.05$, ANOVA tests by Duncan's indicate significant differences.

However, there was a marked difference between the two enzymes activities and their expression levels at the mycelia stage. GGT performs different functions in peptide transferase reaction and hydrolysis reaction under different conditions [16]. Lecsl has already been demonstrated to have one active center involved in the binding of the two substrates, S-methyl-L-cysteine sulfoxide and L-cysteine, with both cysteine sulfoxide lyase and cysteine desulfurase activities [19]. In addition, there are many factors affecting enzyme activity, such as pH, temperature, metal ions and so on. The two enzymes activities were reported to be stimulated by Na^+, K^+, Mg^{2+} and Ca^{2+} ions [18]. The environment conditions are very different between the mycelia stage and the four fruiting body stages, since the former belongs to vegetative growth and the latter belong to reproductive growth. These observations suggested that there might be a sort of regulatory mechanism that activated the two enzymes during the fruiting body stages while stayed inactive at the mycelial stage, which we failed to detect in this study.

2.4. Endogenous Formaldehyde Content in L. edodes

Figure 5 showed the content of endogenous formaldehyde in *L. edodes* at different growth phases. Compared with the mycelia stage, the endogenous formaldehyde content increased significantly ($p < 0.05$) at the four fruiting body stages, reached the maximum at the immature fruiting body stage and slightly decreased at the mature fruiting body stage. The trend of formaldehyde content at these four stages accorded with the findings of Huang et al. and Li et al. [25,28], which ranged from 13 to 89 mg/kg (dry weight) at all five stages. Mason et al. determined the formaldehyde content as 8–24 mg/kg in fresh shiitake mushrooms [29].

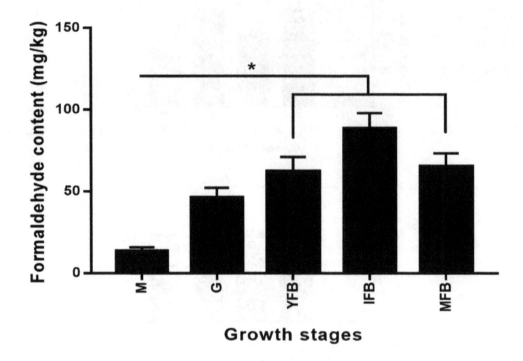

Figure 5. Endogenous formaldehyde content of *L. edodes* strain W1 at different growth stages. M, mycelia; G, grey; YFB, young fruiting body; IFB, immature fruiting body; MFB, mature fruiting body. Error bars indicate standard deviation for three independent experiments. *$p < 0.05$, ANOVA tests by Duncan's indicate significant differences.

The endogenous formaldehyde content at different growth stages showed a similar change trend to that of C-S lyase enzyme activity. The formaldehyde content of *L. edodes* during the drying has been reported to range between 150 and 400 mg/kg (dry weight) [26]. The increase of the two enzyme activities in the drying process also led to a significant increase in the endogenous formaldehyde content. Xu et al. indicated that GGT and C-S lyase were involved in formaldehyde formation and their activities were positively correlated with formaldehyde content [26]. Although the activities of the GGT and C-S Lyase were higher in IFB than in the other four stages, the differences were not significant. The endogenous formaldehyde was found to be produced from oxidative decomposition of the folate backbone and creates a benign 1C unit that can sustain essential metabolism in human cells [30]. Additionally, *L. edodes* also contains folic acid. Therefore, whether there are other enzymes and metabolic pathways involved in the generation of endogenous formaldehyde in *L. edodes* needs to be further studied.

2.5. Correlation Analysis

The effects of GGT and C-S lyase on the generation of endogenous formaldehyde in *Lentinula edodes* at different growth stages were intuitively determined by correlation analysis (Table 1). The formaldehyde content of *L. edodes* showed a positive and significant ($p < 0.01$) correlation (R) with the expression level of *Csl* and Lecsl and the activity of C-S lyase and GGT (0.746, 0.805, 0.867 and 0.768, respectively), while a negative relationship with the expression level of *Ggtl* and Leggt (−0.699 and −0.787; $p < 0.01$).

Table 1. Correlations (R) of the formaldehyde contents in *L. edodes* with *Ggtl* expression levels, *Csl* expression levels, Leggt expression levels, Lecsl expression levels, GGT enzyme activities and C-S lyase enzyme activities at different growth stages.

Properties	Formaldehyde Content
Ggtl expression levels (mRNA)	−0.699 **
Csl expression levels (mRNA)	0.746 **
Leggt expression levels (Protein)	−0.787 **
Lecsl expression levels (Protein)	0.805 **
GGT enzyme activities	0.768 **
C-S lyase enzyme activities	0.867 **

** significant at 0.01 level.

Japanese researchers pointed out that the formaldehyde content of *L. edodes* was stable during the growth process. However, the formaldehyde content after the drying process showed 3–4-fold increase. For example, in the dried shiitake mushrooms, the formaldehyde content ranged from 100 to 230 mg/kg, in contrast to 8–24 mg/kg in fresh ones [14]. Xu et al. indicated that the enzyme activities of GGT and C-S lyase were much higher under high temperature (>45 °C) than under 25 °C. These results demonstrated that the activation of the two key enzymes promoted reactions, leading to the production of a large amount of formaldehyde in *L. edodes* [26], which was well supported by our results in this study. This is the first report to show that the mRNA and protein expression levels of C-S lyase had significant and positive effects on the endogenous formaldehyde content of mushrooms.

Although the mRNA and protein expression levels of GGT were shown to be negatively correlated with the formaldehyde content, both GGT and C-S lyase were proved to be indispensable for the generation of endogenous formaldehyde in *L. edodes*. As previously reported, only the joint action of the two enzymes could promote the generation of endogenous formaldehyde [18], and GGT was the rate-limiting enzyme in the synthesis process of endogenous formaldehyde in *L. edodes* [29]. Our results showed that the activities of both of GGT and C-S lyase played a positive role in endogenous formaldehyde generation, implying the crucial effects of GGT in this process. GGT was also reported to be implicated in the transfer of amino acids across the cellular membrane and in metabolism of glutathione to cysteine by cleaving the glutamyl amide bond to preserve intracellular homeostasis by oxidative stress [31,32]. Besides, the transcription and function of genes are not synchronized in time and space. The presence of *Ggtl* homologous genes was also reported [33]. Moreover, compared with C-S Lyase, GGT has a much more complex structure and function. Despite the negative correlation of *Ggtl* and Leggt expression levels, we could not neglect their effects on the endogenous formaldehyde content in the mushroom. For a better control on the generation of endogenous formaldehyde in *L. edodes*, further studies should focus on the expression regulation of *Ggtl* and *Csl* at the transcription level.

Our study did not involve the influence of other potential metabolic pathways on the generation of endogenous formaldehyde, and whether other enzymes are implicated in the flavor metabolism pathways also needs to be investigated in future studies.

3. Materials and Methods

3.1. Fungal Strain and Culture Conditions

A dikaryotic strain of basidiomycete *Lentinula edodes* strain W1 (preserved in the Institute of Applied Mycology, Huazhong Agricultural University, Wuhan, China) was used in this study [34]. The *L. edodes* samples were obtained at five different stages: mycelia (used as control) and four fruiting body stages (grey, young fruiting body, immature fruiting body and mature fruiting body). Briefly, the mycelia were cultivated on 25 mL CYM liquid medium (2% glucose, 0.2% yeast extracts, 0.2% peptone, 0.1% K_2HPO_4, 0.05% $MgSO_4$ and 0.046% KH_2PO_4) in a conical flask and collected after growth of 12 days. Next, a conventional fruiting treatment was conducted as previously described [35]. The samples of grey (5–10 mm in cap diameter), young fruiting body (15–20 mm in cap diameter), immature fruiting body (with partial veil not ruptured) and mature fruiting body (with partial veil entirely ruptured) were harvested separately during fruiting treatment (Figure 6) [36]. The collected mushroom samples were immediately frozen in liquid nitrogen and stored at −80 °C for further use. All samples were collected in three biological replications.

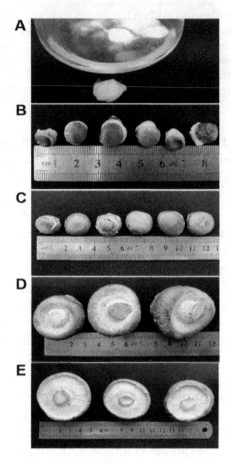

Figure 6. Five growth stages of *L. edodes* strain W1. (**A**) Mycelia. (**B**) Grey (5–10 mm in cap diameter). (**C**) Young fruiting body (15–20 mm in cap diameter). (**D**) Immature fruiting body (with partial veil not ruptured). (**E**) Mature fruiting body (with partial veil entirely ruptured).

3.2. RNA Isolation and Real-Time Quantitative PCR

Total RNA was isolated using RNAiso plus (TaKaRa, Kusatsu, Japan) according to the manufacturer's instructions [37]. The total RNA concentration and purity were detected using a Nano Drop 2000 spectrophotometer (Thermo Scientific, Wilmington, DE, USA; 2.0 < A260/A280 < 2.2). The integrity of RNA was checked by electrophoresis on 1% agarose gel, and the three bands of 28S, 18S and 5S could be clearly observed (Supplementary Figure S1).

Then, 20 μL cDNA was synthesized from 1 μg of total RNA using the HiScript II Q RT SuperMix for qPCR (+ gDNA wiper) kit (Vazyme Biotech, Nanjing, China) according to the manufacturer's instructions. Next, the cDNA was two-fold diluted with double-distilled water and stored at −20 °C for quantitative RT-PCR analysis. Specific primers were designed for quantitative RT-PCR analysis of the tested genes, such as *Ggtl,* encoding Leggt (γ-glutamyl transpeptidase); *Csl,* encoding Lecsl (*L. edodes* C-S lyase, *L. edodes* genome Gene ID: LE01Gene02830) and β-actin gene (*Actinl,* encoding *L. edodes* β-actin, *L. edodes* genome Gene ID: LE01Gene01050; Supplementary Table S1) [33].

Quantitative RT-PCR was performed using a CFX Connect real-time PCR system (BIO-RAD). Each reaction consisted of 0.4 μL each of the forward and reverse primers (10 μM), 1 μL of two-fold diluted cDNA, 5 μL of 2 × Taq Master Mix (Vazyme Biotech, Nanjing, China) and 3.2 μL of double-distilled water. The qRT-PCR was performed at 95 °C for 3 min, followed by 40 cycles of 95 °C for 20 s, 60 °C for

30 s, 72 °C for 30 s and then maintaining at 72 °C for 10 min in a 96-well reaction plate. The specificity and identity of PCR products were verified by melting curve analysis to distinguish specific PCR products from the primer dimmer-caused nonspecific PCR. The existence of a single peak proved each PCR product was specific.

The relative expression was calculated using the $2^{-\Delta\Delta CT}$ method as previously described [38]. The expression of *ActinI* was used as an internal reference [39]. The expressions during the mycelium stage were taken as control. All PCR experiments were performed in three biological and three technical replications (the maximum difference in Ct was 0.5).

3.3. Extraction of Total Protein and Western Blot Analysis

Total protein of *L. edodes* was extracted as previously reported [40]. Briefly, 0.1 g of mycelia or fruiting body powder from each group (three replicates for each group) was mixed with 0.5 mL of extraction buffer (0.5 M Tris-HCl, 50 mM EDTA, 0.1 M NaCl and 40 mM dithiothreitol). The supernatants were collected after extraction for 10 min and centrifugation at 10,000× g for 15 min at 4 °C to remove the insoluble substance. Next, the same volume of saturated Tris-phenol was added to the supernatants, followed by the addition of five volumes of pre-cooled 0.1 M ammonium acetate in methanol to precipitate the protein. After washing with pre-cooled 80% acetone several times, the precipitated proteins were resolubilized and denatured for 10 min in 40 μL solution buffer (7 M urea, 50 mM Tris-HCl, 25 mM EDTA, 10 mM NaCl and 60 mM dithiothreitol). Finally, the pelleted proteins were diluted to 200 μL for further analysis. The concentration of the total protein was tested by the Coomassie Brilliant Blue G250 method [41], and the quality of protein was checked by 10% SDS-PAGE (Supplementary Figure S2) [42].

Western blot was used to analyze the expression of γ-glutamyl transpeptidase (Leggt, EC 2.3.2.2) and S-alkyl-L-cysteine sulfoxide lyase (Lecsl, EC 4.4.1.4) at different growth stages of *L. edodes*. After 50 μg of each protein sample was run on 10% SDS-PAGE gels (Bio-Rad Mini, Hercules, CA, USA), Western blot was performed by standard protocols using 1:200 anti-Leggt and anti-Lecsl polyclonal antibody sera. The antibodies against Leggt and Lecsl were raised by immunizing rabbits with the mixture of purified recombinant protein, which was expressed in *Escherichia coli* BL21 and purified by Ni-NTA Agarose column (Genscript, Nanjing, China) and Freund's adjuvant [43]. The specificity of polyclonal antibodies was detected by Western blot. The results showed that anti-Leggt polyclonal antibody sera had special bands at 68 kDa, 45 kDa and 23 kDa and anti-Lecsl polyclonal antibody sera had special band at 54 kDa, respectively. 1:50,000 horseradish peroxidase conjugated secondary antibody (BOSTER, Wuhan, China). Meanwhile, the β-actin antibody (BOSTER, Wuhan, China) was treated with the same protocol as an internal control [21].

3.4. Enzyme Activity Assays

GGT activity was determined by the transfer rate of γ-glutamyl from γ-glutamyl *p*-nitroanilide (GPNA) as reported by Liu et al. [18]. The mixture including 1 mL crude enzyme extract from *L. edodes*, 1 mL GPNA (3.5 mM) and 3 mL Tris-HCl (0.5 M, pH = 7.6) was incubated at 37 °C for 20 min and the reaction was stopped by adding 3 mL of 1.5 M cold (4 °C) acetic acid. Then, the amount of p-nitroaniline released was measured at 410 nm. The specific activity of GGT was defined as the amount of enzyme that released 1 μmol of p-nitroaniline from the substrate per min per g protein (U/g).

C-S lyase activity was measured as previously described with some modifications [44]. The mixture containing 0.3 mL crude enzyme extract from *L. edodes*, 0.5 mL S-ethyl-L-cysteine sulfoxide and 0.2 mL Tris-HCl (0.5 M, pH = 7.6) was incubated at 37 °C for 5 min. The reaction was terminated by adding 1 mL trichloroacetic acid (TCA, 10%). After supplementation with 1 mL 2,4-dinitrophenylhydrazine (DNPH, 0.1%, *m/v*) were added to the mixture was incubated for 5 min at 25 °C. Finally, 2.5 mL NaOH

(2.5 M) was added to the mixture and incubated for 10 min at 25 °C. The absorbance of DNPH at 520 nm was measured. The specific activity of C-S lyase was expressed as units of enzyme per g of *L. edodes* protein (U/g).

3.5. Determination of Endogenous Formaldehyde Content in L. edodes

Steam distillation was used to extract formaldehyde from *L. edodes* at each growth stage. Each sample was supernatant of 4 g *L. edodes* homogenized with 100 mL Tris-HCl (0.5 M, pH = 7.6) buffer and 10 mL 10% (*v/v*) phosphoric acid aqueous solution in a 250 mL distillation flask. Water vapor was collected into a 150 mL flask, and then immersed in an ice-bath. The distillation process was stopped when 6070 mL of the distillate was collected and made up to 100 mL by deionized water. Formaldehyde in the distillate was derived by adding 1 mL of the distillate, 3.5 mL acetate buffer (0.1 M, pH = 4.0) and 0.5 mL DNPH (3 mg/mL) into a centrifuge tube at 25 °C for 15 min. Then the derived sample was filtered through a 0.22 μm filter for HPLC analysis. The formaldehyde derivative (formaldehyde-DNPH) of each group was separated and determined by a reverse-phase HPLC system (Waters, Milford, MA, USA). The mobile phase was composed of 0.05% acetic acid in acetonitrile and 0.05% acetic acid in water. The injection volume was 20 μL. All samples were detected at 355 nm as previously reported [18].

3.6. Data Analysis

All experimental data were presented as the mean ± standard deviation from at least three independent experiments. The ANOVA tests of statistical significance were performed by Duncan's multiple range tests using SPSS 20.0. p-values of <0.05 and <0.01 were accepted as significant and remarkable significant difference, respectively. The correlations of formaldehyde content with the expression levels of *Ggtl*, *Csl*, Leggt and Lecsl as well as GGT and C-S lyase activities were analyzed separately by Pearson correlation coefficient and trend of data using SPSS 20.0.

4. Conclusions

In this study, we reported for the first time the mRNA and protein expression levels and the activities of GGT and C-S lyase as well as their correlations with the endogenous formaldehyde content in *L. edodes* at different growth stages. The protein expression levels of Leggt and Lecsl were consistent with the mRNA expression levels of *Ggtl* and *Csl*. Additionally, the expression levels of GGT were decreased while those of C-S lyase were increased with the growth and development of *Lentinula edodes*. Furthermore, the enzyme activities and formaldehyde content were found to be the lowest in the mycelium stage. Our results demonstrated that the expression levels of *Csl* and Lecsl as well as the enzyme activities of C-S lyase and GGT were positively correlated with formaldehyde content during the development of *L. edodes*. These findings revealed the role of GGT and C-S lyase in generating endogenous formaldehyde at the molecular level. They also provided a molecular basis for regulating endogenous formaldehyde in the process of *L. edodes* growth.

Author Contributions: X.L., Y.L. and W.H. conceived and designed the experiments. X.L., S.G. and Z.H. prepared the experiment materials. X.L. performed the experiments and analyzed the data. X.L., Y.L. and X.F. wrote the manuscript. Y.B., Y.L. and W.H. provided intellectual input and revised the manuscript. All authors read and approved the final manuscript.

References

1.   Philippoussis, A.; Diamantopoulou, P.; Israilides, C. Productivity of agricultural residues used for the cultivation of the medicinal fungus *Lentinula edodes*. *Int. Biodeter. Biodegr.* **2007**, *59*, 216–219. [CrossRef]

2.   Bruhn, J.N.; Mihail, J.D.; Pickens, J.B. Forest farming of shiitake mushrooms: An integrated evaluation of management practices. *Bioresour. Technol.* **2009**, *100*, 6472–6480. [CrossRef] [PubMed]

3.   Finimundy, T.C.; Dillon, A.J.P.; Henriques, J.A.P.; Ely, M.R. A review on general nutritional compounds and pharmacological properties of the *Lentinula edodes* mushroom. *Food Nutr. Sci.* **2014**, *5*, 1095–1105.

4.   Choi, Y.; Lee, S.M.; Chun, J.; Lee, H.B.; Lee, J. Influence of heat treatment on the antioxidant activities and polyphenolic compounds of Shiitake (*Lentinus edodes*) mushroom. *Food Chem.* **2006**, *99*, 381–387. [CrossRef]

5.   Chen, C.; Ho, C. High-performance liquid chromatographic determination of cyclic sulfur compounds of Shiitake mushroom (*Lentinus edodes* Sing.). *J. Chromatogr. A* **1986**, *356*, 455–459. [CrossRef]

6.   Hiraide, M.; Miyazaki, Y.; Shibata, Y. The smell and odorous components of dried shiitake mushroom, *Lentinula edodes* I: Relationship between sensory evaluations and amounts of odorous components. *J. Wood Sci.* **2004**, *50*, 358–364. [CrossRef]

7.   Yasumoto, K.; Iwami, K.; Mitsuda, H. A new sulfur-containing peptide from *Lentinus edodes* acting as a precursor for lenthionine. *Agric. Biol. Chem.* **1971**, *35*, 2059–2069. [CrossRef]

8.   Yasumoto, K.; Iwami, K.; Mitsuda, H. Enzyme-catalized evolution of lenthionine from lentinic acid. *Agric. Biol. Chem.* **1971**, *35*, 2070–2080. [CrossRef]

9.   Yamazaki, H.; Ogasawara, Y.; Sakai, C.; Yoshiki, M.; Makino, K.; Kishi, T.; Kakiuchi, Y. Formaldehyde in *Lentinus edodes* (in giapponese). *J. Food Hyg. Soc. Jpn.* **1980**, *21*, 165–170. [CrossRef]

10. Yasumoto, K.; Iwami, K.; Mitsuda, H. Enzymatic formation of shiitake aroma from nonvolatile precursor (s)-lenthionine from lentinic acid. *Mushroom Sci.* **1976**, *9*, 371–383.

11. Weng, X.; Chon, C.H.; Jiang, H.; Li, D. Rapid detection of formaldehyde concentration in food on a polydimethylsiloxane (PDMS) microfluidic chip. *Food Chem.* **2009**, *114*, 1079–1082. [CrossRef]

12. Huang, T.; Hong, L.; Yuan, X.; Yan, L.; Gang, Z. Preparation and characterization of a novel absorber for formaldehyde. *Proc. Int. Conf. Biol. Eng. Pharm. (BEP 2016)* **2016**, in press.

13. Tashkov, W. Determination of formaldehyde in foods, biological media and technological materials by headspace gas chromatography. *Chromatographia* **1996**, *43*, 625–627. [CrossRef]

14. Mason, D.; Sykes, M.; Panton, S.; Rippon, E. Determination of naturally-occurring formaldehyde in raw and cooked Shiitake mushrooms by spectrophotometry and liquid chromatography-mass spectrometry. *Food Addit. Contam.* **2004**, *21*, 1071–1082. [CrossRef] [PubMed]

15. Okada, S.; Iga, S.; Isaka, H. Studies on formaldehyde observed in edible mushroom shiitake, *Lentinus edodes* (Berk.) Sing (in giapponese). *J. Hyg. Chem.* **1972**, *18*, 353–357.

16. Tate, S.S.; Meister, A. γ-Glutamyl transpeptidase from kidney. *Methods Enzym.* **1985**, *113*, 400–419.

17. Kuettner, E.B.; Hilgenfeld, R.; Weiss, M.S. The active principle of garlic at atomic resolution. *J. Biol. Chem.* **2002**, *277*, 46402–46407. [CrossRef]

18. Liu, Y.; Yuan, Y.; Lei, X.Y.; Yang, H.; Ibrahim, S.A.; Huang, W. Purification and characterisation of two enzymes related to endogenous formaldehyde in *Lentinula edodes*. *Food Chem.* **2013**, *138*, 2174–2179. [CrossRef]

19. Liu, Y.; Lei, X.Y.; Chen, L.F.; Bian, Y.B.; Yang, H.; Ibrahim, S.A.; Huang, W. A novel cysteine desulfurase influencing organosulfur compounds in *Lentinula edodes*. *Sci. Rep.* **2015**, *5*, 10047. [CrossRef]

20. Gao, S.; Wang, G.Z.; Huang, Z.; Lei, X.; Bian, Y.; Liu, Y.; Huang, W. Selection of reference genes for qRT-PCR analysis in *Lentinula edodes* after hot-air drying. *Molecules* **2018**, *24*, 136. [CrossRef]

21. Rani, N.; Nowakowski, T.; Zhou, H.; Godshalk, S.E.; Lisi, V.; Kriegstein, A.; Kosik, K. A primate lncRNA mediates notch signaling during neuronal development by sequestering miRNA. *Neuron* **2016**, *90*, 1174–1188. [CrossRef] [PubMed]

22. Okada, T.; Suzuki, H.; Wada, K.; Kumagai, H.; Fukuyama, K. Crystal structure of the γ-glutamyltranspeptidase precursor protein from *Escherichia coli* structural changes upon autocatalytic processing and implications for the maturation mechanism. *J. Biol. Chem.* **2007**, *282*, 2433–2439. [CrossRef] [PubMed]

23. Boanca, G.; Sand, A.; Barycki, J.J. Uncoupling the enzymatic and autoprocessing activities of *helicobacter pylori* -Glutamyltranspeptidase. *J. Biol. Chem.* **2006**, *281*, 19029–19037. [CrossRef] [PubMed]

24. Martin, M.N.; Slovin, J.P. Purified gamma-glutamyl transpeptidases from tomato exhibit high affinity for glutathione and glutathione S-conjugates. *Plant Physiol.* **2000**, *122*, 1417–1426. [CrossRef]

25. Huang, J.; Luo, H.; Li, J. Gene cloning of γ-Glutamyltranspeptidase and its relationship to endogenous formaldehyde in shiitake mushroom (*Lentinus edodes*). *Adv. J. Food Sci. Technol.* **2016**, *12*, 579–587. [CrossRef]

26. Xu, L.; Fang, X.; Wu, W.; Chen, H.; Mu, H.; Gao, H. Effects of high-temperature pre-drying on the quality of air-dried shiitake mushrooms (*Lentinula edodes*). *Food Chem.* **2019**, *285*, 406–413. [CrossRef]

27. Huang, Z.; Lei, X.; Feng, X.; Gao, S.; Wang, G.; Bian, Y.; Huang, W.; Liu, Y. Identification of a heat-inducible element of cysteine desulfurase gene promoter in *Lentinula edodes*. *Molecules* **2019**, *24*, 2223. [CrossRef]

28. Li, J.; Song, J.; Huang, J.; Wu, N.; Zhang, L.; Jiang, T. Study on key enzymes of endogenous formaldehyde metabolism and it's content in shiitake mushrooms (*Lentinus Edodes*). *J Chin. Inst. Food Sci. Technol.* **2013**, *13*, 213–218.

29. Aberkane, H.; Frank, P.; Galteau, M.; Wellman, M. Acivicin induces apoptosis independently of gamma-glutamyltranspeptidase activity. *Biochem. Biophys. Res. Commun.* **2001**, *285*, 1162–1167. [CrossRef]

30. Burgos-Barragan, G.; Wit, N.; Meiser, J.; Dingler, F.A.; Pietzke, M.; Mulderrig, L.; Pontel, L.B.; Rosado, I.V.; Brewer, T.F.; Cordell, R.L.; et al. Mammals divert endogenous genotoxic formaldehyde into one-carbon metabolism. *Nature* **2017**, *548*, 549–554. [CrossRef]

31. Pompella, A.; Franzini, M.; Emdin, M.; Passino, C.; Paolicchi, A. Gamma-glutamylaransferase activity in human atherosclerotic plaques: Origin, prooxidant effects and potential roles in progression of disease. *Atheroscler. Supp.* **2007**, *8*, 95. [CrossRef]

32. Lim, J.S.; Yang, J.H.; Chun, B.Y.; Kam, S.; Jr, J.D.; Lee, D.H. Is serum gamma-glutamyltransferase inversely associated with serum antioxidants as a marker of oxidative stress? *Free Radic. Bio. Med.* **2004**, *37*, 1018–1023. [CrossRef] [PubMed]

33. Chen, L.; Gong, Y.; Cai, Y.; Liu, W.; Zhou, Y.; Xiao, Y.; Xu, Z.; Liu, Y.; Lei, X.; Wang, G. Genome sequence of the edible cultivated mushroom *Lentinula edodes* (Shiitake) reveals insights into lignocellulose degradation. *PLoS ONE* **2016**, *11*, e0160336. [CrossRef] [PubMed]

34. Kirk, P.M.; Cannon, P.F.; Minter, D.W.; Stalpers, J.A. Dictionary of the Fungi. *Mycol. Res.* **2009**, *113*, 908–910.

35. Gong, W.; Xu, R.; Xiao, Y.; Zhou, Y.; Bian, Y. Phenotypic evaluation and analysis of important agronomic traits in the hybrid and natural populations of *Lentinula edodes*. *Sci. Hortic.* **2014**, *179*, 271–276. [CrossRef]

36. Leung, G.S.W.; Zhang, M.; Xie, W.J.; Kwan, H.S. Identification by RNA fingerprinting of genes differentially expressed during the development of the basidiomycete *Lentinula edodes*. *Mol. Gen. Genet.* **2000**, *262*, 977–990. [CrossRef]

37. Wang, G.Z.; Ma, C.J.; Luo, Y.; Zhou, S.S.; Zhou, Y.; Ma, X.L.; Cai, Y.L.; Yu, J.J.; Bian, Y.B.; Gong, Y.H.; et al. Proteome and transcriptome reveal involvement of heat shock proteins and indoleacetic acid metabolism process in *Lentinula edodes* thermotolerance. *Cell. Physiol. Biochem.* **2018**, *50*, 1617–1637. [CrossRef]

38. Pfaffl, M.W. A new mathematical model for relative quantification in real-time RT-PCR. *Nucleic Acids Res.* **2001**, *29*, e45. [CrossRef]

39. Masaru, N.; Maki, K.; Hisayuki, W.; Machiko, O.; Kumiko, S.; Toshikazu, T.; Katsuhiro, K.; Toshitsugu, S. Important role of fungal intracellular laccase for melanin synthesis: Purification and characterization of an intracellular laccase from *Lentinula edodes* fruit bodies. *Microbiology* **2003**, *149*, 2455–2462.

40. Cai, Y.; Gong, Y.; Liu, W.; Hu, Y.; Chen, L.; Yan, L.; Zhou, Y.; Bian, Y. Comparative secretomic analysis of lignocellulose degradation by *Lentinula edodes* grown on microcrystalline cellulose, lignosulfonate and glucose. *J. Proteom.* **2017**, *163*, 92–101. [CrossRef]

41. Bradford, M.M. A rapid method for the quantitation of microgram quantities of protein utilizing the principle of protein-dye binding. *Anal. Biochem.* **1976**, *72*, 248–254. [CrossRef]

42. Laemmli, U.K. Cleavage of structural proteins during the assembly of the head of bacteriophage T4. *Nature* **1970**, *227*, 680–685. [CrossRef] [PubMed]

43. Wachino, J.; Shibayama, K.; Suzuki, S.; Yamane, K.; Mori, S.; Arakawa, Y. Profile of Expression of *Helicobacter pylori*γ-Glutamyltranspeptidase. *Helicobacter* **2010**, *15*, 184–192. [CrossRef]

44. Yasumoto, K.; Iwami, K. S-Substituted l-cysteine sulfoxide lyase from shiitake mushroom. *Methods Enzym.* **1987**, *143*, 434–439.

10

Fungi-Mediated Biotransformation of the Isomeric Forms of the Apocarotenoids Ionone, Damascone and Theaspirane

Stefano Serra * and Davide De Simeis

C.N.R. Istituto di Chimica del Riconoscimento Molecolare, Via Mancinelli 7, 20131 Milano, Italy; dav.biotec01@gmail.com
* Correspondence: stefano.serra@cnr.it or stefano.serra@polimi.it.

Abstract: In this work, we describe a study on the biotransformation of seven natural occurring apocarotenoids by means of eleven selected fungal species. The substrates, namely ionone (α-, β- and γ-isomers), 3,4-dehydroionone, damascone (α- and β-isomers) and theaspirane are relevant flavour and fragrances components. We found that most of the investigated biotransformation reactions afforded oxidized products such as hydroxy- keto- or epoxy-derivatives. On the contrary, the reduction of the keto groups or the reduction of the double bond functional groups were observed only for few substrates, where the reduced products are however formed in minor amount. When starting apocarotenoids are isomers of the same chemical compound (e.g., ionone isomers) their biotransformation can give products very different from each other, depending both on the starting substrate and on the fungal species used. Since the majority of the starting apocarotenoids are often available in natural form and the described products are natural compounds, identified in flavours or fragrances, our biotransformation procedures can be regarded as prospective processes for the preparation of high value olfactory active compounds.

Keywords: biotransformation; oxidation; apocarotenoids; flavours; fungi; ionone; damascone; theaspirane

1. Introduction

In Nature, the oxidative degradation of the conjugated tetraterpene carotenoids (C_{40}) produces a plethora of smaller derivatives, called apocarotenoids [1], which possess a range of different chemical structures and biological activities. Among these natural products, compounds having thirteen carbon atoms in their frameworks are relevant flavours or fragrances and their manufacturing represents an important economic resource for chemical companies [2]. The combination of the great diversity of the carotenoids chemical structures with the different possible degradation pathways, gives rise to a huge number of flavours and fragrances.

It is worth noting that the primary odorous C_{13} apocarotenoids, namely ionone, damascone and theaspirane isomers (Figure 1), are cyclohexene derivatives and the possibility of three different positions of the double bond, the presence of a stereogenic center in position 6 (carotenoids numbering) and the eventual structural rearrangements, can give rise to a large number of isomers. In addition, the latter volatile compounds are often accompanied by structural related apocarotenoids having further oxygen atoms in their chemical framework.

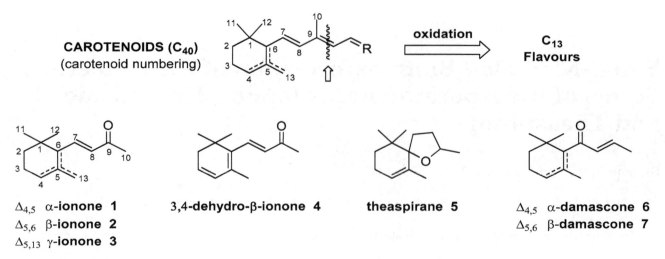

Figure 1. The formation of apocarotenoids through degradation of carotenoids and the seven C_{13} apocarotenoids (compounds **1–7**) selected as substrates for the fungal biotransformation investigated in the present study.

The first consequence of the introduction of a hydroxy- or keto- functional group on these compounds is the decrease of the volatility and the increase of the so-called 'substantivity', namely the long lasting odour of a substance having low vapor pressure. The aforementioned compounds have been recognized as components of different natural flavours. For example 3-hydroxy and 3-keto-α-ionone, 4-hydroxy- and 4-keto-β-ionone and hydroxy-β-damascone isomers have been identified in curry tree [3], eucalyptus honey [4], saffron [5], black tea [6] and tobacco [7,8] respectively, whereas 3-keto-theaspirane (also known under the trade name theaspirone) is the character impact compound of the black tea flavour [9].

All these derivatives are present in vegetables in very minute amounts and the extraction is not a viable process for their production. Consequently, they are currently obtained by chemical synthesis and are not commercially available in their natural form. Since flavours possessing 'natural' status are usually hundreds times as expensive as their synthetic counterparts, any new procedure that provides these compounds in their high value form can be very profitable.

In recent years, some new biocatalytic processes have provided a reliable access route to the most common C_{13} apocarotenoids, such as α- and β-ionone. In addition, the genetic engineering of both carotenoids biosynthesis and carotenoids cleavage pathways in the same microbial host [10] has laid the foundation for the large-scale production of the C_{13} apocarotenoids in natural form.

According to the European and USA legislation, the biotransformation of a natural precursor is a 'natural method' of synthesis [11]. Therefore, we singled out the compounds **1–7** as prospective natural precursors to be used as starting materials for the biotechnological production of different natural flavours.

From a biochemical standpoint, both prokaryotes and eukaryotes are able to degrade the carotenoid frameworks. In spite of this fact, only a limited number of biotransformations of the compounds of type **1–7** have been reported to date. Although the first description of an ionone isomer biotransformation goes back to 1950, when the oxidation of β-ionone in rabbit was investigated [12], only a limited number of studies on this topic took place in the following years. These researches were based mainly on the fungi and bacteria-mediated chemical transformations whereas the exploitation of some specific oxido-reductases were described only recently. In particular, *Aspergillus niger* [13,14], *Lasiodiplodia theobromae* [15], *Cunninghamella blakesleeana* [16], *Botrytis cinerea* [17], *Aspergillus awamori* [18], *Pleurotus sapidus* [19], *Mortierella isabellina* [20] and different *Streptomyces* strains [21] have proved to be active biocatalysts for the transformation of these kind of compounds. Concerning the use of isolated enzymes or the exploitation of a specific enzymatic activities, both cytochrome P450 monooxygenases [22,23] and engineered whole cell biocatalysts expressing mutant P450 monooxygenases [24], were used for the oxidation of different ionone isomers.

It is worth nothing that the biotransformation of some substrates such as γ-ionone, 3,4-dehydro-β-ionone and α-damascone hasn't been investigated yet. This paucity of scientific studies is often due to the difficult availability of some apocarotenoids that can be either substrates or products of the biotransformations. For example, γ-ionone is a rare natural isomer of ionone and can be obtained in high isomeric purity only through demanding multistep syntheses [25–28]. Therefore is not surprising that the fungus-mediated transformation of this compound hasn't been studied yet. In addition, whole cell biotransformations usually afford very complex mixtures of products whose chemical identification, for example by GC or HPLC analysis, require the availability of the corresponding reference standards. The aforementioned compounds are often not commercially available and have to be prepared by specific and multistep chemical syntheses, thus hampering to perform a proper study on the apocarotenoid's biotransformation.

Taking advantage of our previous experience on the stereoselective synthesis of ionone and damascone isomers [2,25–32], we decided to set up a comprehensive study on the biotransformation of the seven natural substrates described above by means of eleven selected fungal species belonging to the three more relevant phylums, namely ascomycota, zygomycota and basidiomycota. More specifically, we selected *Aspergillus niger* and *Mortierella isabellina* because these microorganisms have been already used for the biotransformation of some ionone isomers [13,14,20]. The remaining nine strains were singled out among the plethora of the microorganisms described in the literature based on their prospective biotransformation abilities. In effect, *Nigrospora oryzae*, different *Penicillium* species, *Rhizopus stolonifer*, *Curvularia lunata* and *Fusarium culmorum* have been successful employed in the biotransformation of terpenoids and steroids [33–35] whereas *Geotrichum candidum* was used for the oxidation of the cyclohexanone derivatives [36]. Finally, we selected also the yeasts *Yarrowia lipolytica* and *Xanthophyllomyces dendrorhous* since they are microorganisms of primary interest in the industrial synthesis/degradation of lipids [37] and carotenoids [38], respectively.

The results obtained by our work, beside confirming and extending those described by some previous researches, give new insights on the ability of fungi in the biotransformation of apocarotenoids, establishing their prospective utility for flavour production.

2. Results and Discussion

As described in the introduction, each one of the selected apocarotenoid isomers was incubated with a growing culture of each one of the eleven fungal strains. After a defined period of time (see Experimental Section) the crude biotransformation mixtures were derivatized (by acetylation) and analysed by GC-MS. To this end, several reference standard compounds were prepared by chemical synthesis and then were used for the unambiguous identification of the compounds formed in the biotransformation experiments. In spite of our efforts, a number of these derivatives were not identified. Therefore, in order to spot the most relevant biochemical transformations that each fungal strain is able to perform, we carried out the chromatographic isolation of the unknown metabolites that were formed in relevant amount or that make up of the main part of the transformed derivatives. The structures of these compounds were then elucidated through their comprehensive chemical characterization. More specifically, the combined use of ^{1}H-NMR, ^{13}C-NMR, DEPT experiments, GC-MS and ESI-MS spectroscopy allowed us to identify some compounds that haven't been correlated with any biotransformation experiment yet or that haven't been described in the literature until now.

We first investigated the transformation of the ionone isomers 1–3. Although only α- and β-ionone are very common in Nature, we deemed that it would be very interesting to also study the reactivity of the rare γ-isomer. In effect, even if the latter three compounds differ only for the position of the cyclohexene double bond, they possess very different reactivity from each other. Therefore, the identification of the derivatives obtained through biotransformation can help in understanding the chemical processes involved in these fungi-mediated reactions.

As collectively described in Figure 2, the main part of the identified metabolites are the result of four different biochemical reactions, namely the oxidation of the methylene functional groups, the reduction of the conjugated double bond, the epoxidation of the 4,5-double bond and the reduction of the keto group. The investigated fungal strains are able to perform both single step reactions and sequential multi-steps transformations, in turn deriving from the combination of the aforementioned four chemical reactions.

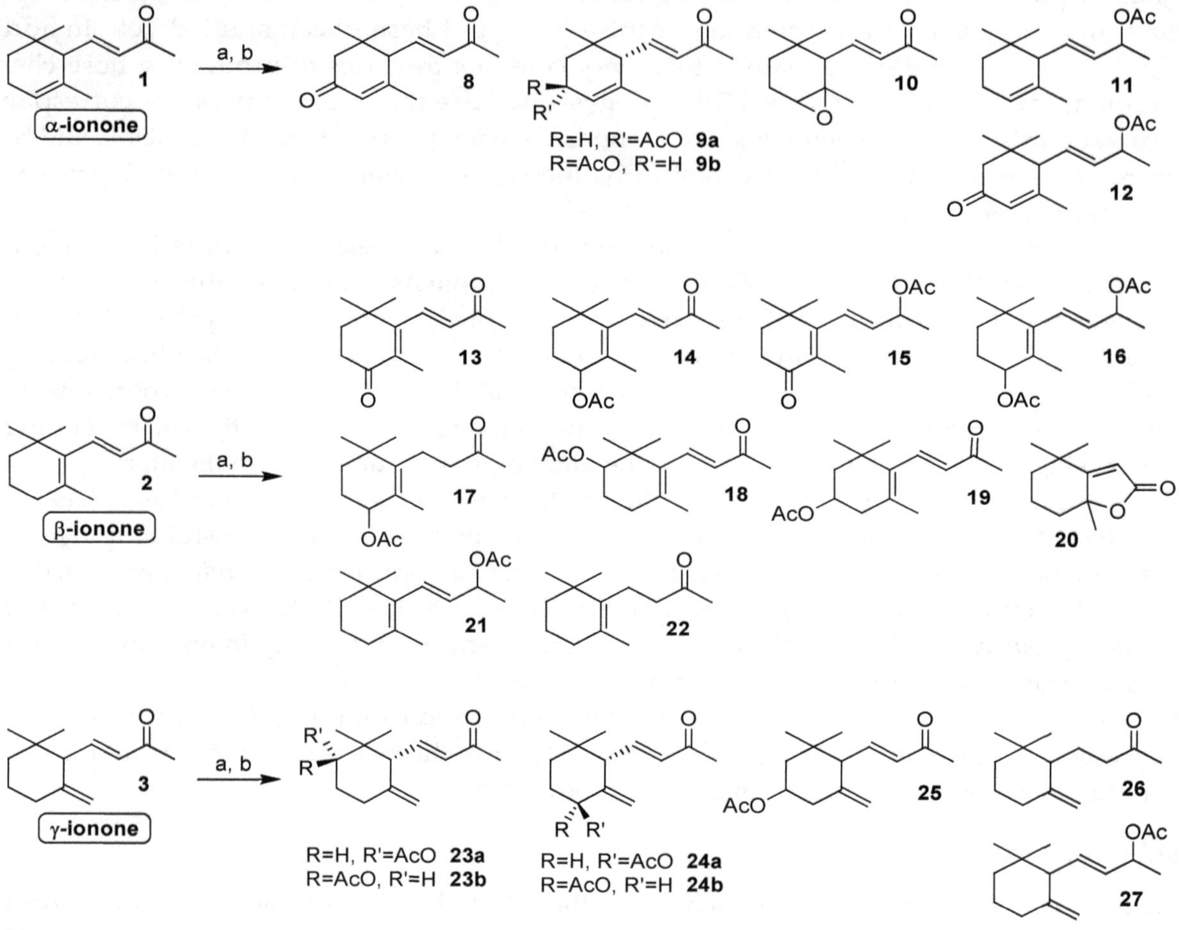

Figure 2. Compounds obtained through biotransformation of α-, β- and γ-ionone by means of the fungal strains employed in the present study. *Reagents and conditions*: (**a**) fungal strain, malt extract medium (MEM), aerobic conditions, 140 rpm, 20 or 25 °C, 8–20 days; (**b**) Ac_2O/pyridine (Py), 4-dimethylaminopyridine (DMAP) catalyst, room temperature (rt), 6 h.

Overall, ionone isomers 1–3 were converted into derivatives 8–27 in relative amounts described in Table 1. A thorough perusal of these data shows considerable differences in the reactivity of the three isomers. α-Ionone 1 was oxidized almost exclusively at the activated allylic methylene functional group, with the exception of *Geotrichum candidum* that is completely inactive and of *Curvularia lunata* and *Fusarium culmorum* that are also able to oxidize the 4,5-double bond to give the corresponding epoxy-ionone 10. The relative ratio of the obtained 3-keto-α-ionone 8 and of the diastereoisomeric 3-hydroxy-α-ionones 9a and 9b changes significantly based on the fungal strain used. Overall, the microorganisms that better performed the oxidation of 1 were *Aspergillus niger*, *Nigrospora oryzae* and *Fusarium culmorum*.

The global amount of the 3-oxidized metabolites obtained using the latter fungal strains took account of the 74%, 69% and 58% of the crude biotransformation mixture, respectively. On the contrary, none of the tested strains showed notable reductive activity on α-ionone, as confirmed by the modest formation of derivatives 11 and 12.

Table 1. Results of the fungi-mediated biotransformation of α-, β- and γ-ionone isomers.

Substrate	Biotransformation Products	Fungal Strains and Distribution of the Biotransformation Products [1]										
		A. niger	N. oryzae	G. candidum	Y. lipolytica	P. roqueforti	R. stolonifer	P. corylophilum	M. isabellina	C. lunata	X. dendrorhous	F. culmorum
α-ionone (1)	1	21	2	100	87	82	85	84	62	-	88	24
	8	13	27	-	-	4	2	4	1	2	-	10
	9a	33	7	-	1	3	3	4	9	7	2	21
	9b	28	25	-	1	10	1	6	12	-	1	25
	10	-	-	-	-	-	-	-	-	10	-	3
	11	-	-	-	8	-	-	-	3	-	2	-
	12	-	10	-	-	-	-	-	-	7	-	2
	N.D. [2]	5	31	-	3	1	9	2	13	74	7	15
β-ionone (2)	2	45	15	76	82	87	66	61	84	16	90	40
	13	1	-	5	4	1	3	3	1	-	-	4
	14	39	26	8	1	8	14	16	7	15	3	11
	15	-	-	-	-	-	-	1	-	-	-	-
	16	-	-	-	-	-	-	-	-	23	-	3
	17	-	26	-	-	-	-	-	-	-	-	-
	18	11	-	-	-	-	-	8	-	-	-	-
	19	-	-	-	-	2	-	-	-	-	-	-
	20	-	-	1	-	-	-	-	-	-	-	-
	21	-	2	-	11	-	-	3	1	12	-	3
	22	-	5	-	1	-	-	-	-	-	-	-
	N.D. [2]	4	26	10	1	2	17	8	7	34	7	39
γ-ionone (3)	3	-	6	100	43	83	100	44	85	26	78	21
	23a	35	-	-	-	5	-	14	-	-	-	-
	23b	-	-	-	-	-	-	-	-	-	-	-
	24	2	-	-	-	-	-	-	2	8	-	13
	25	49	4	-	20	-	-	19	-	-	-	10
	26	-	2	-	16	-	-	6	-	-	8	7
	27	-	2	-	-	1	-	-	2	20	5	-
	N.D. [2]	14	86	-	21	11	-	17	11	46	9	49

[1] Percentage of the compound detected by GC–MS analysis of the biotransformation mixture, after extraction and chemical acetylation; [2] N.D. = not determined: the value indicates the overall percentage of the compounds obtained by biotransformation whose chemical structure wasn't assigned.

A more complex outcome were observed when β-ionone **2** was used as substrate. In this case, even if the latter ketone was oxidized mainly at the activated allylic methylene functional group (compounds **13–17**), the biotransformations also afforded the 2-hydroxy-β-ionone derivative **18** (*Aspergillus niger* and *Penicillium corylophilum*) and a little amount of 3-hydroxy-β-ionone derivative **19** (*Penicillium roqueforti*). In addition, the partial oxidation of the ionone side chain is also possible, as shown by the detection of trace (1%) of dihydroactinodiolide **20** in the biotransformation mixture of *Geotrichum candidum*. Interestingly, all the tested fungal strains left unaffected the C(13) methyl group of both α- and β-ionone isomers. In effect, we observed neither the formation of 13-hydroxy-derivatives nor the presence of the epoxy-megastigmaen-9-one isomers that can arise from the intramolecular 1,4-addition of the hydroxy-group to the conjugated double bond [31].

On the contrary, the reduction of the conjugated double bond and of the keto group are chemical transformations of major significance in fungal β-ionone biotransformation. The abovementioned reactions can proceed also on intermediates deriving from the oxidation of the activated position 4 of the β-ionone framework, affording a number of oxidized-reduced derivatives (**15–17**) besides compounds obtained by simple reduction (i.e., **21** and **22**).

Completely different results were observed for the biotransformation of γ-ionone where the oxidation of the allyl methylene group, to afford derivative **24**, appeared to be a path of minor relevance. On the contrary, the oxidation of the positions 2 and 3 of the latter ionone isomer was efficiently performed by *Aspergillus niger* that produced compounds **23a** and **25** in high yield. *Penicillium corylophilum* is also able to convert γ-ionone into the derivative **23a** but for this microorganism, as well as for *Yarrowia lipolytica*, the reductive steps are more relevant than the oxidative ones. In effect, the reduction of the conjugated double bond and of the carbonyl functional group afforded the compound **26** and **27**, respectively. It is worth noting that *Aspergillus niger*, *Penicillium roqueforti* and *Penicillium corylophilum* produced diastereoselectively *cis*-2-hydroxy-γ-ionone (**23a**) and none of the microorganisms tested afforded its diastereoisomer, namely *trans*-2-hydroxy-γ-ionone (**23b**). Differently, the fungi *Nigrospora oryzae*, *Curvularia lunata* and *Fusarium culmorum* transformed efficiently γ-ionone, but the main part of the biotransformation reaction consisted in a mixture of unknown compounds, most likely deriving by the extensive oxidative degradation of the ionone framework.

The second group of the investigated apocarotenoids regards 3,4-dehydro-β-ionone (**4**) and theaspirane (**5**, Figure 3).

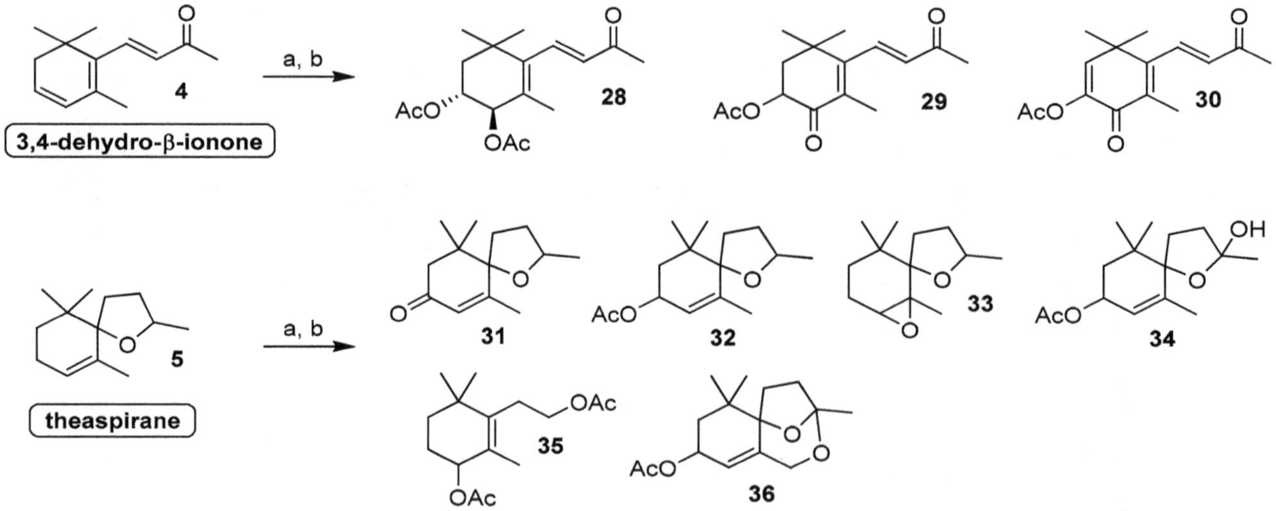

Figure 3. Compounds obtained through biotransformation of 3,4-dehydro-β-ionone and theaspirane by means of the fungal strains employed in the present study. *Reagents and conditions*: (**a**) fungal strain, MEM, aerobic conditions, 140 rpm, 20 or 25 °C, 14–20 days; (**b**) Ac₂O/Py, DMAP catalyst, rt, 6 h.

Although these compounds have different chemical structures, they both showed high reactivity and the results of their biotransformation experiments were described together. More specifically, we observed that all the investigated fungal strains were not able to oxidize the allylic positions of compound **4**. We identified as biotransformation products only compounds **28–30** (Table 2). These three ionone derivatives originated from the oxidation of the conjugated diene functional group. Most likely the first step is the epoxidation of the 3,4-double bond followed by the opening of the oxirane ring by addition of a molecule of water. This two step mechanism could justify the formation of the compound **28**, possessing *trans* relative configuration of the two contiguous hydroxy groups, as the major diastereoisomer. In addition, the following oxidation of the alcohol functional groups gave rise to the keto derivatives. More specifically, the oxidation at position 4 or at both position 4 and 3 gave compounds **29** and **30**, respectively.

Otherwise, when a spirocyclic ether group replaces the conjugated carbonyl group, our selected fungal strains become able to oxidize the theaspirane framework both at the allylic positions and at the methine carbon linked to the ether oxygen atom. Overall, the main part of the compounds obtained by biotransformation of **5** derive from allylic oxidation. In particular, the experiments performed using *Aspergillus niger*, *Rhizopus stolonifer* and *Fusarium culmorum* showed a total content of the compounds **31** and **32** that accounts for at least 60% of the reaction mixtures. This result is noteworthy as theaspirone **31** is a relevant natural flavour and the corresponding alcohol (**32** is the acetylated derivative) can be regarded as its direct precursor. In effect, the preparation of **31** in natural form could be possible by means of oxidation of the allyl alcohol, for example using a biocatalytic transformation involving alcohol dehydrogenases.

Jointly with allylic oxidation, we observed also the oxidation of the ether functional group. Hence, the latter moiety was transformed into the hemiketal and ketal groups as well in a completely rearranged framework, as demonstrated by the isolation and characterization of compounds **34**, **36** and **35**, respectively. Most likely, the latter derivatives are the result of a multistep oxidation reaction. The transformation of the allyl methylene and methyl groups proceeds faster than the oxidation of ether functional group, which is finally converted into a hemiketal functional group. Accordingly, *Nigrospora oryzae* and *Mortierella isabellina* completely oxidized theaspirane (**5**) to give two biotransformation mixtures containing 28% and 50% of compound **34**, respectively. It is worth mentioning that the latter compound is the direct precursor of 8,9-dehydrotheaspirone [32], a relevant apocarotenoid flavour identified in white-fleshed nectarines [39].

The biotransformation experiments performed using *Curvularia lunata* gave results of more complex interpretation affording a plethora (60% of the mixture) of undetermined compounds beside the 3%, 8%, 12% and 9% of derivatives **32**, **34**, **35** and **36**, respectively. Compound **35** could result from a multistep transformation comprising both of oxidation and transposition reactions. The C(9) oxidation is responsible for the formation of the hemiketal functional group that is in equilibrium with the less stable open hydroxy-ketone form. The Baeyer-Villiger oxidation of the latter ketone functional group could explain the formation of the primary alcohol acetate moiety whereas the 1,3-allyl transposition of the tertiary hydroxy group could take account of the unexpected position of the C(4) acetate group in compound **35**. Similarly, the formation of compound **36** in *Curvularia lunata*–mediated biotransformation could be explained by the hydroxylation of the carbons placed in position 3, 9 and 13 of the ionone framework.

The subsequent reaction of the obtained primary alcohol group with the hemiketal functional group should afford the more thermodynamically stable tricyclic ketal **36**. It is worth nothing that a natural compound having an identical ketal structure, but devoid of the hydroxy group at the C(4), is an aroma component of quince brandy [40]. The comparison of the NMR data measured for **36** with those reported for the natural flavour, allowed us to assign the above described chemical structure suggesting a new synthetic approach to this chemical framework by fungal biotransformation of theaspirane.

Table 2. Results of the fungi-mediated biotransformation of 3,4-dehydro-β-ionone and theaspirane.

Substrate	Biotransformation Products	Fungal Strains and Distribution of the Biotransformation Products[1]										
		A. niger	N. oryzae	G. candidum	Y. lipolytica	P. roqueforti	R. stolonifer	P. corylophilum	M. isabellina	C. lunata	X. dendrorhous	F. culmorum
3,4-dehydro-β-ionone (4)	4	48	-	45	87	45	80	76	71	52	6	-
	28	25	69	18	-	32	7	22	8	26	26	35
	29	3	2	17	1	9	3	-	7	8	22	20
	30	2	3	12	1	6	2	-	5	6	21	17
	N.D.[2]	22	26	8	3	8	8	2	9	8	25	28
theaspirane (5)	5	20	-	56	82	31	-	22	-	8	46	-
	31	9	12	11	4	19	5	16	-	-	3	28
	32	51	25	15	1	28	62	34	-	3	28	34
	33	-	-	-	-	2	-	-	-	-	2	2
	34	-	28	-	-	-	6	-	50	8	-	9
	35	-	-	-	-	-	-	-	-	12	-	-
	36	-	-	-	-	-	-	-	-	9	-	-
	N.D.[2]	20	35	18	1	20	27	28	50	60	21	27

[1] Percentage of the compound detected by GC-MS analysis of the biotransformation mixture, after extraction and chemical acetylation; [2] N.D. = not determined: the value indicates the overall percentage of the compounds obtained by biotransformation whose chemical structure wasn't assigned.

A completely different reactivity was observed in the biotransformation of the α- and β-damascone isomers **6** and **7** (Figure 4). The latter apocarotenoids are isomers of α- and β-ionone respectively, as each one damascone isomer is interconvertible into the corresponding ionone isomer by 1,3-shift of the enone moiety.

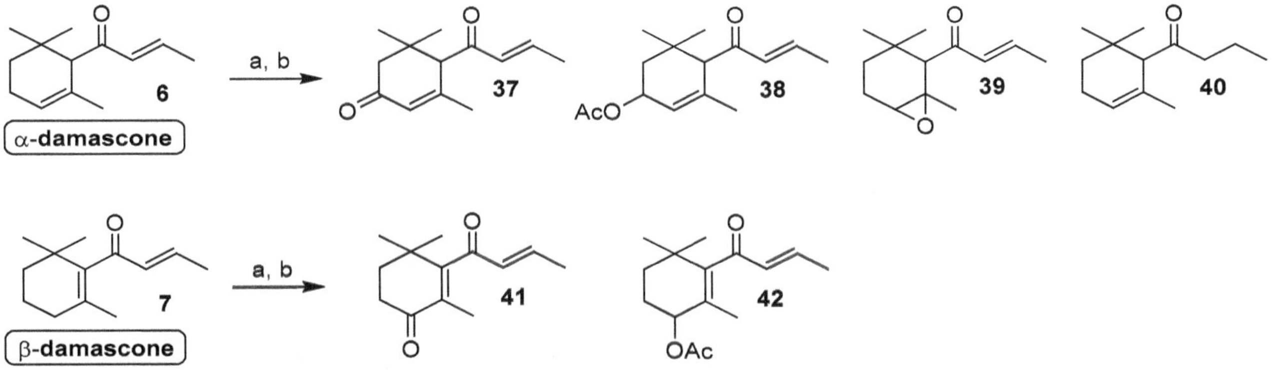

Figure 4. Compounds obtained through biotransformation of α- and β-damascone by means of the fungal strains employed in the present study. *Reagents and conditions*: (**a**) fungal strain, MEM, aerobic conditions, 140 rpm, 20–25 °C, 14–20 days; (**b**) Ac$_2$O/Py, DMAP catalyst, rt, 6 h.

Despite the structural similarity of these compounds, none of the investigated fungal strains was able to reduce the carbonyl functional group present in the damascone framework. In addition, the fungi-mediated oxidative transformations of damascone isomers are restricted mainly to the allyl methylene functional groups, as indicated by the formation of derivative **37** and **38** from α-damascone and derivatives **41** and **42** from β-damascone (Table 3). The other position of the damascone framework were unaffected with the exception of the 4,5-double bond that was oxidized by both *Mortierella isabellina* and *Xanthophyllomyces dendrorhous* to produce a very minor amount of the epoxy-α-damascone **39**. Similarly, we did not record any reductive reactions with the exception of the conjugated double bond of the α-damascone isomer that was reduced by *Penicillium corylophilum* to give a small amount of compound **40**.

Overall, using the described fungal strains, we observed that the damascone isomers are less reactive than the corresponding ionone isomers. This effect is more pronounced for the β-isomer where only *Nigrospora oryzae* and *Fusarium culmorum* produced a significant amount of the corresponding 4-hydroxydamascone, identified as acetyl derivative **42** (23 and 22%, respectively).

Our results seem in contrast to those reported in a recent study on the fungal biotransformation of β-damascone [20], where a different *Mortierella isabellina* strain provides 4-hydroxydamascone in a much higher yield. The recorded differences between these experimental data could be justified considering both the substrate concentrations and the different biocatalytic activity among fungal strains belonging to the same species.

In particular the substrate concentration seem to be the most relevant factor as apocarotenoid derivatives show significant toxicity for many fungal strains and high concentration of these compounds could inhibit their growth as well as their biocatalytic activity. We performed all the investigated biotransformation experiments using a substrate concentration of about 2.5 g/L whereas, in the above-mentioned work, the β-damascone concentration was set to 0.1 g/L. Our choice is justified by the need of devising a preparative protocol for fungi-mediated ionone biotransformation. Since the synthesis of this kind of flavours, in natural form, can show industrial significance only working with substrate concentrations superior to 1 g/L, we set the above indicated concentration for all experiments. As a consequence, it is reasonable that our selected fungal strains could have transformed better both damascone as well as ionone and theaspirane isomers if they have been used in lower concentration.

Table 3. Results of the fungi-mediated biotransformation of α- and β-damascone isomers.

Substrate	Biotransformation Products	Fungal Strains and Distribution of the Biotransformation Products [1]										
		A. niger	N. oryzae	G. candidum	Y. lipolytica	P. roqueforti	R. stolonifer	P. corylophilum	M. isabellina	C. lunata	X. dendrorhous	F. culmorum
α-damascone (6)	6	56	89	55	93	100	100	24	30	96	39	56
	37	20	5	10	-	-	-	-	22	-	10	5
	38	5	2	10	-	-	-	14	17	3	21	30
	39	-	-	-	-	-	-	-	2	-	3	-
	40	-	-	-	-	-	-	13	-	-	-	-
	N.D.[2]	19	4	25	7	-	-	49	29	1	27	9
β-damascone (7)	7	98	70	95	95	100	88	95	95	92	95	57
	41	-	2	-	1	-	1	-	-	-	-	4
	42	1	23	1	1	-	2	-	2	7	1	22
	N.D.[2]	1	5	4	4	-	9	5	3	1	4	17

[1] Percentage of the compound detected by GC-MS analysis of the biotransformation mixture, after extraction and chemical acetylation; [2] N.D. = not determined: the value indicates the overall percentage of the compounds obtained by biotransformation whose chemical structure wasn't assigned.

3. Materials and Methods

3.1. Materials and General Methods

All air and moisture sensitive reactions were carried out using dry solvents and under a static atmosphere of nitrogen. All solvents and reagents were of commercial quality and were purchased from Sigma-Aldrich (St. Louis, MO, USA). A large number of reference standard compounds were synthesized in our laboratory and were used for the unambiguous identification of the compounds formed in the biotransformation experiments. α-Ionone, γ-ionone and α-damascone were used in racemic form. Commercial theaspirane consists of an equimolar mixture of racemic diastereoisomers. γ-Ionone (**3**) and 3,4-dehydro-β-ionone (**4**) were prepared starting from α-ionone, according to the procedures previously described by us [26–28].

The keto derivatives: 3-keto-α-ionone (**8**), 4-keto-β-ionone (**13**), 3-keto-α-damascone (**37**), 4-keto-β-damascone (**41**), theaspirone (**31**), 3-keto-α-ionol acetate (**12**) and 4-keto-β-ionol acetate (**15**) were prepared by oxidation of α-ionone, β-ionone, α-damascone, β-damascone, theaspirane, α-ionol acetate (**11**) and β-ionol acetate (**21**), respectively. The oxidation reactions were performed using TBHP/MnO$_2$ as oxidant according to our previously reported procedure [41].

The diastereoisomeric forms of 4,5-epoxy-α-ionone (**10**), 4,5-epoxy-α-damascone (**39**), 4,5-epoxy-theaspirane (**33**) as well as 5,6-epoxy-β-ionone and 5,6-epoxy-β-damascone were prepared by epoxidation of α-ionone, α-damascone, theaspirane, β-ionone and β-damascone, respectively, using m-chloroperbenzoic acid and CH$_2$Cl$_2$ as solvent.

α-7,8-Dihydroionones, β-7,8-dihydroionone (**22**) and γ-7,8-dihydroionone (**26**) were prepared by reduction of α-, β- and γ-ionone respectively, using hydrogen and Ni Raney as catalyst for α-ionone [42] and Bu$_3$SnH and (Ph$_3$P)$_2$PdCl$_2$ as catalyst for β- and γ-ionone [25,26]. α-8,9-Dihydrodamascone (**40**) was prepared by reduction of α-damascone using NaBH$_4$ in methanol. β-8,9-Dihydrodamascone was prepared by addition of propylmagnesium bromide to β-cyclocitral followed by oxidation of the resulting carbinol using Dess-Martin periodinane [43].

β-Ionol acetate (**21**, racemic), α-ionol acetate (**11**, as a mixture of two racemic diastereoisomers), γ-ionol acetates (**27**, as a mixture of two racemic diastereoisomers) and 3-acetoxy-theaspirane (**32**) (as a mixture of four racemic diastereoisomers) were prepared by chemical acetylation (Ac$_2$O/Py) of the corresponding alcohols, which were in turn obtained through the reduction of α-, β-, γ-ionone and theaspirone, respectively, using NaBH$_4$ in methanol.

cis-2-Acetoxy-α-ionone, and cis-2-acetoxy-γ-ionone (**23a**) were prepared starting from 2,8,8-trimethyl-6-oxabicyclo[3.2.1]oct-2-en-7-one and 8,8-dimethyl-2-methylene-6-oxabicyclo[3.2.1]octan-7-one (kaharana lactone) respectively, according to the synthetic procedure developed by Audran [44]. The latter lactones were in turn prepared from racemic cis-2-hydroxy-α-cyclogeraniol and cis-2-hydroxy-γ-cyclogeraniol [45] by oxidation using BAIB and TEMPO as catalyst. In addition, the partial reduction of the two isomeric lactones afforded the corresponding lactols, whose condensation with acetone [46] followed by acetylation (Ac$_2$O/Py) of the crude reaction mixtures, gave the cis/trans mixtures of acetoxy-α-ionone and acetoxy-γ-ionone, respectively, that were used as GC-MS reference standards for the identification of the corresponding trans isomers.

Racemic 2-acetoxy-β-ionone (**18**) was prepared starting from 2-hydroxy-β-cyclogeraniol [45] by selective oxidation of the primary alcohol functional group by MnO$_2$ in CHCl$_3$, condensation with acetone [46] and acetylation (Ac$_2$O/Py) of the obtained hydroxy-ionone derivative.

Samples of 3-acetoxy-α-ionone (**9**) (2:1 cis/trans mixture), of 4-acetoxy-γ-ionone (**24**) (4:1 cis/trans mixture) and of 3-acetoxy-β-ionone (**19**) were prepared starting from α-ionone according to the procedure described by Tu [47], by Serra [42] and Khachik [48], respectively. A sample of 3-acetoxy-α-damascone (**38**) (1:1 cis/trans mixture) was prepared starting from ethyl 3-hydroxy-α-cyclogeraniate according to the procedure described by Takei [49].

4-Acetoxy-β-ionone (**14**) and 4-acetoxy-β-damascone (**42**) were prepared starting from 4,5-epoxy-α-ionone and 4,5-epoxy-α-damascone, respectively, by means of NaOMe mediated transposition followed by chemical acetylation (Ac$_2$O/Py) of the obtained allyl alcohols. A sample of 4-acetoxy-β-ionol acetate (**16**) (1:1 mixture of diastereoisomers) was prepared from 4-keto-β-ionol acetate by reduction with NaBH$_4$ in methanol followed by chemical acetylation (Ac$_2$O/Py).

A sample of 4-acetoxy-β-7,8-dihydroionone acetate (**17**) was prepared from 4-hydroxy-β-ionone by reduction with Ph$_3$SiH followed by chemical acetylation (Ac$_2$O/Py), according to the procedure described by Pascual [50].

3,4-Diacetoxy-β-ionone (**28**) (*cis/trans* mixture), was prepared from 3,4-dehydro-β-ionone according to the procedure described by Buschor [51]. The oxidation of 4-keto-β-ionone with IBDA in methanol [52] afforded 3-hydroxy-4-keto-β-ionone that was further oxidated using oxygen in presence of *t*BuOK [53] to give 3,4-diketo-β-ionone. The acetylation (Ac$_2$O/Py) of the latter two compounds afforded 3-acetoxy-4-keto-β-ionone (**29**) and 3-acetoxy-4-keto-2,3-dehydro-β-ionone (**30**). Racemic dihydroactinodiolide (**20**) and 2-hydroxy-2,6,10,10-tetramethyl-1-oxaspiro[4.5]dec-6-en-8-yl acetate (**34**) were prepared as described previously [32,54].

A comprehensive characterization of the above described reference standards is reported in the Supplementary Materials section.

3.2. Analytical Methods and Characterization of the Products Deriving from the Biotransformation Experiments

The crude biotransformation mixtures obtained according to the procedures described below were then acetylated by treatment with pyridine/acetic anhydride (2 mL of a 2:1 mixture) and catalytic DMAP (10 mg) for 6 hours at rt. The obtained acetylated mixture was analyzed by GC-MS. The compounds whose chemical structure couldn't be assigned only by GC-MS analysis were isolated from the biotransformation mixtures by means of chromatographic separation and then characterized by NMR analysis and GC-MS or ESI-MS analysis.

^{1}H- and ^{13}C-NMR spectra and DEPT experiments were recorded/performed at 400, 100 and 100 MHz, respectively, in CDCl$_3$ solutions at rt using an AC-400 spectrometer (Bruker, Billerica, MA, USA); ^{13}C spectra are proton decoupled; chemical shifts in ppm rel to internal SiMe$_4$ (=0 ppm).

TLC: silica gel 60 F_{254} plates (Merck, Kenilworth, NJ, USA). Column chromatography: silica gel.

Melting points were measured on a Reichert apparatus, equipped with a Reichert microscope, and are uncorrected.

Mass spectrum were recorded on a ESQUIRE 3000 PLUS spectrometer equipped with an ESI detector (Bruker, Billerica, MA, USA) or by GC-MS analyses.

GC-MS analyses: *HP-6890* gas chromatograph equipped with a 5973 mass detector, using a HP-5MS column (30 m × 0.25 mm, 0.25 μm film thickness; Hewlett Packard, Palo Alto, CA, USA) with the following temp. program: 60° (1 min)—6°/min—150° (1 min)—12°/min—280° (5 min); carrier gas, He; constant flow 1 mL/min; split ratio, 1/30; t_R given in min: t_R(**1**) 16.23, t_R(**2**) 17.57, t_R(**3**) 16.53, t_R(**4**) 17.53, t_R(**5**) 13.40 and 13.77, t_R(**6**) 15.44, t_R(**7**) 15.91, t_R(**8**) 20.42, t_R(**9a**) 21.41, t_R(**9b**) 21.61, t_R(**10**) 18.63 and 18.72, t_R(**11**) 17.78, t_R(**12**) 21.52, t_R(**13**) 20.61, t_R(**14**) 21.82, t_R(**15**) 22.03, t_R(**16**) 22.56, t_R(**17**) 21.59, t_R(**18**) 22.18, t_R(**19**) 22.23, t_R(**20**) 18.61, t_R(**21**) 18.55, t_R(**22**) 16.51, t_R(**23a**) 21.51, t_R(**24a**) 21.32, t_R(**24b**) 21.60, t_R(**25**) 21.79, t_R(**26**) 15.67, t_R(**27**) 17.47, 17.58, 18.01 and 18.08, t_R(**28**) 24.10, t_R(**29**) 23.75, t_R(**30**) 23.92, t_R(**31**) 19.02 and 19.19, t_R(**32**) 20.17, 20.33, 20.50 and 20.61, t_R(**33**) 15.14 and 15.50, t_R(**34**) 19.37 and 19.67, t_R(**35**) 21.78, t_R(**36**) 21.64, t_R(**37**) 20.15, t_R(**38**) 21.17 and 21.30, t_R(**39**) 17.84, t_R(**40**) 14.62, t_R(**41**) 20.91, t_R(**42**) 20.93, t_R(7,8-dihydro-α-ionone) 15.95, t_R(7,8-dihydro-β-damascone) 20.95, t_R(*cis*-2-acetoxy-α-ionone) 21.42, t_R(*trans*-2-acetoxy-α-ionone) 21.36.

3.3. Microorganisms and Biotransformation Experiments

Geotrichum candidum (DSM 10452), Yarrowia lipolytica (DSM 8218), Rhizopus stolonifer (DSM 855), Xanthophyllomyces dendrorhous (DMS 5626), Curvularia lunata (CBS 215.54), Mortierella isabellina (CBS 167.60), Aspergillus niger (CBS 626.26) were purchased from the DSMZ (Braunschweig, Germany) or CBS-KNAW (Utrecht, The Netherlands) collections.

Penicillium corylophilum (MUT 5838), *Nigrospora oryzae* (MUT 5844), *Penicillium roqueforti* (MUT 5856) and *Fusarium culmorum* (MUT 5855) were isolated as axenic cultures in our laboratory, then identified by the Mycotheca Universitatis Taurinensis (MUT) of the University of Turin and finally deposited in the same institution under the collection number given in brackets.

All the biotransformations were carried out in triplicate and the presented results are the average of three experimental runs.

3.3.1. Representative Procedures for Biotransformations

The experimental conditions used for the biotransformations are based on the type of microorganism used. Here is described a general procedure depending on the different morphological features regarding the various active grow mycelia. The main ones could be classified in yeast-shape mycelia (*Xanthophyllomyces dendrorhous*, *Geotrichum candidum* and *Yarrowia lipolytica*) and spore-forming mycelia (*Aspergillus niger*, *Rhizopus stolonifer*, *Curvularia lunata*, *Penicillium corylophilum*, *Nigrospora oryzae*, *Penicillium roqueforti*, *Mortierella isabellina* and *Fusarium culmorum*). In the first case, a small amount of the active mycelia grew previously in a petri dish, was suspend in 1 mL of sterile water and then inoculated in a 100 mL conical Pyrex flask containing 40 mL of Malt Extract Medium (MEM) for 2 days at 25 °C and 140 rpm (with exception of *Xanthophyllomyces dendrorhous* that was grown at 20 °C). After this period, the cells were centrifuged 3 minutes, (rt, 3220·g) and collected removing the media. The cells (approx. 600 mg wet-weight) were suspended in 3 mL of sterile water than 350 μL of the same suspension were used for inoculating each biotransformation flask containing 40 mL of MEM. In order to ensure aerobic conditions, the flasks were sealed with cellulose plugs. The microorganism was leave to growth for 2 days and then was treated with a solution of 100 mg of substrate dissolved in 400 μL of DMSO. Generally, after 14 days from the substrate injection using the growing condition described above, the reaction media was filtered under vacuum through a celite pad then was extracted 3 times with ethyl acetate. The organic phase was separated, dried on Na_2SO_4 and the solvent removed at reduced pressure to give the crude biotransformation mixture.

In the case of the spore-forming mycelia, the spore were collected from a sporulated surface cultures and suspended in 3 mL of sterile water. After that, 350 μL of the same suspension were used for inoculating each biotransformation flask containing 40 mL of MEM. The subsequent steps are the same of that described previously. The unique exception was carried out for *Penicillium corylophilum*, in the case of γ-ionone. The toxicity of the compound forced us to keep its concentration lower than the others (7.5 mM) and to block the biotransformation earlier (8 days). After this period, the most important products are degraded. In the case of *Fusarium culmorum* the biotransformation was blocked after 20 days instead of 14 days because the activity of the fungus did not stop in the prefixed time.

3.3.2. Preparative Biotransformations and Chemical Characterization of Compounds 23a, 25, 28, 35 and 36

For the main part of the strains tested, the GC-MS analysis of the crude biotransformation mixtures indicated the presence of different compounds whose chemical structures could not be assigned only on the basis of our reference standards. The unidentified peaks taking account of less than 5% of the overall percentage of the compounds obtained by biotransformation were collectively indicated as 'not determined'. Otherwise, compounds **23a**, **25**, **28**, **35** and **36** were isolated from the fermentation broths by extraction and chromatographic separation and then submitted to chemical characterization. Different reasons prompted us to undertake the isolation procedure. First of all, these compounds were a relevant part of the biotransformation mixture and their MS fragmentations clearly indicated a chemical structure deriving from the corresponding starting materials. The compounds **25** was not one of the references standard available from our laboratory. Hence, we identified this compound only after its isolation and chemical characterization. Both the diastereoisomeric forms of the compound **25** have been described in the literature [55] but only low resolution ^{1}H-NMR data was reported. Therefore, we were not able to assign the relative configuration to the compound obtained by biotransformation. Differently, compound **36** is completely new and its analytic data has not described yet. Concerning compound **28**, we observed that its diastereoisomeric forms (*trans* and *cis* isomers) have the same retention time by GC-MS analysis. As a consequence, the isolation of compound **28** from the biotransformation mixture followed by its NMR analysis was necessary in order to understand what was the main isomer formed. Finally, the case of compound **35** is singular. It is the only compound obtained by biotransformation that was formed through a Baeyer-Villiger oxidation. Consequently, acetate **35** was completely unexpected and the proper reference standard was not synthesized.

Hereafter we reported the procedure for the preparative biotransformation experiments allowing the isolation of compounds **23a, 25, 28, 35** and **36** as well as their main analytic data.

According to the procedure described before for the preparation of the inoculum of spore forming mycelia, *Aspergillus niger*, *Nigrospora oryzae* and *Curvularia lunata*, were inoculated in three 1 L conical pyrex flasks containing 400 mL of MEM. The microorganisms were left to grow at 25 °C and 140 rpm for 2 days. Hence the cultures of *Aspergillus niger*, *Nigrospora oryzae* and *Curvularia lunata*, were treated with a solution of 1 g of γ-ionone, 3,4-dehydro-β-ionone and theaspirane, respectively, each one dissolved in 3 mL of DMSO. After 14 days from the substrate injection, using the growing condition described above, the reaction media was filtered through a celite pad, the filter was washed with ethyl acetate and the filtrate was extracted 3 times with the same solvent. The combined organic phases were separated, were washed with brine, dried on Na_2SO_4 and the solvent was removed under reduced pressure. The residue was then acetylated by treatment with pyridine/acetic anhydride (10 mL of a 2:1 mixture) and catalytic DMAP (10 mg) for 6 h at rt. The acetylating mixture (Py/Ac_2O) was then removed under reduced pressure and the resulting oil was purified by chromatography using *n*-hexane/AcOEt mixture as eluent.

The biotransformation of γ-ionone performed using *Aspergillus niger* allowed isolating 0.21 g (18% yield) of compound **23a** and 0.36 g (28% yield) of compound **25** as a single diastereoisomeric form (configuration not determined):

cis-2-Hydroxy-γ-ionone acetate (**23a**) = *(1SR,3RS)-2,2-Dimethyl-4-methylene-3-((E)-3-oxobut-1-en-1-yl) cyclohexyl acetate*. ^{1}H-NMR: δ = 6.97 (dd, J = 15.8, 9.9 Hz, 1H), 6.10 (d, J = 15.8 Hz, 1H), 4.88 (s, 1H), 4.74 (dd, J = 9.2, 4.0 Hz, 1H), 4.61 (s, 1H), 2.65 (d, J = 9.9 Hz, 1H), 2.41 (dt, J = 14.0, 5.4 Hz, 1H), 2.37–2.05 (m, 1H) 2.28 (s, 3H), 2.07 (s, 3H), 1.93–1.83 (m, 1H), 1.72–1.57 (m, 1H), 0.91 (s, 6H). ^{13}C-NMR δ = 197.9 (C), 170.4 (C), 146.0 (C), 145.5 (CH), 132.9 (CH) 111.1 (CH_2), 77.8 (CH), 55.9 (CH), 39.0 (C), 31.2 (CH_2), 27.6 (CH_2), 27.3 (Me), 26.2 (Me), 21.1 (Me), 17.8 (Me). GC-MS (EI): m/z (%) = 250 [M$^+$] (12), 235 [M$^+$ − Me] (1), 208 (13), 190 (45), 175 (40), 165 (36), 147 (100), 131 (23), 122 (39), 109 (96), 91 (34), 79 (35), 71 (12).

3-Hydroxy-γ-ionone acetate (**25**) = *(E)-3,3-Dimethyl-5-methylene-4-(3-oxobut-1-en-1-yl)cyclohexyl acetate*. ^{1}H-NMR: δ = 6.82 (dd, J = 15.8, 10.1 Hz, 1H), 6.13 (d, J = 15.8 Hz, 1H), 4.97 (s, 1H), 4.95–4.86 (m, 1H), 4.63 (s, 1H), 2,73 (dd, J = 12.6, 4.9 Hz, 1H), 2.58 (d, J = 10.1 Hz, 1H), 2.29 (s, 3H), 2.15–2.01 (m, 1H), 2.03 (s, 3H), 1.85 (dd, J = 12.6, 4.5 Hz, 1H), 1.45 (t, J = 12.1 Hz, 1H), 0.95 (s, 3H), 0.92 (s, 3H). ^{13}C-NMR: δ = 197.8 (C), 170.3 (C), 145.3 (CH), 144.5 (C), 134.0 (CH), 112.5 (CH_2), 69.8 (CH), 56.0 (CH), 45.5 (CH_2), 41.1 (CH_2), 35.8 (C), 30.4 (Me), 27.3 (Me), 21.4 (Me), 21.3 (Me). GC-MS (EI): m/z (%) = 250 [M$^+$] (<1), 235 [M$^+$ − Me] (<1), 190 (28), 175 (29), 157 (14), 147 (100), 131 (32), 119 (20), 105 (39), 91 (27), 79 (15), 69 (14), 55 (8).

The biotransformation of 3,4-dehydro-β-ionone performed using *Nigrospora oryzae* allowed isolating 0.56 g (34% yield) of compound **28** as a 5:1 mixture of *trans/cis* isomers:

trans-3,4-Dihydroxy-β-ionone diacetate (**28**) = *(1RS,2RS)-3,5,5-Trimethyl-4-((E)-3-oxobut-1-en-1-yl)cyclohex-3-ene-1,2-diyl diacetate*. ^{1}H-NMR: δ = 7.10 (dq, J = 16.4, 0.9 Hz, 1H), 6.11 (d, J = 16.4 Hz, 1H), 5.51 (d, J = 7.8 Hz, 1H), 5.17–5.09 (m, 1H), 2.28 (s, 3H), 2.06 (s, 3H), 2.00 (s, 3H), 1.88–1.66 (m, 2H), 1.62 (br s, 3H), 1.17 (s, 3H), 1.07 (s, 3H). ^{13}C-NMR: δ = 197.8 (C), 170.7 (C), 170.4 (C), 141.0 (CH), 140.2 (C), 133.7 (CH),

128.5 (C), 74.7 (CH), 70.3 (CH), 41.4 (CH$_2$), 36.4 (C), 29.7 (Me), 27.7 (Me), 27.5 (Me), 21.1 (Me), 20.8 (Me), 16.5 (Me). MS (ESI): 331.2 (M$^+$ + Na).

The biotransformation of theaspirane performed using *Curvularia lunata* allowed isolating 95 mg (7% yield) of compound **35** and 65 mg (5% yield) of compound **36**:

2-(3-Acetoxy-2,6,6-trimethylcyclohex-1-en-1-yl)ethyl acetate (**35**). ^{1}H-NMR: δ = 5.13 (t, *J* = 4.6 Hz, 1H), 4.05 (dd, *J* = 9.4, 7.5 Hz, 2H), 2.46–2.37 (m, 2H), 2.06 (br s, 6H), 1.92–1.79 (m, 1H), 1.74–1.52 (m, 2H), 1.66 (s, 3H), 1.44–1.33 (m, 1H), 1.08 (s, 3H), 1.00 (s, 3H). ^{13}C-NMR: δ = 171.0 (C), 171.0 (C), 139.5 (C), 128.2 (C), 72.3 (CH), 63.3 (CH$_2$), 35.1 (C), 34.6 (CH$_2$), 28.2 (Me), 28.0 (CH$_2$), 26.8 (Me), 25.3 (CH$_2$), 21.3 (Me), 21.0 (Me), 16.8 (Me). GC-MS (EI): *m/z* (%) = 268 [M$^+$] (<1), 226 (13), 208 [M$^+$ − AcOH] (16), 166 (40), 148 (17), 133 (100), 120 (28), 110 (36), 91 (16), 79 (8).

3,6,6-Trimethyl-1,4,5,6,7,8-hexahydro-3H-3,5a-epoxybenzo[c]oxepin-8-yl acetate (**36**). ^{1}H-NMR: δ = 5.43–5.36 (m, 1H), 5.31 (s, 1H), 4.49 (dt, *J* = 14.2, 2.1 Hz, 1H), 4.20 (d, *J* = 14.2 Hz, 1H), 2.30–1.40 (m, 6H), 2.02 (s, 3H), 1.47 (s, 3H), 1.11 (s, 3H), 1.01 (s, 3H). ^{13}C-NMR: δ = 170.5 (C), 138.7 (C), 116.3 (CH), 105.3 (C), 86.0 (C), 68.4 (CH), 63.7 (CH$_2$), 40.7 (CH$_2$), 35.3 (C), 34.3 (CH$_2$), 30.5 (CH$_2$), 24.6 (Me), 24.4 (Me), 23.3 (Me), 21.2 (Me). GC-MS (EI): *m/z* (%) = 266 [M$^+$] (1), 236 (50), 224 (8), 207 [M$^+$ − AcO] (59), 195 (36), 178 (27), 164 (84), 153 (100), 136 (61), 121 (59), 107 (62), 91 (53), 79 (29).

4. Conclusions

Our work provides some relevant findings. First, we demonstrated that fungi are able to perform different biotransformations on the isomeric forms of the C$_{13}$ apocarotenoids ionone, theaspirane and damascone. With respect to the eleven strains tested, we observed that the most common chemical transformations are oxidation reactions that afford oxygenated products such as hydroxy- keto- or epoxy-derivatives. On the contrary, the reduction of the keto groups or the reduction of the double bond functional groups are less relevant transformations, occurring for few substrates and yielding a minority amount of products.

A very significant feature of our study concern the prospective applicability of the fungi-mediated biotransformation of apocarotenoids for the synthesis of high value natural flavours. Since some ionone, damascone and theaspirane isomers are available in natural form and the biotransformation of a natural precursor is considered a 'natural method' of synthesis, the flavours obtained by means of the fungi-mediated reactions described above possess the natural status and could be commercialized accordingly.

Finally, we would like to highlight that our microbial biotransformations allow the preparation of many derivatives whose synthesis, using the classical chemical reactions, is very difficult. For example, different fungal strains proved to be able to oxidize some inactivated positions of the ionone or theaspirane framework, such as the position 2 and 3 of the β- and γ-ionone and the methine carbon linked to the theaspirane oxygen atom. By means of these microbial capabilities, we isolated one new compound (compound **36**) and we devised a new biocatalytic procedure for the synthesis of 2-hydroxy-γ-ionone, 3-hydroxy-γ-ionone, 3,4-dihydroxy-β-ionone and 2,6,10,10-tetramethyl-1-oxa-spiro[4.5]dec-6-ene-2,8-diol (identified as its monoacetate **34**), which are natural apocarotenoids or their direct precursors.

Author Contributions: S.S. and D.D.S. equally contributed to the conceptualization of this study. S.S. and D.D.S. equally contributed to conceive, design and perform the experiments as well as to analyze the data. S.S. wrote the paper.

Acknowledgments: The authors thank Cariplo Foundation for supporting this study.

References

1. Walter, M.H.; Strack, D. Carotenoids and their cleavage products: Biosynthesis and functions. *Nat. Prod. Rep.* **2011**, *28*, 663–692. [CrossRef] [PubMed]

2. Serra, S. Recent advances in the synthesis of carotenoid-derived flavours and fragrances. *Molecules* **2015**, *20*, 12817–12840. [CrossRef] [PubMed]

3. Ma, Q.-G.; Wang, Y.-G.; Liu, W.-M.; Wei, R.-R.; Yang, J.-B.; Wang, A.-G.; Ji, T.-F.; Tian, J.; Su, Y.-L. Hepatoprotective sesquiterpenes and rutinosides from *Murraya koenigii* (L.) Spreng. *J. Agric. Food Chem.* **2014**, *62*, 4145–4151. [CrossRef] [PubMed]

4. Schievano, E.; Morelato, E.; Facchin, C.; Mammi, S. Characterization of markers of botanical origin and other compounds extracted from unifloral honeys. *J. Agric. Food Chem.* **2013**, *61*, 1747–1755. [CrossRef] [PubMed]

5. Li, C.-Y.; Wu, T.-S. Constituents of the stigmas of *Crocus sativus* and their tyrosinase inhibitory activity. *J. Nat. Prod.* **2002**, *65*, 1452–1456. [CrossRef] [PubMed]

6. Ina, K.; Etō, H. 3-keto-β-ionone in the essential oil from black tea. *Agric. Biol. Chem.* **1971**, *35*, 962–963. [CrossRef]

7. Fujimori, T.; Kasuga, R.; Matsushita, H.; Kaneko, H.; Noguchi, M. Neutral aroma constituents in burley tobacco. *Agric. Biol. Chem.* **1976**, *40*, 303–315. [CrossRef]

8. Bolt, A.J.N.; Purkis, S.W.; Sadd, J.S. A damascone derivative from *Nicotiana tabacum*. *Phytochemistry* **1983**, *22*, 613–614. [CrossRef]

9. Sato, S.; Sasakura, S.; Kobayashi, A.; Nakatani, Y.; Yamanishi, T. Flavor of black tea. Part VI. Intermediate and high boiling components of the neutral fraction. *Agric. Biol. Chem.* **1970**, *34*, 1355–1367. [CrossRef]

10. Beekwilder, J.; van Rossum, H.M.; Koopman, F.; Sonntag, F.; Buchhaupt, M.; Schrader, J.; Hall, R.D.; Bosch, D.; Pronk, J.T.; van Maris, A.J.A.; et al. Polycistronic expression of a β-carotene biosynthetic pathway in *Saccharomyces cerevisiae* coupled to β-ionone production. *J. Biotechnol.* **2014**, *192*, 383–392. [CrossRef] [PubMed]

11. Serra, S.; Fuganti, C.; Brenna, E. Biocatalytic preparation of natural flavours and fragrances. *Trends Biotechnol.* **2005**, *23*, 193–198. [CrossRef] [PubMed]

12. Prelog, V.; Meier, H.L. Untersuchungen über organextrakte und harn. 18. Mitteilung. Über die biochemische oxydation von β-jonon im tierkörper. *Helv. Chim. Acta* **1950**, *33*, 1276–1284. [CrossRef]

13. Mikami, Y.; Watanabe, E.; Fukunaga, Y.; Kisaki, T. Formation of 2S-hydroxy-β-ionone and 4-hydroxy-β-ionone by microbial hydroxylation of β-ionone. *Agric. Biol. Chem.* **1978**, *42*, 1075–1077. [CrossRef]

14. Mikami, Y.; Fukunaga, Y.; Arita, M.; Kisaki, T. Microbial transformation of β-ionone and β-methylionone. *Appl. Environ. Microbiol.* **1981**, *41*, 610–617. [PubMed]

15. Krasnobajew, V.; Helmlinger, D. Fermentation of fragrances: Biotransformation of β-ionone by *Lasiodiplodia theobromae*. *Helv. Chim. Acta* **1982**, *65*, 1590–1601. [CrossRef]

16. Hartman, D.A.; Pontones, M.E.; Kloss, V.F.; Curley, R.W.; Robertson, L.W. Models of retinoid metabolism: Microbial biotransformation of α-ionone and β-ionone. *J. Nat. Prod.* **1988**, *51*, 947–953. [CrossRef] [PubMed]

17. Schoch, E.; Benda, I.; Schreier, P. Bioconversion of α-damascone by *Botrytis cinerea*. *Appl. Environ. Microbiol.* **1991**, *57*, 15–18.

18. Kakeya, H.; Sugai, T.; Ohta, H. Biochemical preparation of optically active 4-hydroxy-β-ionone and its transformation to (*S*)-6-hydroxy-α-ionone. *Agric. Biol. Chem.* **1991**, *55*, 1873–1876. [CrossRef]

19. Weidmann, V.; Kliewer, S.; Sick, M.; Bycinskij, S.; Kleczka, M.; Rehbein, J.; Griesbeck, A.G.; Zorn, H.; Maison, W. Studies towards the synthetic applicability of biocatalytic allylic oxidations with the lyophilisate of *Pleurotus sapidus*. *J. Mol. Catal. B Enzym.* **2015**, *121*, 15–21. [CrossRef]

20. Gliszczyńska, A.; Gładkowski, W.; Dancewicz, K.; Gabryś, B.; Szczepanik, M. Transformation of β-damascone to (+)-(*S*)-4-hydroxy-β-damascone by fungal strains and its evaluation as a potential insecticide against aphids *Myzus persicae* and lesser mealworm *Alphitobius diaperinus* Panzer. *Catal. Commun.* **2016**, *80*, 39–43. [CrossRef]

21. Lutz-Wahl, S.; Fischer, P.; Schmidt-Dannert, C.; Wohlleben, W.; Hauer, B.; Schmid, R.D. Stereo- and regioselective hydroxylation of α-ionone by *Streptomyces* strains. *Appl. Environ. Microbiol.* **1998**, *64*, 3878–3881.

22. Maurer, S.C.; Schulze, H.; Schmid, R.D.; Urlacher, V. Immobilisation of P450 BM-3 and an NADP+ cofactor recycling system: Towards a technical application of heme-containing monooxygenases in fine chemical synthesis. *Adv. Synth. Catal.* **2003**, *345*, 802–810. [CrossRef]

23. Litzenburger, M.; Bernhardt, R. Selective oxidation of carotenoid-derived aroma compounds by CYP260B1 and CYP267B1 from *Sorangium cellulosum* So ce56. *Appl. Microbiol. Biotechnol.* **2016**, *100*, 4447–4457. [CrossRef] [PubMed]

24. Venkataraman, H.; Beer, S.B.A.d.; Geerke, D.P.; Vermeulen, N.P.E.; Commandeur, J.N.M. Regio- and stereoselective hydroxylation of optically active α-ionone enantiomers by engineered cytochrome P450 BM3 mutants. *Adv. Synth. Catal.* **2012**, *354*, 2172–2184. [CrossRef]

25. Fuganti, C.; Serra, S.; Zenoni, A. Synthesis and olfactory evaluation of (+)- and (−)-gamma-ionone. *Helv. Chim. Acta* **2000**, *83*, 2761–2768. [CrossRef]

26. Serra, S.; Fuganti, C.; Brenna, E. Synthesis, olfactory evaluation, and determination of the absolute configuration of the 3,4-didehydroionone stereoisomers. *Helv. Chim. Acta* **2006**, *89*, 1110–1122. [CrossRef]

27. Serra, S.; Fuganti, C.; Brenna, E. Two easy photochemical methods for the conversion of commercial ionone alpha into regioisomerically enriched gamma-ionone and gamma-dihydroionone. *Flavour Fragr. J.* **2007**, *22*, 505–511. [CrossRef]

28. Serra, S. An expedient preparation of enantio-enriched ambergris odorants starting from commercial ionone alpha. *Flavour Fragr. J.* **2013**, *28*, 46–52. [CrossRef]

29. Brenna, E.; Fuganti, C.; Serra, S.; Kraft, P. Optically active ionones and derivatives: Preparation and olfactory properties. *Eur. J. Org. Chem.* **2002**, 967–978. [CrossRef]

30. Serra, S.; Fuganti, C. Synthesis of the enantiomeric forms of alpha- and gamma-damascone starting from commercial racemic alpha-ionone. *Tetrahedron Asymmetry* **2006**, *17*, 1573–1580. [CrossRef]

31. Brenna, E.; Fuganti, C.; Serra, S. Synthesis and olfactory evaluation of the enantiomerically enriched forms of 7,11-epoxymegastigma-5(6)-en-9-one and 7,11-epoxymegastigma-5(6)-en-9-ols isomers, identified in passiflora edulis. *Tetrahedron Asymmetry* **2005**, *16*, 1699–1704. [CrossRef]

32. Serra, S.; Barakat, A.; Fuganti, C. Chemoenzymatic resolution of *cis*- and *trans*-3,6-dihydroxy-alpha-ionone. Synthesis of the enantiomeric forms of dehydrovomifoliol and 8,9-dehydrotheaspirone. *Tetrahedron Asymmetry* **2007**, *18*, 2573–2580. [CrossRef]

33. Mazur, M.; Grudniewska, A.; Wawrzeńczyk, C. Microbial transformations of halolactones with *p*-menthane system. *J. Biosci. Bioeng.* **2015**, *119*, 72–76. [CrossRef] [PubMed]

34. Simeo, Y.; Sinisterra, J.V. Biotransformation of terpenoids: A green alternative for producing molecules with pharmacological activity. *Mini-Rev. Org. Chem.* **2009**, *6*, 128–134. [CrossRef]

35. Bhatti, H.N.; Khera, R.A. Biological transformations of steroidal compounds: A review. *Steroids* **2012**, *77*, 1267–1290. [CrossRef]

36. Carballeira, J.D.; Álvarez, E.; Sinisterra, J.V. Biotransformation of cyclohexanone using immobilized *Geotrichum candidum* NCYC49: Factors affecting the selectivity of the process. *J. Mol. Catal. B Enzym.* **2004**, *28*, 25–32. [CrossRef]

37. Bankar, A.V.; Kumar, A.R.; Zinjarde, S.S. Environmental and industrial applications of *Yarrowia lipolytica*. *Appl. Microbiol. Biotechnol.* **2009**, *84*, 847. [CrossRef]

38. Johnson, E.A. *Phaffia rhodozyma*: Colorful odyssey. *Int. Microbiol.* **2003**, *6*, 169–174. [CrossRef]

39. Knapp, H.; Weigand, C.; Gloser, J.; Winterhalter, P. 2-hydroxy-2,6,10,10-tetramethyl-1-oxaspiro[4.5]dec-6-en-8-one: Precursor of 8,9-dehydrotheaspirone in white-fleshed nectarines. *J. Agric. Food Chem.* **1997**, *45*, 1309–1313. [CrossRef]

40. Näf, R.; Velluz, A.; Decorzant, R.; Näf, F. Structure and synthesis of two novel ionone-type compounds identified in quince brandy (*Cydonia oblonga* Mil.). *Tetrahedron Lett.* **1991**, *32*, 753–756. [CrossRef]

41. Serra, S. MnO_2/TBHP: A versatile and user-friendly combination of reagents for the oxidation of allylic and benzylic methylene functional groups. *Eur. J. Org. Chem.* **2015**, *2015*, 6472–6478. [CrossRef]

42. Serra, S.; Lissoni, V. First enantioselective synthesis of marine diterpene ambliol-A. *Eur. J. Org. Chem.* **2015**, *2015*, 2226–2234. [CrossRef]

43. Dess, D.B.; Martin, J.C. Readily accessible 12-I-5 oxidant for the conversion of primary and secondary alcohols to aldehydes and ketones. *J. Org. Chem.* **1983**, *48*, 4155–4156. [CrossRef]

44. Audran, G.; Galano, J.M.; Monti, H. Enantioselective synthesis and determination of the absolute configuration of natural (−)-elegansidiol. *Eur. J. Org. Chem.* **2001**, 2293–2296. [CrossRef]

45. Serra, S.; Gatti, F.G.; Fuganti, C. Lipase-mediated resolution of the hydroxy-cyclogeraniol isomers: Application to the synthesis of the enantiomers of karahana lactone, karahana ether, crocusatin C and gamma-cyclogeraniol. *Tetrahedron Asymmetry* **2009**, *20*, 1319–1329. [CrossRef]

46.   Kaiser, R.; Lamparsky, D. Inhaltsstoffe des *Osmanthus*-absolues. 1. Mitteilung: 2,5-epoxy-megastigma-6,8-dien. *Helv. Chim. Acta* **1978**, *61*, 373–382. [CrossRef]

47.   Tu, V.A.; Kaga, A.; Gericke, K.-H.; Watanabe, N.; Narumi, T.; Toda, M.; Brueckner, B.; Baldermann, S.; Mase, N. Synthesis and characterization of quantum dot nanoparticles bound to the plant volatile precursor of hydroxy-apo-10′-carotenal. *J. Org. Chem.* **2014**, *79*, 6808–6815. [CrossRef]

48.   Khachik, F.; Chang, A.-N. Synthesis of (3*S*)- and (3*R*)-3-hydroxy-β-ionone and their transformation into (3*S*)- and (3*R*)-β-cryptoxanthin. *Synthesis* **2011**, *2011*, 509–516. [CrossRef]

49.   Takei, Y.; Mori, K.; Matsui, M. Synthesis of a stereoisomeric mixture of 3-hydroxy-α-damascone. *Agric. Biol. Chem.* **1973**, *37*, 2927–2928. [CrossRef]

50.   Pascual, A.; Bischofberger, N.; Frei, B.; Jeger, O. Photochemical reactions. 149th communication. Photochemistry of 7,8-dihydro-4-hydroxy-β-ionone and derivatives. *Helv. Chim. Acta* **1988**, *71*, 374–388. [CrossRef]

51.   Buschor, D.J.; Eugster, C.H. Synthese der (3*S*,4*R*,3′*S*,4′*R*)- und (3*S*,4*R*,3′*S*,4′*S*)crustaxanthine sowie weiterer verbindungen mit 3,4-dihydroxy-β-endgruppen. *Helv. Chim. Acta* **1990**, *73*, 1002–1021. [CrossRef]

52.   Irie, H.; Matsumoto, R.; Nishimura, M.; Zhang, Y. Synthesis of (±)-heritol, a sesquiterpene lactone belonging to the aromatic cadinane group. *Chem. Pharm. Bull.* **1990**, *38*, 1852–1856. [CrossRef]

53.   Cooper, R.D.G.; Davis, J.B.; Leftwick, A.P.; Price, C.; Weedon, B.C.L. Carotenoids and related compounds. Part XXXII. Synthesis of astaxanthin, phoenicoxanthin, hydroxyechinenone, and the corresponding diosphenols. *J. Chem. Soc. Perkin Trans. 1* **1975**, 2195–2204. [CrossRef]

54.   Serra, S.; Piccioni, O. A new chemo-enzymatic approach to the stereoselective synthesis of the flavors tetrahydroactinidiolide and dihydroactinidiolide. *Tetrahedron Asymmetry* **2015**, *26*, 584–592. [CrossRef]

55.   Oritani, T.; Yamamoto, H.; Yamashita, K. Synthesis of (±)-4′-hydroxy-γ-ionylideneacetic acids, fungal biosynthetic intermediates of abscisic acid. *Agric. Biol. Chem.* **1990**, *54*, 125–130. [CrossRef]

11

Ultrasonic Processing Induced Activity and Structural Changes of Polyphenol Oxidase in Orange (*Citrus sinensis* Osbeck)

Lijuan Zhu [1,2]**, Linhu Zhu** [1,2]**, Ayesha Murtaza** [1,2]**, Yan Liu** [1,2]**, Siyu Liu** [3]**, Junjie Li** [1,2]**, Aamir Iqbal** [1,2]**, Xiaoyun Xu** [1,2]**, Siyi Pan** [1,2] **and Wanfeng Hu** [1,2,]*****

[1]   College of Food Science and Technology, Huazhong Agricultural University, No.1, Shizishan Street, Hongshan District, Wuhan 430070, China; lijuanzhu2019@hotmail.com (L.Z.); orion_zhu@webmail.hzau.edu.cn (L.Z.); ayeshamurtaza@webmail.hzau.edu.cn (A.M.); y20170022@mail.ecust.edu.cn (Y.L.); junjieli2019@hotmail.com (J.L.); aamirraoiqbal@webmail.hzau.edu.cn (A.I.); xiaoyunxu88@gmail.com (X.X.); drpansiyi.hzau.edu@outlook.com (S.P.)

[2]   Key Laboratory of Environment Correlative Dietology, Huazhong Agricultural University, Ministry of Education, Wuhan 430070, China

[3]   Key Laboratory of Structural Biology, School of Chemical Biology & Biotechnology, Peking University Shenzhen Graduate School, Shenzhen 518055, China; siyuliu@pku.edu.cn

*   Correspondence: wanfenghu@mail.hzau.edu.cn.

Academic Editor: Stefano Serra

Abstract: Apart from non-enzymatic browning, polyphenol oxidase (PPO) also plays a role in the browning reaction of orange (*Citrus sinensis* Osbeck) juice, and needs to be inactivated during the processing. In this study, the protein with high PPO activity was purified from orange (*Citrus sinensis* Osbeck) and inactivated by ultrasonic processing. Fluorescence spectroscopy, circular dichroism (CD) and Dynamic light scattering (DLS) were used to investigate the ultrasonic effect on PPO activity and structural changes on purified PPO. DLS analysis illustrated that ultrasonic processing leads to initial dissociation and final aggregation of the protein. Fluorescence spectroscopy analysis showed the decrease in fluorescence intensity leading to the exposure of Trp residues to the polar environment, thereby causing the disruption of the tertiary structure after ultrasonic processing. Loss of α-helix conformation leading to the reorganization of secondary structure was triggered after the ultrasonic processing, according to CD analysis. Ultrasonic processing could induce aggregation and modification in the tertiary and secondary structure of a protein containing high PPO activity in orange (*Citrus sinensis* Osbeck), thereby causing inactivation of the enzyme.

Keywords: browning reaction; polyphenol oxidase; ultrasonic processing; structural changes; aggregation

1. Introduction

The browning of citrus fruits during storage and juice processing often leads to undesirable flavor and nutritional loss in the final products. The reason for this browning is usually attributed to non-enzymatic browning caused by ascorbic acid degradation [1]. The enzymatic browning, which is catalyzed by polyphenol oxidase (PPO), is often ignored with regard to browning reaction in citrus products. This may be due to the presence of a high level of ascorbic acid in citrus fruits, which could reduce the colored quinone to colorless phenol, and therefore prevent the browning reaction from happening [2]. However, in long-term storage, citrus products tend to brown gradually when ascorbic acid is oxidized during storage, and the existing PPO may play its role in and be responsible for the final browning [3,4].

In previous studies, protein with high PPO activity was found in *Satsuma mandarine* juice [3,5]. In this study, PPO with high activity was also found in *Citrus sinensis* Osbeck. These enzymes accumulate in citrus peel, could easily mix with citrus juice during processing and thus catalyze the phenols to quinones, which further polymerize to generate the melanin pigments [6,7]. These colored compounds negatively affect the nutritional and organoleptic qualities, and consequently lower the marketability of citrus products [3].

Conventional methods such as thermal treatments and chemical reagents are mostly used to inhibit the browning [7]. However, thermal processing could even activate the enzyme [3,8], and may also cause loss of quality. Chemical reagents may bring safety problems to the products. Ultrasonic processing is an innovative, non-thermal technology which could retain the quality of food products at mild conditions [9]. Low-frequency, high-intensity ultrasonic processing may effectively enhance the shelf life of the juice product with minimal damage to its quality [10]. The effects of ultrasonic processing on food processing include cavitation bubbles, vibration on shear strength, and temporary generation of spots of extreme physical phenomena, as well as generation of free radicals through sonolysis of water [11]. The ultrasonic energy in liquid causes the formation of cavitation bubbles due to changes in pressure. The collision of these bubbles leads to an increase in temperature (5000 K) and pressure (1000 atm), which may generate turbulence and extreme shear force in the cavitation zone [10].

Numerous studies were conducted to investigate the effect of ultrasonic processing on enzyme inactivation during fruit and vegetable processing [7]. Up to now, literature about PPO inactivation in orange (*Citrus sinensis* Osbeck) juice has rarely been covered. The aim of the current study was to explore the effect of high-intensity ultrasonic processing on the inactivation of PPO and the structural changes of the enzyme through circular dichroism, fluorescence spectral analysis and dynamic light scattering analysis. These structural analyses may explain the mechanism of enzyme inactivation in orange (*Citrus sinensis* Osbeck) juice during ultrasonic processing.

2. Results and Discussion

2.1. Purity and Molecular Weight

As shown in Figure 1, the purified protein showed only one band after staining with R-250 Coomassie Brilliant Blue by non-denaturing (native) polyacrylamide gel electrophoresis (PAGE), which demonstrated that the protein was electrophoretically pure. Elution profile of protein extraction are shown in Figure S1. The protein purification fold and yield are showed in Table S1. The gel stained with pyrogallol showed the same protein band in the same position, confirming that the protein had PPO activity. As shown in Figure S2, it can also be confirmed that the protein had PPO activity. The protein band coincided approximately with the known protein marker of 100 kDa on the SDS and native PAGE. These results showed that the PPO might be a monomer revealing one band during electrophoresis [12]. Literature demonstrated that PPO from present molecular weight varied from 30 kDa to 128 kDa [13–15].

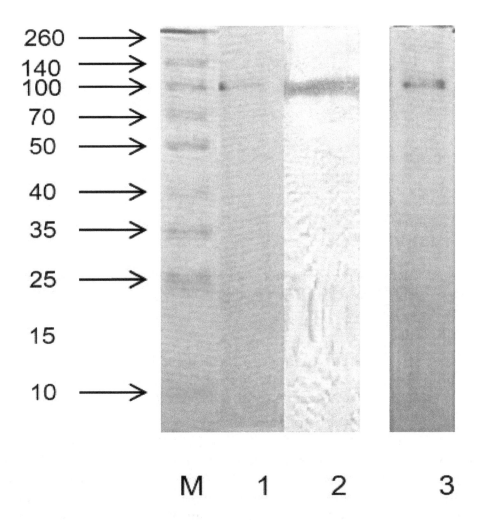

Figure 1. Electrophoresis pattern of the marker (M) and purified enzyme (1—sodium salt (SDS)-Polyacrylamide gel electrophoresis (PAGE) dyed with Coomassie Blue R-250, 2—native PAGE dyed with catechol, 3—native PAGE dyed with Coomassie Blue R-250).

2.2. Effect of Ultrasonication on PPO Activity

The PPO activities of the ultrasonicated protein solutions are presented in Figure 2A,B, respectively. As can be seen from Figure 2A, the PPO activity was very high. Ultrasonic processing at lower duration stimulated the activation of PPO. With constant power of 20 W/mL at different times (5, 10, 15, 20 and 30 min) in ultrasonic processing, the relative enzyme activity changed. The PPO activity increased with the increase of time up to 15 min and decreased with a further increase in time. As seen in Figure 2B, with constant ultrasonication for 20 min under different powers, the relative enzyme activity changed. With increasing ultrasonic power, a peak of the relative enzyme activity was observed at 20 W/mL. At above 20 W/mL, the relative enzyme activity began to reduce. Under an ultrasonic power of 40 W/mL, the enzyme activity was restrained; in particular, PPO activity was reduced by 48% under ultrasonic processing at 50 W/mL for 20 min.

These results indicate that low-intensity (20 W/mL or lower) ultrasonic processing stimulates the PPO activity, but high-intensity ultrasonic processing can inhibit the PPO activity. Previous studies demonstrated the accordant effect of ultrasonic processing on PPO activity [7,10,16]. The effect of ultrasonication on the PPO activity may be attributed to the ability of ultrasonic processing to break down the molecular aggregates. Ultrasonic processing at low intensity may dissociate protein monomers which results in the exposure of the active site to the substrate, causing the activation of the PPO enzyme [16]. However, the catalytic center was destroyed under strong ultrasonic processing and more treatment time, which might cause higher levels of denaturation and inactivation of PPO protein.

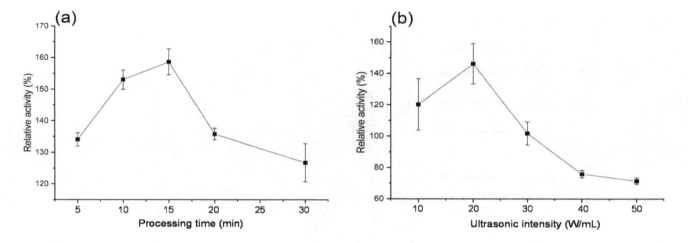

Figure 2. Relative activities of ultrasonic-processed purified PPO at 20 W/mL for 5, 10, 15, 20 and 30 min (**a**); processed for 20 min at 10, 20, 30, 40 and 50 W/mL (**b**).

2.3. Particle Size Distribution

The particle size distributions (PSD) for the untreated and ultrasonicated samples are shown in Figure 3a,b. The untreated protein has a peak value of 49.1% in number fraction at 142 nm with a relatively narrow span of 122 nm to 164 nm (42 nm), which indicates that the native protein aggregate was monodispersed in the aqueous systems [5].

The effect of ultrasonic time (5, 10, 15, 20 and 30 min at 20 W/mL) on the PSD pattern of purified protein is shown in Figure 3a. With ultrasonication for 5 min, the peak particle diameter did not change obviously, but the diameter span increased and the minimum particle diameter decreased. With increased ultrasonic time, the peak particle size and diameter span increased. The maximum peak particle size (255 nm) and diameter span (620.7 nm) were obtained by increasing ultrasonic duration to 30 min. The influence of increasing ultrasonic intensity (10 W/mL to 50 W/mL for 20 min) on PPO is shown in Figure 3b. With the intensity of 10 and 20 W/mL, the peak particle diameter did not obviously change, but the diameter span increased. When the ultrasonic intensity increased to 30 W/mL, the particles started aggregation to attain higher particle diameter. The maximum peak diameter (342 nm) and diameter span (863.7 nm) was obtained after ultrasonic processing at 50 W/mL for 20 min.

In general, after ultrasonic processing, the PSD patterns of protein exhibited remarkable changes with wider spans of larger and smaller particle diameters, suggesting the polydispersion and aggregation of protein particles [17]. The native protein aggregate was monodispersed in the aqueous systems because of hydrogen bonding, hydrophobic interaction and electrostatic interaction. A previous study found that ultrasonic processing caused the cleavage of agglomerates, hydrophobic interaction, and van der Waal's forces among protein molecules, thereby inducing the structural changes in enzyme protein [18]. In this case, the particle sizes were decreased because of the shear forces due to cavitation [18,19]. However, as the ultrasonic processing duration and intensity increased, some particles started aggregation. Similarly, Gulseren et al. [20] found that ultrasonic processing at a high duration of 40 min could cause the formation of large aggregates in bovine serum albumin protein. This aggregation might be due to the electrostatic and hydrophobic non-covalent interactions among protein particles [20]. The surface hydrophobicity decreased after prolonged sonication on reconstituted whey protein concentrate, which is a sign of protein aggregation [21]. Under the low-intensity or short-term ultrasonic processing, the enzyme exhibited dissociation because of the forces caused by cavitation. However, the aggregation increased with the increased intensity and time due to noncovalent interactions.

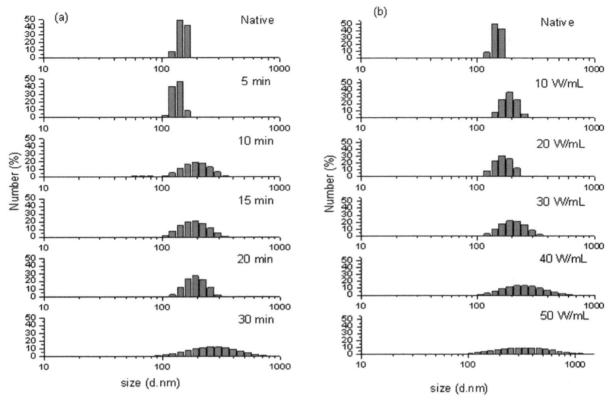

Figure 3. Particle size distributions of native and ultrasonic-processed PPO at 20 W/mL for 5, 10, 15, 20 and 30 min (**a**); processed for 20 min at 10, 20, 30, 40 and 50 W/mL (**b**).

2.4. Fluorescence Spectroscopic Analysis

The fluorescence property of the native and ultrasonicated purified enzyme was studied by fluorescence spectroscopy. It is a valuable technique to investigate the transition in the tertiary structure of proteins because the fluorescence from the tryptophan of amino acid is sensitive to the polarity of its local environment [16,22]. As shown in Figure 4, the λ_{max} of the native protein was 334 nm with an intensity of 39.74, signifying that Trp residues in enzymes were located in the nonpolar hydrophobic environment. Figure 4a shows the effect of the ultrasonic duration (5, 10, 15, 20 and 30 min at 20 W/mL) on the intrinsic fluorescence spectrum of protein. The ultrasonically treated enzyme showed a red-shift of 0.6–1.2 nm in λ_{max}, with a gradual decrease in fluorescence intensity following the increasing treatment duration. This finding indicated the exposure of Trp residues to the polar environment, thereby disrupting the tertiary structure. More exposure of fluorophores to the polar environment might cause the release and transfer of energy, which consequently leads to the quenching of fluorescence intensity [23,24].

Figure 4B illustrates the effect of increasing ultrasonic intensities (10 W/mL to 50 W/mL) on the fluorescence spectra of purified protein. The purified protein showed red-shifts of 0.2–2.0 nm in λ_{max} when treated with low ultrasonic intensities of 10, 20 and 30 W/mL. In the meantime, the fluorescence intensity gradually decreased with increasing ultrasonic intensity. The fluorophores were exposed to a more polar environment due to the polydispersity of protein aggregate after ultrasonic processing as described above, which caused the red-shift and consequently led to the quenching of fluorescence intensity. However, the extreme ultrasonic processing at 50 W/mL induced aggregation and may bury the exposed fluorophores inside the molecules, thereby decreasing the λ_{max}. PPO structural change is responsible for the activity change. The changes in fluorescence intensity and λ_{max} indicated a possible change in PPO's tertiary structure, which ultimately led to the activity reduction of PPO [25]. The result was similar to the observation that the fluorescence intensity of mushroom tyrosinase decreased in an aqueous system following mild thermal and supercritical CO_2 treatments [24].

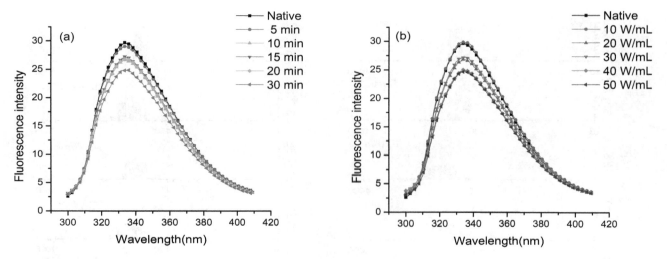

Figure 4. Fluorescence spectra of native and ultrasonic-processed PPO at 20 W/mL for 5, 10, 15, 20 and 30 min (**a**); processed for 20 min at 10, 20, 30, 40 and 50 W/mL (**b**).

2.5. Circular Dichroism Spectroscopy Analysis

The secondary structures of the native and ultrasonicated enzyme were analyzed through CD spectroscopy. The CD spectra of protein are shown in Figure 5a,b. The native protein showed a positive peak at 193 nm with two double-negative slots (208 and 222 nm), which were considered as typical α-helix conformation in the secondary structures [16,26,27]. As shown in Figure 5a,b, the negative peak (208 nm) increased with the negative peak decrease at 222 nm after ultrasonic processing. As the ultrasonic time and intensity increased, the change grew and the two double-negative slots gradually disappeared. These changes show that the ultrasonic processing triggered α-helix conformation of secondary structure loss [3,16,26–28]. The contents based on the CONTIN algorithms of the protein secondary structure of the native and ultrasonicated protein [24] are shown in Table 1. At the high ultrasonic intensity of 40 W/mL, the α-helix conformation remarkably decreased, while β-turn contents were increased. A similar study was reported by Liu et al. [10], where ultrasonic processing at high intensity caused a loss of α-helix conformation in protein structure. The decrease in α-helix was found to be correlated with the enzymatic activity of PPO molecules [29]. Ultrasonic processing induced the changes in molecular interaction, thus leading to changes in secondary structure and eventually causing the loss of PPO activity.

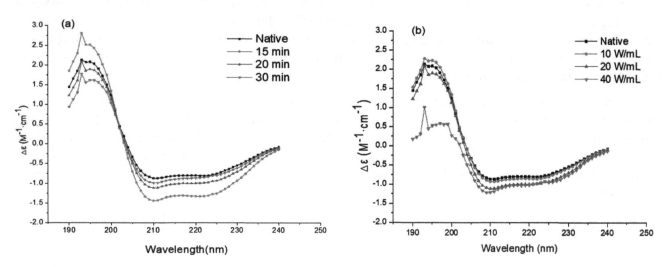

Figure 5. Circular Dichroism (CD) spectra of native and ultrasonic-processed PPO at 20 W/mL for 15, 20 and 30 min (**a**); processed for 20 min at 10, 20 and 40 W/mL (**b**).

Table 1. Secondary structure contents of native and ultrasonic-processed PPO.

Samples	α-Helix	β-Sheet	β-Turn	Random Coil
Native	76.20%	0.00%	23.60%	0.10%
10 W/mL/20 min	76.00%	0.00%	24.00%	0.00%
20 W/mL/20 min	53.30%	0.00%	29.30%	17.50%
40 W/mL/20 min	32.80%	0.00%	33.60%	33.60%
20 W/mL/15 min	54.60%	0.00%	26.50%	18.90%
20 W/mL/20 min	53.30%	0.00%	29.30%	17.50%
20 W/mL/30 min	51.80%	0.00%	31.10%	17.20%

3. Materials and Methods

3.1. Materials and Chemicals

Fresh orange (*Citrus sinensis* Osbeck) used in this study was procured from a local market (Chongqing, China). All chemicals used were of analytical grade.

3.2. Extraction and Partial Purification of Protein

Orange tissue (250 g) was homogenized in 700 mL of Tris-HCl buffer (0.5 mol/L) containing 10% polyvinylpyrrolidone (PVPP). The homogenate liquid was stored for 8 h at 4 °C and then subjected to centrifugation at 2057× g for 10 min using a centrifuge machine (Eppendorf centrifuge 5804 R, Eppendorf, Hamburg, Germany). The resultant supernatant was fractionated with 25% solid ammonium sulfate to remove impurities. The process was repeated using 90% ammonium sulfate saturation to precipitate proteins with PPO activity. The precipitate was re-dissolved in Tris-HCl (0.5 mol/L) and dialyzed against the same buffer for 34 h. The buffer was changed every hour during the whole process of dialysis. Ultrafilter was used to concentrate the crude extract of protein. Then, DEAE Sepharose Fast Flow and Sepracryl S-200 columns were used to purify the crude protein [5]. The fractions containing the highest activity of PPO were selected, concentrated and stored for further analysis.

3.3. Electrophoresis Assay

Native PAGE was carried out on preparative 12% polyacrylamide gels using the method described by Davis [30] with slight modifications. After running, the gels were stained with 0.1 mol/L Catechol (50 mL) and Coomassie Brilliant Blue R-250. The gels were analyzed for activity and estimation of molecular weight. The molecular weight and subunit of purified PPO enzyme were determined by SDS–PAGE [31].

3.4. Protein Content

Protein content was determined according to the Bradford method [32]. The protein solution was stained with Coomassie Brilliant Blue G-250 and the absorption peak was observed at 595 nm wavelength [33]. The absorption value of A595 was directly proportional to the protein concentration. The standard curve of bovine serum protein (BSA) was used as a standard protein. The concentration of sample protein was calculated according to the standard curve.

3.5. Ultrasonic Processing

An ultrasonic processor (JY92-2D, Ningbo, China) containing a titanium probe of diameter 0.636 cm was used to sonicate 10 mL of protein in 25 mL of centrifugal tubes, surrounded by ice to maintain the low temperature. All ultrasonic processing was definitely below 30 °C. The protein samples were treated at a low frequency of 20 kHz with the pulse duration of 5 s on and 5 s off setting, to investigate the effect of ultrasonic time (10, 15, 20 and 30 min at 20 W/mL) and intensity (10,

20 and 40 W/mL for 30 min) on the protein. The ultrasonic-processed samples were stored at 4 °C for further analysis.

3.6. PPO Activity Assay

The activity of the PPO enzyme was measured by determining the increasing rate of absorbance per minute at 420 nm using an Eppendorf Bio-spectrometer (Eppendorf, BioSpectrometer kinetic, Hamburg, Germany). The protein solution of 0.5 mL was mixed with 0.1 mol/L catechol (1 mL) and 0.1 mol/L Tris-HCl buffer (1 mL), then the absorbance of the resultant mixture was measured at 420 nm [5]. The relative activity (RA) of the PPO enzyme was measured according to the following equation.

$$\text{Relative PPO activity} = \frac{PPO\ activity\ of\ ultrasonic\ treated\ ppo}{PPO\ activity\ of\ native\ ppo} \times 100\% \tag{1}$$

3.7. Particle Size Distribution Analysis

Particle size determination was performed using a zetasizer, Nano-ZS device (Malvern Instruments, Malvern, Worcestershire, UK). The purified protein solutions prepared in Tris–HCl buffers (50 mmol/L) were subjected to scattered light wavelength (532 nm) and 15° laser reflection angle at a measuring temperature of 25 °C. The size measurements were taken as the mean of five readings.

3.8. CD Spectral Measurement

The purified protein solutions were subjected to CD spectra measurement with a spectropolarimeter (JASCO J-1500, Tokyo, Japan). The samples of treated protein solutions were prepared in Tris–HCl buffer (50 mmol/L) using this buffer as a blank. The secondary structure of protein solutions was determined in the range of far-ultraviolet (196–260 nm) with the scanning speed of 120 nm/min and the bandwidth of 1 nm. Data for CD spectra were presented as changes in the molar extinction coefficient ($\Delta\varepsilon$, M^{-1} cm^{-1}). The contents of secondary structure were calculated from the CD spectra by the estimation software of Spectra Manager (JASCO, Japan) [5].

3.9. Fluorescence Spectral Measurement

Intrinsic fluorescence spectral measurements were recorded using a fluorescence spectrophotometer F-4600 (Hitachi, Tokyo, Japan). Tris–HCl (50 mmol/L) was used to prepare the protein solution for fluorescence assay. The samples were measured at an emission wavelength of 350 nm to obtain the maximum excitation wavelength and then scanned at this excitation wavelength to record the emission spectra. The E_m and E_x slits were set as 5 nm; scan speed was set as 200 nm/min with a response time of 0.1 s [34,35].

4. Conclusions

Low-intensity ultrasonic processing stimulated the activation of PPO but exposure to high-intensity ultrasonic processing exhibited an inactivation effect on the PPO enzyme. The increasing power under ultrasonic processing induced the polydispersity of the protein, as well as the change of the secondary and tertiary structures of the protein. The loss of α-helical contents after ultrasonic processing led to the reorganization of the secondary structure of the protein. High-intensity ultrasonic processing caused the exposure of Trp residues to a more polar environment, thereby leading to the quenching of fluorescence intensity and subsequently decreasing the suitability of the catalytic center of the enzyme for reaction with substrates.

Author Contributions: Data curation, A.M.; formal analysis, Y.L.; project administration, X.X. and S.P.; software, S.L.; supervision, W.H.; visualization, J.L. and A.I.; writing—original draft, L.Z. (Lijuan Zhu); writing—review and editing, L.Z. (Linhu Zhu).

References

1. Kacem, B.; Matthews, R.F.; Crandall, P.G.; Cornell, J.A. Nonenzymatic browning in aseptically packaged orange juice and orange drinks. Effect of amino acids, deaeration, and anaerobic storage. *J. Food Sci.* **1987**, *52*, 1665–1667. [CrossRef]

2. Landi, M.; Degl'innocenti, E.; Guglielminetti, L.; Guidi, L. Role of ascorbic acid in the inhibition of polyphenol oxidase and the prevention of browning in different browning-sensitive Lactuca sativa var. capitata (L.) and Eruca sativa (Mill.) stored as fresh-cut produce. *J. Sci. Food Agric.* **2013**, *93*. [CrossRef]

3. Huang, N.; Cheng, X.; Hu, W.; Pan, S. Inactivation, aggregation, secondary and tertiary structural changes of germin-like protein in Satsuma mandarine with high polyphenol oxidase activity induced by ultrasonic processing. *Biophys. Chem.* **2015**, *197*, 18–24. [CrossRef]

4. Zhao, G.Y.; Li, B. Studies on the occurrence of non-enzymatic browning during the storage of cloudy apple juice. *Proc. 2007 Int. Conf. Agric. Eng.* **2007**, *32*, 634–643.

5. Cheng, X.; Huang, X.; Liu, S.; Tang, M.; Hu, W.; Pan, S. Characterization of germin-like protein with polyphenol oxidase activity from Satsuma mandarine. *Biochem. Biophys. Res. Commun.* **2014**, *449*, 313–318. [CrossRef]

6. Vamos-Vigyázó, L. Polyphenol Oxidase and Peroxidase in Fruits and Vegetables. *CRC Crit. Rev. Food Sci. Nutr.* **1981**, *15*, 49–127. [CrossRef]

7. Jang, J.H.; Moon, K.D. Inhibition of polyphenol oxidase and peroxidase activities on fresh-cut apple by simultaneous treatment of ultrasound and ascorbic acid. *Food Chem.* **2011**, *124*, 444–449. [CrossRef]

8. Murtaza, A.; Muhammad, Z.; Iqbal, A.; Ramzan, R.; Liu, Y.; Pan, S.; Hu, W. Aggregation and Conformational Changes in Native and Thermally Treated Polyphenol Oxidase From Apple Juice (Malus domestica). *Front. Chem.* **2018**, *6*, 1–10. [CrossRef] [PubMed]

9. Fonteles, T.V.; Costa, M.G.M.; de Jesus, A.L.T.; de Miranda, M.R.A.; Fernandes, F.A.N.; Rodrigues, S. Power ultrasound processing of cantaloupe melon juice: Effects on quality parameters. *Food Res. Int.* **2012**, *48*, 41–48. [CrossRef]

10. Liu, S.; Liu, Y.; Huang, X.; Yang, W.; Hu, W.; Pan, S. Effect of ultrasonic processing on the changes in activity, aggregation and the secondary and tertiary structure of polyphenol oxidase in oriental sweet melon (Cucumis melo var. makuwa Makino). *J. Sci. Food Agric.* **2017**, *97*, 1326–1334. [CrossRef] [PubMed]

11. Illera, A.E.; Sanz, M.T.; Benito-Román, O.; Varona, S.; Beltrán, S.; Melgosa, R.; Solaesa, A.G. Effect of thermosonication batch treatment on enzyme inactivation kinetics and other quality parameters of cloudy apple juice. *Innov. Food Sci. Emerg. Technol.* **2018**, *47*, 71–80. [CrossRef]

12. Yang, C.; Fujita, S.; Nakamura, N. Purification and Characterization of Polyphenol Oxidase from Banana (Musa sapientum L.) Pulp. *J. Agric. Food Chem.* **2000**, *48*, 2732–2735. [CrossRef] [PubMed]

13. Wititsuwannakul, D.; Chareonthiphakorn, N.; Pace, M.; Wititsuwannakul, R. Polyphenol oxidases from latex of Hevea brasiliensis: Purification and characterization. *Phytochemistry* **2002**, *61*, 115–121. [CrossRef]

14. Xu, J.; Zheng, T.; Meguro, S.; Kawachi, S. Purification and characterization of polyphenol oxidase from Henry chestnuts (Castanea henryi). *J. Wood Sci.* **2004**, *50*, 260–265. [CrossRef]

15. Cheng, X.F.; Zhang, M.; Adhikari, B. The inactivation kinetics of polyphenol oxidase in mushroom (Agaricus bisporus) during thermal and thermosonic treatments. *Ultrason. Sonochem.* **2013**, *20*, 674–679. [CrossRef]

16. Ma, H.; Huang, L.; Jia, J.; He, R.; Luo, L.; Zhu, W. Effect of energy-gathered ultrasound on Alcalase. *Ultrason. Sonochem.* **2011**, *18*, 419–424. [CrossRef]

17. Li, R.; Wang, Y.; Hu, W.; Liao, X. Changes in the activity, dissociation, aggregation, and the secondary and tertiary structures of a thaumatin-like protein with a high polyphenol oxidase activity induced by high pressure CO2. *Innov. Food Sci. Emerg. Technol.* **2014**, *23*, 68–78. [CrossRef]

18. Jambrak, A.R.; Mason, T.J.; Lelas, V.; Paniwnyk, L.; Herceg, Z. Effect of ultrasound treatment on particle size and molecular weight of whey proteins. *J. Food Eng.* **2014**, *121*, 15–23. [CrossRef]

19. Bi, X.; Hemar, Y.; Balaban, M.O.; Liao, X. The effect of ultrasound on particle size, color, viscosity and polyphenol oxidase activity of diluted avocado puree. *Ultrason. Sonochem.* **2015**, *27*, 567–575. [CrossRef]

20. Gülseren, I.; Güzey, D.; Bruce, B.D.; Weiss, J. Structural and functional changes in ultrasonicated bovine serum albumin solutions. *Ultrason. Sonochem.* **2007**, *14*, 173–183. [CrossRef] [PubMed]

21. Chandrapala, J.; Zisu, B.; Palmer, M.; Kentish, S.; Ashokkumar, M. Effects of ultrasound on the thermal and structural characteristics of proteins in reconstituted whey protein concentrate. *Ultrason. Sonochem.* **2011**, *18*, 951–957. [CrossRef] [PubMed]

22. Vivian, J.T.; Callis, P.R. Mechanisms of tryptophan fluorescence shifts in proteins. *Biophys. J.* **2001**, *80*, 2093–2109. [CrossRef]

23. Carvalho, A.S.L.; Ferreira, B.S.; Neves-Petersen, M.T.; Petersen, S.B.; Aires-Barros, M.R.; Melo, E.P. Thermal denaturation of HRPA2: pH-dependent conformational changes. *Enzyme Microb. Technol.* **2007**, *40*, 696–703. [CrossRef]

24. Hu, W.; Zhang, Y.; Wang, Y.; Zhou, L.; Leng, X.; Liao, X.; Hu, X. Aggregation and homogenization, surface charge and structural change, and inactivation of mushroom tyrosinase in an aqueous system by subcritical/supercritical carbon dioxide. *Langmuir* **2011**, *27*, 909–916. [CrossRef] [PubMed]

25. Benito-román, Ó.; Sanz, M.T.; Melgosa, R.; de Paz, E.; Escudero, I. Studies of polyphenol oxidase inactivation by means of high pressure carbon dioxide (HPCD). *J. Supercrit. Fluids* **2019**, *147*, 310–321. [CrossRef]

26. Kelly, S.M.; Jess, T.J.; Price, N.C. How to study proteins by circular dichroism. *Biochim. Biophys. Acta - Proteins Proteomics* **2005**, *1751*, 119–139. [CrossRef] [PubMed]

27. Greenfield, N.J. Applications of circular dichroism in protein and peptide analysis. *TrAC - Trends Anal. Chem.* **1999**, *18*, 236–244. [CrossRef]

28. Barteri, M.; Diociaiuti, M.; Pala, A.; Rotella, S. Low frequency ultrasound induces aggregation of porcine fumarase by free radicals production. *Biophys. Chem.* **2004**, *111*, 35–42. [CrossRef] [PubMed]

29. Yu, Z.L.; Zeng, W.C.; Lu, X.L. Influence of ultrasound to the activity of tyrosinase. *Ultrason. Sonochem.* **2013**, *20*, 805–809. [CrossRef] [PubMed]

30. Davis, B.J. Disc Electrophoresis ? II Method and Application to Human Serum Proteins. *Ann. N. Y. Acad. Sci.* **1964**, *121*, 404–427. [CrossRef] [PubMed]

31. Aydemir, T. Partial purification and characterization of polyphenol oxidase from artichoke (Cynara scolymus L.) heads. *Food Chem.* **2004**, *87*, 59–67. [CrossRef]

32. Bradford, M.M. A Rapid and Sensitive Method for the quantitation of Mocrigram Quantities of Protein Utilizing the Principle of Protein-Dye Binding. *Anal. Biochem.* **1976**, *72*, 248–254. [CrossRef]

33. Liu, S.; Murtaza, A.; Liu, Y.; Hu, W.; Xu, X.; Pan, S. Catalytic and Structural Characterization of a Browning-Related Protein in Oriental Sweet Melon (Cucumis Melo var. Makuwa Makino). *Front. Chem.* **2018**, *6*, 1–11. [CrossRef] [PubMed]

34. Murtaza, A.; Iqbal, A.; Linhu, Z.; Liu, Y.; Xu, X.; Pan, S.; Hu, W. Effect of high-pressure carbon dioxide on the aggregation and conformational changes of polyphenol oxidase from apple (Malus domestica) juice. *Innov. Food Sci. Emerg. Technol.* **2019**. [CrossRef]

35. Iqbal, A.; Murtaza, A.; Muhammad, Z.; Elkhedir, A.; Tao, M.; Xu, X. Inactivation, Aggregation and Conformational Changes of Polyphenol Oxidase from Quince (Cydonia oblonga Miller) Juice Subjected to Thermal and High-Pressure Carbon Dioxide Treatment. *Molecules* **2018**, *23*, 1743. [CrossRef] [PubMed]

12

Chemical Modification of Sweet Potato β-amylase by Mal-mPEG to Improve its Enzymatic Characteristics

Xinhong Liang, Wanli Zhang, Junjian Ran, Junliang Sun *, Lingxia Jiao, Longfei Feng and Benguo Liu

School of Food Science, Henan Institute of Science and Technology, Xinxiang 453003, China; liangxinhong2005@163.com (X.L.); zwl6996468@126.com (W.Z.); ranjunjian@126.com (J.R.); jiaolingxia@163.com (L.J.); flf18738380903_007@163.com (L.F.); zzgclbg@126.com (B.L.)
* Correspondence: sjl@hist.edu.cn.

Academic Editor: Stefano Serra

Abstract: The sweet potato β-amylase (SPA) was modified by 6 types of methoxy polyethylene glycol to enhance its specific activity and thermal stability. The aims of the study were to select the optimum modifier, optimize the modification parameters, and further investigate the characterization of the modified SPA. The results showed that methoxy polyethylene glycol maleimide (molecular weight 5000, Mal-mPEG5000) was the optimum modifier of SPA; Under the optimal modification conditions, the specific activity of Mal-mPEG5000-SPA was 24.06% higher than that of the untreated SPA. Mal-mPEG5000-SPA was monomeric with a molecular weight of about 67 kDa by SDS-PAGE. The characteristics of Mal-mPEG5000-SPA were significantly improved. The K_m value, V_{max} and Ea in Mal-mPEG5000-SPA for sweet potato starch showed that Mal-mPEG5000-SPA had greater affinity for sweet potato starch and higher speed of hydrolysis than SPA. There was no significant difference of the metal ions' effect on Mal-mPEG5000-SPA and SPA.

Keywords: sweet potato β-amylase (SPA) 2; methoxy polyethylene glycol maleimide (Mal-mPEG) 3; chemical modification 4; enzymatic characteristics

1. Introduction

β-amylase (E.C. 3.2.1.2) is an exo-type saccharifying enzyme that can act on the α-1,4 glucosidic bonds and cleave off maltose units at the non-reducing end of starch molecules [1]. The cleave-off process is accompanied by Walden inversion that turns the product from α-maltose into β-maltose [2,3]. β-amylase exists widely in higher plants such as sweet potato, barley, wheat, and soybea [4], and is mainly applied in the industrial process including food, fermentation, textiles, pharmaceuticals, etc. [5,6]. Sweet potato β-amylase (SPA) is an important component of protein in sweet potato tubers next only to sporamin, and is primarily obtained by extraction and separation from the waste water of the sweet potato starch production [7]. As a bio-active biomacromolecule, SPA's biological activity and thermal stability are among the key factors that limit its application in the food industry.

Chemical modification of molecules is an effective means to increase enzymatic stability and biological activity. It can also effectively prolong the half-life of enzymes. Polyethylene glycol (PEG) is a good non-irritant amphipathic organic solvent without immunogenicity, antigenicity and toxicity [8,9]. The studies made by Abuehowski et al. as early as 1977 showed that proteins modified by PEG were of greater efficacy than the unmodified ones [10]. Meanwhile, through enzyme modification, the antigenicity of particular enzymes can be reduced or eliminated, and the enzymatic stability can be enhanced. Therefore, the techniques of PEG-modified proteins have been developing rapidly [11,12]. Presently, over 10 types of PEG-modified proteins have been certified by the US Food and Drug

Administration [13]. However, due to deficiencies such as frequent occurrence of crosslinking and agglomeration of -OH at both ends of PEG, its usage in protein modification is limited.

With similar properties to PEG, methoxy polyethylene glycol (mPEG) also works as a protein modifier with the active hydroxyl group at one end of PEG being blocked by the methoxy group. Modification through mPEG can change the relevant biological characteristics of enzymes or proteins, including hydrophobicity, surface charge, stability and water solubility [13–15]. It is reported that the activity, thermal stability and pH stability of enzymes modified by mPEG can be enhanced. Using mPEG to modify neutral protease changed the thermal stability of the enzyme, and the modified enzyme showed higher affinity for substrate [16]. Fang et al. [17] modified phospholipase C by using mPEG-succinimidyl succinate ester. The results indicated an increase by three times in the catalytic efficiency of the modified phospholipase C, and also higher thermal stability. Daba et al. [18] adopted glutaraldehyde (GA), mPEG chlorotriazine and trinitro-benzene-sulfonic acid (TNBS) to modify β-amylase in malt. The result indicated that mPEG chlorotriazine enhanced the enzymatic activity and thermal stability of β-amylase in malt. However, there are very few reports about mPEG modified SPA.

Methoxy polyethylene glycol N-hydroxylsuccinimide ester (NHS-mPEG5000, NHS-mPEG20000), Methoxy polyethylene glycol tosylate (Ts-mPEG5000, Ts-mPEG5000, Ts-mPEG10000, Ts-mPEG20000) and Methoxy polyethylene glycol maleimide (Mal-mPEG5000) were adopted in this study for chemical modification on SPA to screen the optimum modifier. The response surface method was applied to optimize the molar ratio of the optimum modifier to SPA, as well as the modification temperature, pH value and other parameters. The enzymatic properties of the modification enzyme under optimal parameters were studied to improve its catalytic activity and thermal stability, so as to help lay the foundation for its industrial application.

2. Results and Discussion

2.1. Screening of Modifiers

The change in SPA specific activity modified by 6 different modifiers was shown in Figure 1. Compared with the untreated SPA, Mal-mPEG5000, NHS-mPEG5000 and Ts-mPEG5000 could significantly increase SPA's specific activity ($p < 0.05$) in the same molar ratio. The specific activity of Mal-mPEG5000-SPA reached the maximum value at $(2.081 \pm 0.050) \times 10^4$ U/mg. There was a significant difference between Mal-mPEG5000 and other modifiers ($p < 0.05$). However, there was no significant difference between Ts-mPEG10000, Ts-mPEG20000 and NHS-mPEG20000 ($p > 0.05$). Therefore, Mal-mPEG5000 was selected as the optimal modifier for the next experiments.

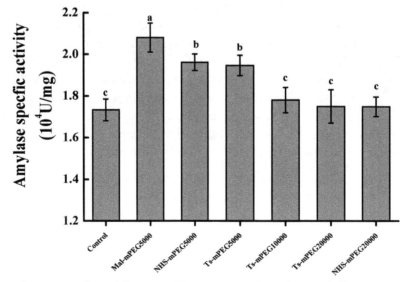

Figure 1. Effect of 6 types of modifiers on SPA activity. Significant difference in each column are expressed as different superscript letters ($p < 0.05$).

2.2. Effect of the Molar Ratio on Modification

The results from Figure 2 indicated that as the molar ratio rose from 1:1 to 1:4, increasingly high SPA specific activity was found along with the increase of Mal-mPEG5000 concentration. The enzymatic specific activity reached the maximum $(2.143 \pm 0.050) \times 10^4$ U/mg when the molar ratio was 1:4, an increase by 23.66% compared with that of the untreated SPA. As the concentration of Mal-mPEG5000 increased further, and the molar ratio increased to 1:6, the enzymatic specific activity decreased to $(2.016 \pm 0.051) \times 10^4$ U/mg, and there was a significant difference between the molar ratio of 1:4 and 1:6 ($p < 0.05$). Therefore, the optimal response molar ratio of SPA to Mal-mPEG5000 was concluded to be 1:4.

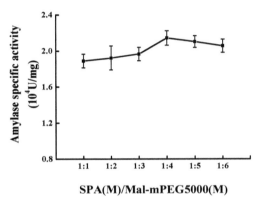

Figure 2. Effect of the molar ratio of SPA to Mal-mPEG5000 on SPA activity.

2.3. Effect of Temperature on Modification

The results from Figure 3 indicated that the Mal-mPEG5000-SPA specific activity gradually increased as the temperature rose from 25 °C to 55 °C. The specific activity reached the maximum $(2.131 \pm 0.059) \times 10^4$ U/mg at 55 °C. As the temperature increased to 65 °C, the specific activity dropped to $(1.801 \pm 0.055) \times 10^4$ U/mg, and decreased by 15.49%, and there was a significant difference between 55 °C and 65 °C ($p < 0.05$). Therefore, the optimal modification temperature was concluded to be 55 °C.

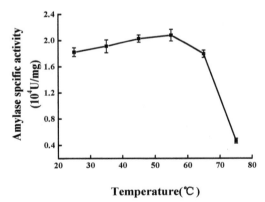

Figure 3. Effect of modification temperature on modification.

2.4. Effect of pH on Modification

The results from Figure 4 indicated that the Mal-mPEG5000-SPA specific activity gradually increased as the pH moved from 3 to 6. The specific activity reached the maximum $(2.155 \pm 0.046) \times 10^4$ U/mg at pH 6.0. As pH increased to 7.0, the specific activity was $(1.925 \pm 0.045) \times 10^4$ U/mg and decreased by 10.69%, and there was a significant difference between pH 6.0 and 7.0 ($p < 0.05$). Therefore, the optimal modification pH was concluded to be 6.0.

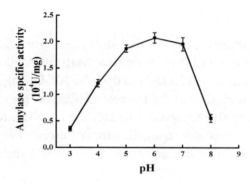

Figure 4. Effect of modification pH values on modification.

2.5. Effect of Time on Modification

The results from Figure 5 indicated that the Mal-mPEG5000-SPA specific activity increased as the modification time rose from 5 min to 10 min. The specific activity reached the maximum (2.142 ± 0.059) $\times 10^4$ U/mg at 10 min. As the modification time further increased, no significant change was noted in the enzymatic specific activity ($p = 0.05$). The ANVOA results showed no significant difference of enzymatic specific activity between 10 min and 30 min. Therefore, taking into account of practical application, the optimal modification time was concluded to be 10 min.

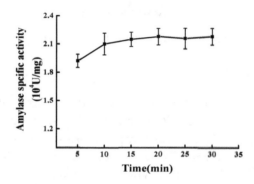

Figure 5. Effect of modification time on modification.

2.6. Optimizing the Modification Procedure

According to the effect of the molar ratio, modification temperature, pH and time, the time has little effect on modification, and modification pH (A), the molar ratio (B) and modification temperature (C) were selected as three factors to optimize the modification procedure by central composite design (CCD). With modification time set at 10 min, and Mal-mPEG5000-SPA specific activity (Y) as the response value, Box-Behnken design principles were followed to conduct a test of three factors and three levels. Results were shown in Table 1.

Table 1. Experimental design and results for the activity of Mal-mPEG5000-SPA.

Runs	Coded Levels			Response Value
	pH	Molar Ratio	Temperature (°C)	($\times 10^4$ U/mg)
1	−1	−1	0	1.7763
2	0	−1	−1	1.8858
3	0	0	0	2.1592
4	0	0	0	2.1481
5	1	0	−1	1.8142
6	0	1	−1	1.8667
7	0	−1	1	1.8758
8	1	−1	0	1.9097

Table 1. *Cont.*

| Runs | Coded Levels | | | Response Value |
	pH	Molar Ratio	Temperature (°C)	($\times 10^4$ U/mg)
9	0	0	0	2.1323
10	−1	1	0	1.8620
11	−1	0	1	1.8226
12	0	0	0	2.1597
13	0	0	0	2.1326
14	0	1	1	1.9097
15	1	0	1	1.9097
16	1	1	0	1.8619
17	−1	0	−1	1.8362

Multiple regression analysis was adopted to the experimental data by using the software Design-Expert V8.0.6. The response, Y (Mal-mPEG5000-SPA specific activity), was selected as the test variables, and by the second order, a polynomial equation was developed (Equation (1)).

$$Y = 2.15 + 0.025A + 0.00528B + 0.016C - 0.033AB + 0.027AC + 0.0011BC - 0.17A^2 - 0.13B^2 - 0.13C^2 \quad (1)$$

The analysis of variance (ANOVA) for the response Y, the Mal-mPEG5000-SPA specific activity, was shown in Table 2. The regression model was highly significant ($p < 0.01$), while the lack of fit was not significant ($p = 0.317 > 0.05$) and the value of the determination coefficient (R^2) was 0.996, which indicated the goodness of fit of the regression model [19]. Based on the analysis of experimental data in Tables 1 and 2, the regression model demonstrated a high correlation, and can be used as theoretical prediction for the enzymatic specific activity of Mal-mPEG5000 modified SPA. The significance test on the regression model suggested that the effect on enzymatic specific activity was as follows: Modification pH > modification temperature > Molar ratio of SPA to Mal-mPEG5000.

Table 2. Analysis of variance (ANOVA) for the experimental results.

Source	Sum of Squares	df	Mean Square	F-Value	*p*-Value	Significance
Model	0.3	9	0.034	145.34	<0.0001	**
A	4.91×10^{-3}	1	4.91×10^{-3}	21.05	0.0025	**
B	2.23×10^{-4}	1	2.23×10^{-4}	0.96	0.3606	
C	1.95×10^{-3}	1	1.95×10^{-3}	8.36	0.0233	*
AB	4.48×10^{-3}	1	4.48×10^{-3}	19.23	0.0032	**
AC	2.98×10^{-3}	1	2.98×10^{-3}	12.76	0.0091	**
BC	4.62×10^{-4}	1	4.62×10^{-4}	1.98	0.2022	
A2	0.12	1	0.12	507.79	<0.0001	**
B2	0.067	1	0.067	288.2	<0.0001	**
C2	0.074	1	0.074	319.52	<0.0001	**
Residual	1.63×10^{-3}	7	2.33×10^{-4}			
Lack of fit	8.97×10^{-4}	3	2.99×10^{-4}	1.63	0.317	
Pure error	7.35×10^{-4}	4	1.84×10^{-4}			
Cor total	0.31	16				
R2	0.996					

* indicate significant ($p < 0.05$); ** indicate highly significant ($p < 0.01$).

In order to further comprehend the interaction between the parameters, the response surfaces were obtained using Equation (2), which were plotted between two independent variables and the other independent variable was set at the zero-coded level. The analysis on the response was shown in Figure 6a–c.

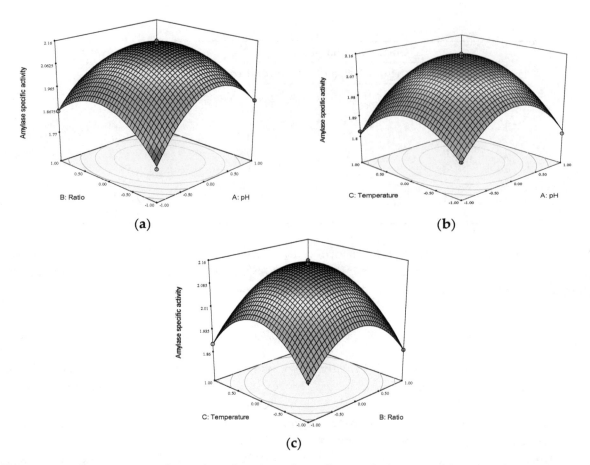

Figure 6. Response surface plot showing the effects of the variables on the activity of Mal-mPEG5000-SPA. The three independent variables set were modification pH (**a**), the molar ratio of SPA to Mal-mPEG5000 (**b**) and modification temperature (**c**).

Figure 6a indicated that with the increase of the modification pH and the molar ratio, the Mal-mPEG5000-SPA specific activity first increased, and then declined, meanwhile, steep response surfaces and oval contour line were found, which indicated significant interaction of the pH value and the molar ratio.

Figure 6b suggested that with the increase of modification pH and temperature, the Mal-mPEG5000-SPA specific activity firstly increased, and then decreased, and the oval contour line was noticed, which demonstrated a significant interaction of modification pH value temperature.

Figure 6c revealed that with the increase of the molar ratio and modification temperature, the Mal-mPEG5000-SPA specific activity increased firstly before it declined, and the occurrence of circular contour line showed no significant interaction of the molar ratio and modification temperature.

Through the analysis by the software Design-Expert, the optimized combination for maximum Mal-mPEG5000-SPA specific activity was determined through canonical analysis of the response surfaces and contour plots as A = 6.08, B = 4.04 and C = 58.85 °C, i.e., the predicted value of Mal-mPEG5000-SPA specific activity of 2.247×10^4 U/mg. under modification pH 6.08, the molar ratio 1:4.04, and modification temperature 58.85 °C. Considering the practicality of the validation test, the optimum parameters were corrected into modification pH 6.0, the molar ratio 1:4, and modification temperature 58 °C. The specific activity of Mal-mPEG5000-SPA was determined under the corrected parameters to be $(2.150 \pm 0.055) \times 10^4$ U/mg, an increase by 24.06% than that of the untreated one. The validation test showed that the optimization results from the response surface method were reliable, and adopting response surface method to optimize the modification process of Mal-mPEG5000 on SPA was feasible.

2.7. Separation and Purification of Mal-mPEG5000-SPA

Mal-mPEG5000-SPA was separated and purified by AKTA purifier^{TM}10 with Superdex^{TM}75 gel column (Figure 7a–c). Separated by Superdex^{TM}75 gel column, there were three protein peaks: S_1, S_2 and S_3 (Figure 7a), and the eluents in the collection tubes corresponding to the peak position were collected and determined for enzymatic specific activity. The specific activity of S_1 and S_2 was $(1.245 \pm 0.047) \times 10^4$ U/mg and $(0.533 \pm 0.036) \times 10^4$ U/mg respectively as shown in Figure 7c. As the peak of SPA elution was noted around 17 min according to the result of the pre-test, the elution peak S_1 was collected, centrifuged and concentrated for further purification.

Mal-mPEG5000-SPA was separated for the second time using Superdex^{TM}75 gel column, and there were elution peaks S_{11} and S_{12} (Figure 7b). The specific activity of S_{11} and S_{12} was determined to be $(1.422 \pm 0.057) \times 10^4$ U/mg and $(0.061 \pm 0.009) \times 10^4$ U/mg respectively (Figure 7d). According to the separation principles of gel column chromatography, different positions of the peaks suggest different molecular weight of the protein under each peak. The position of SPA peak was around 17 min, while the position of Mal-mPEG5000-SPA was about 7 min, which indicated change in molecular weight as SPA was modified by Mal-mPEG5000.

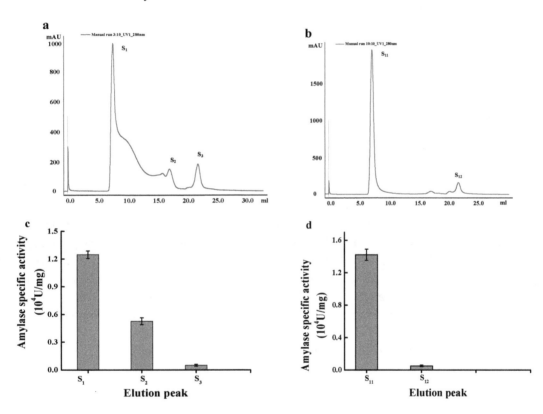

Figure 7. Separation curve of Mal-mPEG5000-SPA by gel column Superdex^{TM}75 (**a**)Elution profile of Mal-mPEG5000-SPA by Superdex^{TM}75 for the first time; (**b**) Elution profile of Mal-mPEG5000-SPA by Superdex^{TM}75 for the second time; (**c**) The specific activity of the elution fractions from Superdex^{TM}75 for the first time; (**d**) The specific activity of the elution fractions from Superdex^{TM}75 for the second time.

SDS-PAGE gel electrophoresis was carried out on the collected elution peaks S_1 and S_{11}. The results were shown in Figure 8a,b. An obvious band of S_1 was found above the SPA band as indicated in Figure 8a. The band was of Mal-mPEG5000-SPA, whose molecular weight was about 67 kDa compared with the standard marker. Figure 8b demonstrated the electrophoretogram of elution peak S_{11}. A single band was noted after separation twice by Superdex^{TM}75, and Mal-mPEG5000-SPA was obtained through purification.

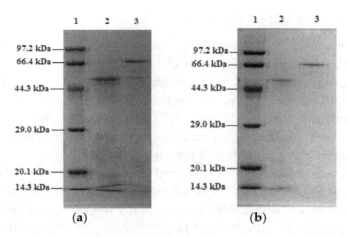

Figure 8. SDS-PAGE electrophoresis spectra of Mal-mPEG5000-β-SPA (**a**) separation by Superdex™75 for the first time; Lane 1: marker proteins; Lane 2: SPA; Lane 3: Mal-mPEG5000-SPA by Superdex™75. (**b**) separation by Superdex™75 for the second time; Lane 1: marker proteins; Lane 2: SPA; Lane 3: Mal-mPEG5000-SPA by Superdex™75.

2.8. Optimum Temperature and Thermal Stability of Mal-mPEG5000-SPA and SPA

According to Figure 9a, the maximum specific activity of SPA was found to be $(1.733 \pm 0.050) \times 10^4$ U/mg at 55 °C, while the maximum of Mal-mPEG5000-SPA was $(2.121 \pm 0.058) \times 10^4$ U/mg at 45 °C, an increase by 22.38% compared to the untreated one. Moreover, the activities of Mal-mPEG5000-SPA at 50 °C, 55 °C and 60 °C were determined as $(2.074 \pm 0.060) \times 10^4$ U/mg, $(2.057 \pm 0.062) \times 10^4$ U/mg, and $(2.031 \pm 0.064) \times 10^4$ U/mg respectively, all significantly higher than that of untreated SPA ($p < 0.05$). These results indicated that the enzymatic specific activity of modification enzyme was significantly improved after Mal-mPEG5000 modification.

Figure 9b showed the thermal stability of Mal-mPEG5000-SPA and SPA. No significant difference in SPA specific activity at 20 °C and 25 °C for 1 h was noted. However, when the temperature rose to 30 °C, the specific activity of SPA was significantly reduced ($p < 0.05$), which indicated high sensitiveness of SPA to temperature. No significant difference in the specific activity of Mal-mPEG5000-SPA from 20 °C to 45 °C for 1 h was found. But when the temperature rose to 50 °C, the specific activity of Mal-mPEG5000-SPA started to decline ($p < 0.05$), which suggested significant improvement of the thermal resistance of Mal-mPEG5000-SPA. These results showed that hydrolyzed starch by Mal-mPEG5000-SPA has a wider range of application in the food industry, which can significantly enhance enzymatic efficiency to reduce production costs.

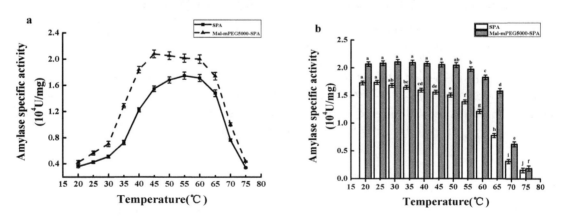

Figure 9. Temperature profiles of Mal-mPEG5000- SPA and SPA. (**a**) Optimum enzymolysis temperature; (**b**) Thermal stability. Significant difference in each column are expressed as different superscript letters ($p < 0.05$).

2.9. Optimum pH and pH Stability of Mal-mPEG5000-SPA and SPA

Figure 10a showed that Mal-mPEG5000-SPA and SPA both showed relatively high specific activity from pH 5.0 to 7.0. At pH 6.0, the specific activity of Mal-mPEG5000-SPA and SPA reached respective maximum values at $(2.061 \pm 0.051) \times 10^4$ U/mg and $(1.733 \pm 0.050) \times 10^4$ U/mg, and the specific activity was improved by 18.92% after modification.

Figure 10b showed the pH stability of Mal-mPEG5000-SPA and SPA. A significant difference in SPA's specific activity at pH 6.0 and 6.5 was found ($p < 0.05$), while the difference for Mal-mPEG5000-SPA was not significant ($p = 0.05$), which indicated higher adaptability of Mal-mPEG5000-SPA to a pH environment. In addition, the specific activity of Mal-mPEG5000-SPA was constantly higher than that of SPA under pH value from 4.0 to 7.5, which indicated lower susceptibility of Mal-mPEG5000-SPA to pH environment. This means Mal-mPEG5000-SPA could applied more broadly and is more suitable for industrial application.

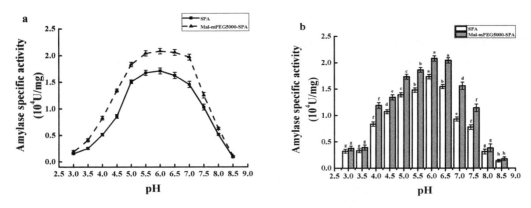

Figure 10. pH profiles of Mal-mPEG5000- SPA and SPA. (**a**) Optimum enzymolysis pH; (**b**) pH stability. Significant difference in each column are expressed as different superscript letters ($p < 0.05$).

2.10. Kinetic Parameters of Mal-mPEG5000-SPA

The results of kinetic parameters of Mal-mPEG5000-SPA were shown in Table 3, the K_m value for sweet potato starch, potato starch, corn starch, soluble starch, amylase and amylopectin by Mal-mPEG5000-SPA hydrolysis was respectively (1.63 ± 0.033) mg/mL, (2.06 ± 0.028) mg/mL, (2.36 ± 0.063) mg/mL, (1.84 ± 0.025) mg/mL, (2.18 ± 0.029) mg/mL and (2.11 ± 0.052) mg/mL. The V_{max} value was respectively (32.06 ± 0.61) mmol/min/mL, (16.23 ± 0.32) mmol/min/mL, (10.66 ± 0.37) mmol/min/mL, (20.88 ± 0.78) mmol/min/mL, (13.35 ± 0.38) mmol/min/mL, (13.89 ± 0.41) mmol/min/mL. The E_a value was respectively (11.07 ± 0.43) kJ/mol, (18.24 ± 1.12) kJ/mol, (26.52 ± 1.21) kJ/mol, (14.71 ± 1.15) kJ/mol, (21.16 ± 1.12) kJ/mol and (21.66 ± 0.8) kJ/mol. These results indicated that Mal-mPEG5000-SPA had the lowest Michaelis constant for sweet potato starch. Therefore, Mal-mPEG5000-SPA was determined to have the strongest binding affinity for sweet potato starch. Next only to sweet potato starch, soluble starch was followed by potato starch, amylase and amylopectin. Compared with the SPA kinetic parameters reported in earlier studies [20], for sweet potato starch hydrolyzed by Mal-mPEG5000-SPA, K_m declined by 12.95%, V_{max} increased by 26.87%, and Ea dropped by 12.63%, which suggested Mal-mPEG5000-SPA showed stronger affinity for sweet potato starch than SPA.

Table 3. Kinetic and activation energy parameters of Mal-mPEG5000-SPA.

Substrates	K_m (mg/mL)	V_{max} (mmol/min/mL)	E_a (kJ/mol)
Sweet potato starch	1.87 ± 0.032 [a]	19.32 ± 0.65 [e]	12.67 ± 0.73 [a]
Potato starch	2.16 ± 0.031 [c]	14.62 ± 0.35 [c]	19.84 ± 1.03 [c]
Corn starch	2.56 ± 0.053 [e]	9.36 ± 0.27 [a]	28.23 ± 1.37 [e]

Table 3. *Cont.*

Substrates	K_m (mg/mL)	V_{max} (mmol/min/mL)	E_a (kJ/mol)
Soluble starch	1.96 ± 0.029 [b]	16.75 ± 0.68 [d]	16.61 ± 0.99 [b]
Amylase	2.32 ± 0.037 [d]	12.85 ± 0.42 [b]	22.26 ± 1.09 [d]
Amylopectin	2.37 ± 0.045 [d]	12.53 ± 0.41 [b]	22.54 ± 0.98 [d]

Significant difference in each column are expressed as different superscript letters ($p < 0.05$).

2.11. Effect of Metal Ions on the Activity of Mal-mPEG5000-SPA

As shown in Table 4, the relative specific activity after addition of Mn^{2+}, K^+, Zn^{2+} and Ca^{2+} was respectively $(143.48 \pm 6.25)\%$, $(115.49 \pm 4.51)\%$, $(111.28 \pm 4.82)\%$ and $(102.88 \pm 4.32)\%$, which demonstrated that Mn^{2+}, K^+, Zn^{2+} and Ca^{2+} had good activation effect. Mg^{2+} showed a slight inhibitive effect on Mal-mPEG5000-SPA, with relative specific activity standing at $(95.25 \pm 4.14)\%$. However, Cu^{2+}, NH_4^+, Fe^{3+}, Al^{3+}, Ba^{2+} and EDTA demonstrated a relatively strong inhibitory effect on Mal-mPEG5000-SPA, with relative specific activity reduced to 55–80%. Meanwhile, under the effect of Cd^{2+}, Hg^+ and Ag^+, the relative specific activity were all below 30%, among which Hg^+ and Ag^+ indicated the strongest inhibitory effect on Mal-mPEG5000-SPA. Compared with the effect of metal ions on SPA reported in earlier studies [21], the effect of metal ions on Mal-mPEG5000-SPA and on SPA showed no significant difference ($p = 0.05$).

Table 4. Effect of metal ions on the activity of Mal-mPEG5000-SPA.

Metal Ions	Relative Activity (%)
Control	100
Ca^{2+}	102.88 ± 4.32 [c]
Mg^{2+}	95.25 ± 4.14 [d]
Cu^{2+}	60.95 ± 3.61 [f]
Zn^{2+}	111.28 ± 4.82 [b]
Mn^{2+}	143.48 ± 6.25 [a]
K^+	115.49 ± 4.51 [b]
NH_4^+	78.89 ± 4.05 [e]
Hg^+	5.21 ± 0.63 [i]
Ag^+	3.28 ± 0.12 [i]
Al^{3+}	45.81 ± 3.52 [g]
Fe^{3+}	80.85 ± 4.62 [e]
Ba^{2+}	59.68 ± 3.92 [f]
Cd^{2+}	22.85 ± 2.06 [h]
EDTA	55.84 ± 3.27 [f]

Significant difference in each column are expressed as different superscript letters ($p < 0.05$).

3. Materials and Methods

3.1. Materials

Sweet potato, molar weight 56.043 kDa, harvested in Xinxiang City, Henan Province, China, its cultivar was Shangshu 19. NHS-mPEG5000, NHS-mPEG20000, Ts-mPEG5000, Ts-mPEG10000, Ts-mPEG20000 and Mal-mPEG5000 were from Nanocs (New York, NY, USA; purity $\geq$ 95%). All other reagents were of analytical reagent grade and used without further purification.

3.2. Separation and Purification of SPA

Using the methods described by Liang et al. [21] with minor modification. SPA was separated and purified with the following steps. Fresh sweet potatoes were washed and sliced into chips. 100 g sweet potatoes were weighed out, and 200 mL distilled water was added for pulverization in a pulverizer

(SQ2002, Shanghai Shuaijia Electronic Technology Company, China) for 1 min. The sample was filtered through 40 mesh sieve, and the resulting filtrate was placed in a refrigerated centrifuge at 4 °C for centrifugation at 4000 rpm for 15 min. The supernatant was obtained, and ammonium sulfate was added to the supernatant to 70% saturation. The resulting solution was stored in a refrigerator at 4 °C for 4 h. The centrifuge parameters were set at 4 °C and 8000 rpm for refrigerated centrifugation for 15 min. The resulting precipitate was collected, and dissolved in a bufferA. Ammonium sulfate was desalinated by using 1 kDa ultrafiltration membrane for 4 h. Protein purifier (AKTA purifier™10, General Electric Company, Boston, Massachusetts, MA, USA) was used to purify the enzyme, and Mono Q anion exchange chromatographic column and Superdex™75 gel column were adopted for separation and purification with detection wavelength set at 280 nm. The buffers for purification were as follows. Buffer A: 20 mM pH 5.8 disodium-hydrogen phosphate-citric acid, Buffer B: 1 mol/L NaCl. The buffers were filtered through a 0.22 μm membrane, ultrasound treated for 20 min, and stored in a refrigerator at 4 °C. Buffer A was used to equilibrate the chromatographic column. The sample was injected after equilibration of buffer to the baseline. The column was washed with 5 column volumes of buffer A, and then eluted with a gradient from 0–100% Buffer B at the flow rate of 1.0 mL/min, and maximum back-pressure of 4 MPa for 30 min. A collector was used to obtain 1 mL fractions. After the elution was complete, the enzymes in the collection tube corresponding to the peak position were collected, frozen and dried in a vacuum freeze dryer (Thermo Savan, Thermo Electron Co., Waltham, MA, USA) and stored at 4 °C for later usage. The separation and purification of mPEG5000-β-SPA: 70% ammonium sulfate was added into the reaction liquid to precipitate. The rest steps were the same as those of the β-amylase separation and purification.

The molecular mass and the purity of the enzyme was conducted by SDS-PAGE according to the Laemmli [22] method, and 15% (w/v) polyacrylamide gel was adopted.

3.3. Protein Content and Enzyme Assay

Protein content was measured as described by Lowry et al. [23], and bovine serum albumin was used as the standard. Enzyme assay was determined by the dinitrosalicylic acid (DNS) method according to Sagu et al. [24] with minor modification. The enzyme assay was performed at 40 °C, pH 5.8, and 1 mg maltose released per hour from 1.1% soluble starch was defined as a unit of enzyme specific activity. In this paper, the enzyme activity was indicated by specific enzyme activity, expressed as U/mg.

3.4. Screening of Modifiers

The molar ratio of SPA to modifiers was determined at 1:4. SPA and 6 different types of modifiers (NHS-mPEG5000, NHS-mPEG20000, Ts-mPEG5000, Ts-mPEG10000, Ts-mPEG20000 and Mal-mPEG5000) were placed in the buffer (disodium-hydrogen phosphate-citric acid acid, pH 6.0), respectively, and then kept in a water-bath thermostatic metal oscillator at 55 °C for 10 min. The reaction mixture was obtained for dialysis. After the dialysis was complete, the enzyme specific activity was measured to obtain the optimal modifier.

3.5. Selection of Relevant Variables and Experimental Design

SPA was modified by Mal-mPEG5000 at the molar ratio of SPA to Mal-mPEG5000 from 1:1 to 1:6, for modification temperature from 25 °C to 75 °C and for modification pH from 3.0 to 8.0. The enzyme specific activity of modified enzyme (expressed as Mal-mPEG5000- SPA) was used as the index for evaluating the effects of the different modification parameters to be optimized via response surface methodology (RSM). A central composite design (CCD) were used with three variables and three levels for optimizing the modification conditions. These three factors were modification pH (A), the molar ratio of SPA to Mal-mPEG5000 (B) and modification temperature (C) (Table 5). A second-order polynomial equation (Equation (2)) for the variables was developed:

$$Y = \alpha_0 + \alpha_1 A + \alpha_2 B + \alpha_3 C + \alpha_{11} A^2 + \alpha_{22} B^2 + \alpha_{33} C^2 + \alpha_{12} AB + \alpha_{13} AC + \alpha_{23} BC \tag{2}$$

Y is the predicted response, α_0 is the intercept; α_1, α_2, α_3, linear coefficients; α_{11}, α_{22}, α_{33}, squared coefficients; and α_{12}, α_{13}, α_{23}, interaction coefficients. Analysis of variance was used to evaluate the model's adequacy and to determine the regression coefficients and their statistical significance. The response surface contour plots showed how the independent variables interacted and how those interactions influenced the overall response [20].

Table 5. Independent variables and levels for central composite design (CCD).

Factors	Levels		
	−1	0	1
Modification pH	5.0	6.0	7.0
Molar ratio of SPA to Mal-mPEG5000	1:3	1:4	1:5
Temperature/°C	45	55	65

3.6. Enzyme Characterization

Influence of temperature on Mal-mPEG5000-SPA and SPA: The optimum temperature of Mal-mPEG5000-SPA was investigated at pH 5.8 over a temperature range of 20–75 °C. SPA and Mal-mPEG5000-SPA was incubated in 20 mM disodium-hydrogen phosphate-citric acid buffer (pH 5.8) during 60 min, respectively. The thermostability of the enzyme residual specific activity were determined under different temperature conditions (20–75 °C).

Influence of pH on Mal-mPEG5000-SPA and SPA: The optimum pH of Mal-mPEG5000-SPA was investigated at 50 °C over a pH range of 3.0–8.5. Mal-mPEG5000-SPA was preincubated for 60 min under 50 °C, the residual specific activity was determined at a pH range of 3.0–8.5 to evaluate the pH stability.

3.7. Effect of Metal Ions on Mal-mPEG5000-SPA and SPA

Incubation of the enzyme was carried out by 10 mM of metal ions salts (chlorides including Ca^{2+}, Mg^{2+}, Cu^{2+}, Zn^{2+}, Mn^{2+}, K^+, NH^{4+}, Hg^+, Al^{3+}, Fe^{3+}, Ba^{2+} and Cd^+), $AgNO_3$ and EDTA at 40 °C for 30 min, and the residual activities were determined respectively. An enzyme that did not contain metal ions was chosen as control (100%).

3.8. Kinetic Constant

The Michaelis constant (K_m) and the maximum velocity (V_{max}) of Mal-mPEG5000-SPA and SPA was defined by Lineweaver-Burk plot. Different concentrations of substrates (1–20 mg/mL) in 20 mM disodium-hydrogen phosphate-citric acid buffer pH 5.8 were used and specific activity was assessed by the DNS method. The values of K_m and V_{max} were calculated respectively based on the double reciprocal plot [25].

3.9. Activation Energy (Ea)

E_a of the enzyme was measured by Arrhenius equation ($\ln k_{cat} = \ln k_0 - Ea/RT$), and the temperature was set from 25 °C to 65 °C. Representing $1/T$ in K on the axis of x and natural log of specific activity on the axis of y, Arrhenius plot was generated. E_a was defined based on the value of the slope.

3.10. Statistical Analysis

All experiments were carried out in triplicate. SPSS 22.0 statistical software (SPSS Inc., Chicago, IL, USA) was used for the analysis of variance, and the significance test ($p < 0.05$) was performed by Duncan new complex method. All analyses were done in triplicate, and the statistical significance was determined using the mean values $\pm$ standard deviation.

4. Conclusions

In this study, we selected Mal-mPEG5000 as the best modifier from six modifiers and optimized the modification parameters by the response surface method. We used column chromatography to isolate and purify Mal-mPEG5000-SPA from the reaction solution between SPA and Mal-mPEG5000. Thereafter, the enzymatic properties were determined and compared in the presence and absence of Mal-mPEG5000. For the results, under the optimal conditions (the molar ratio of Mal-mPEG5000 to SPA, 1:4, modification temperature, 58 °C, pH 6.0), the specific activity of Mal-mPEG5000-SPA was $(2.150 \pm 0.055) \times 10^4$ U/mg, an increase by 24.06% than that of the unmodified SPA. Mal-mPEG5000-SPA was separated and purified, and a single band was noticed and its molecular weight was about 67 kDa. The enzyme properties of Mal-mPEG5000-SPA were significantly improved as the optimum temperature declined from 55 °C to 45 °C, the thermal stability and pH stability were obviously enhanced; the K_m value of sweet potato starch by Mal-mPEG5000-SPA declined by 12.95%, while V_{max} increased by 26.87%, and Ea dropped by 12.63%, which showed that Mal-mPEG5000-SPA had greater affinity for sweet potato starch and higher the speed of hydrolysis than SPA; as for Mal-mPEG5000-SPA, Mn^{2+}, K^+, Zn^{2+} and Ca^{2+} demonstrated activation effect; Mg^{2+} showed a slight inhibitory effect; Cu^{2+}, NH_4^+, Fe^{3+}, Al^{3+}, Ba^{2+}, EDTA, Hg^+ and Ag^+ had a relatively strong inhibitory effect; there was no significant difference of metal ions' effect on Mal-mPEG5000-SPA and SPA. It is concluded that Mal-mPEG5000-SPA will have a wider range of application in beer processing, maltose production and other food industries. Therefore, the studies on SPA application in food industries are theoretically and practically significant.

Author Contributions: J.S. and X.L. conceived and designed the experiments; X.L. and W.Z. performed the experiments; J.R. and L.J. analyzed the data; L.F. and B.L. provided the software and formal analysis.

References

1. Chang, C.T.; Lion, H.Y.; Tang, H.L.; Sung, H.Y. Activation, purification and properties of beta-amylase from sweet potatoes (Ipomoea batatas). *Biotechnol. Appl. Biochem.* **1996**, *126*, 120–128.
2. Dicko, M.H.; Searle-van Leeuwen, M.J.F.; Beldman, G.; Ouedraogo, O.G.; Hilhorst, O.G.; Traore, A.S. Purification and characterization of β-amylase from Curculigo pilosa. *Appl. Microbiol. Biotechnol.* **1999**, *52*, 802–805. [CrossRef]
3. Kaplan, F.; Dong, Y.S.; Guy, C.L. Roles of β-amylase and starch breakdown during temperatures stress. *Physiol. Plant.* **2010**, *126*, 120–128. [CrossRef]
4. Vajravijayan, S.; Pletnev, S.; Mani, N.; Nandhagopal, N.; Gunasekaran, K. Structural insights on starch hydrolysis by plant β-amylase and its evolutionary relationship with bacterial enzymes. *Int. J. Biol. Macromol.* **2018**, *113*, 329–337. [CrossRef] [PubMed]
5. Miao, M.; Li, R.; Huang, C.; Jiang, B.; Zhang, T. Impact of beta-amylase degradation on properties of sugary maize soluble starch particles. *Food Chem.* **2015**, *177*, 1–7. [CrossRef] [PubMed]
6. Mihajlovski, K.R.; Radovanovic, N.R.; Veljovic, D.N.; Siler-Marinkovic, S.S.; Dimitrijevic-Brankovic, S.I. Improved β-amylase production on molasses and sugar beet pulp by a novel strain Paenibacillus chitinolyticus CKS1. *Ind. Crops Prod.* **2016**, *80*, 115–122. [CrossRef]
7. Li, H.S; Oba, K. Major Soluble Proteins of Sweet Potato Roots and Changes in Proteins after Cutting, Infection, or Storage. *J. Agric. Chem. Soc. Jpn.* **1985**, *49*, 737–744.
8. Li, L.; Shang, B.; Hu, L.; Shao, R.; Zhen, Y. Site-specific PEGylation of lidamycin and its antitumor activity. *Acta Pharm. Sin. B* **2015**, *5*, 264–269. [CrossRef] [PubMed]

9. Hsieh, Y.P.; Lin, S.C. Effect of PEGylation on the Activity and Stability of Horseradish Peroxidase and L-*N*-Carbamoylase in Aqueous Phases. *Process Biochem.* **2015**, *50*, 1372–1378. [CrossRef]

10. Abuchowski, A.; van Es, T.; Palczuk, N.C.; Davis, F.F. Alteration of immunological properties of bovine serum albumin by covalent attachment of polyethylene glycol. *J. Biol. Chem.* **1977**, *252*, 3578–3581. [PubMed]

11. Pfister, D.; Morbidelli, M. Process for protein PEGylation. *J. Control. Release Off. J. Control. Release Soc.* **2014**, *180*, 134–149. [CrossRef] [PubMed]

12. Roberts, M.J.; Bentley, M.D.; Harris, J.M. Chemistry for peptide and protein PEGylation. *Adv. Drug Deliv. Rev.* **2002**, *54*, 459–476. [CrossRef]

13. Dozier, J.K.; Distefano, M.D. Site-Specific PEGylation of Therapeutic Proteins. *Int. J. Mol. Sci.* **2015**, *16*, 25831–25864. [CrossRef] [PubMed]

14. Maiser, B.; Baumgartner, K.; Dismer, F.; Hubbuch, J. Effect of lysozyme solid-phase PEGylation on reaction kinetics and isoform distribution. *J. Chromatogr. B* **2015**, *1002*, 313–318. [CrossRef] [PubMed]

15. Xu, Y.; Shi, Y.; Zhou, J.; Yang, W.; Bai, L.; Wang, S.; Jin, X.; Niu, Q.; Huang, A.; Wang, D. Structure-based antigenic epitope and PEGylation improve the efficacy of staphylokinase. *Microb. Cell Fact* **2017**, *16*, 197. [CrossRef] [PubMed]

16. Zhao, S.G.; Fang, L.M.; Wang, L.; Yin, R.C. Preparation and properties of neutral protease modified with monomethoxypolyethylene glycol. *Food Ferment. Ind.* **2014**, *40*, 160–163.

17. Fang, X.; Wang, X.; Li, G.; Zeng, J.; Li, J.; Liu, J. SS-mPEG chemical modification of recombinant phospholipase C for enhanced thermal stability and catalytic efficiency. *Int. J. Biol. Macromol.* **2018**, *111*, 1032–1039. [CrossRef] [PubMed]

18. Daba, T.; Kojima, K.; Inouye, K. Chemical modification of wheat beta-amylase by trinitrobenzenesulfonic acid, methoxypolyethylene glycol, and glutaraldehyde to improve its thermal stability and activity. *Enzyme Microb. Technol.* **2013**, *53*, 420–426. [CrossRef] [PubMed]

19. Bhardwaj, S.K.; Basu, T. Study on binding phenomenon of lipase enzyme with tributyrin on the surface of graphene oxide array using surface plasmon resonance. *Thin Solid Films* **2018**, *645*, 10–18. [CrossRef]

20. Liang, X.; Ran, J.; Sun, J.; Wang, T.; Jiao, Z.; He, H.; Zhu, M. Steam-explosion-modified optimization of soluble dietary fiber extraction from apple pomace using response surface methodology. *CyTA-J. Food* **2018**, *16*, 1–7. [CrossRef]

21. Liang, X.; Zhang, W.; Wang, Y.; Sun, J. Purification and characterization of β-amylase from sweet potato (Baizhengshu 2) tuberous roots. *Res. J. Biotechnol.* **2018**, *13*, 84–91.

22. Laemmli, U.K. Cleavage of structural proteins during the assembly of the head of bacteriophage T4. *Nature* **1970**, *227*, 680–685. [CrossRef] [PubMed]

23. Lowry, B.O.H.; Rosebrough, N.J.; Farr, A.L.; Randall, T.J. Protein measurement with Folin phenol reagent. *J. Biol. Chem.* **1951**, *193*, 265. [PubMed]

24. Sagu, S.T.; Nso, E.J.; Homann, T.; Kapseu, C.; Rawel, H.M. Extraction and purification of beta-amylase from stems of Abrus precatorius by three phase partitioning. *Food Chem.* **2015**, *183*, 144–153. [CrossRef] [PubMed]

25. He, L.; Park, S.-H.; Dang, N.D.H.; Duong, H.X.; Duong, T.P.C.; Tran, P.L.; Park, J.-T.; Ni, L. ; Park., K.-H. Characterization and thermal inactivation kinetics of highly thermostable ramie leaf β-amylase. *Enzyme Microb. Technol.* **2017**, *101*, 17–23. [CrossRef] [PubMed]

13

A New Triterpenoid Glucoside from a Novel Acidic Glycosylation of Ganoderic Acid A via Recombinant Glycosyltransferase of *Bacillus subtilis*

Te-Sheng Chang [1,†], **Chien-Min Chiang** [2,†], **Yu-Han Kao** [1], **Jiumn-Yih Wu** [3], **Yu-Wei Wu** [4,5] and **Tzi-Yuan Wang** [6,*]

[1] Department of Biological Sciences and Technology, National University of Tainan, Tainan 70005, Taiwan; mozyme2001@gmail.com (T.-S.C.); aa0920281529@gmail.com (Y.-H.K.)

[2] Department of Biotechnology, Chia Nan University of Pharmacy and Science, No. 60, Sec. 1, Erh-Jen Rd., Jen-Te District, Tainan 71710, Taiwan; cmchiang@mail.cnu.edu.tw

[3] Department of Food Science, National Quemoy University, Kinmen County 892, Taiwan; wujy@nqu.edu.tw

[4] Graduate Institute of Biomedical Informatics, College of Medical Science and Technology, Taipei Medical University, Taipei 106, Taiwan; yuwei.wu@tmu.edu.tw

[5] Clinical Big Data Research Center, Taipei Medical University Hospital, Taipei 110, Taiwan

[6] Biodiversity Research Center, Academia Sinica, Taipei 115, Taiwan

* Correspondence: tziyuan@gmail.com.

† The two authors contributed equally.

Academic Editor: Stefano Serra

Abstract: Ganoderic acid A (GAA) is a bioactive triterpenoid isolated from the medicinal fungus *Ganoderma lucidum*. Our previous study showed that the *Bacillus subtilis* ATCC (American type culture collection) 6633 strain could biotransform GAA into compound (**1**), GAA-15-*O*-β-glucoside, and compound (**2**). Even though we identified two glycosyltransferases (GT) to catalyze the synthesis of GAA-15-*O*-β-glucoside, the chemical structure of compound (**2**) and its corresponding enzyme remain elusive. In the present study, we identified BsGT110, a GT from the same *B. subtilis* strain, for the biotransformation of GAA into compound (**2**) through acidic glycosylation. BsGT110 showed an optimal glycosylation activity toward GAA at pH 6 but lost most of its activity at pH 8. Through a scaled-up production, compound (**2**) was successfully isolated using preparative high-performance liquid chromatography and identified to be a new triterpenoid glucoside (GAA-26-*O*-β-glucoside) by mass and nuclear magnetic resonance spectroscopy. The results of kinetic experiments showed that the turnover number (k_{cat}) of BsGT110 toward GAA at pH 6 (k_{cat} = 11.2 min^{-1}) was 3-fold higher than that at pH 7 (k_{cat} = 3.8 min^{-1}), indicating that the glycosylation activity of BsGT110 toward GAA was more active at acidic pH 6. In short, we determined that BsGT110 is a unique GT that plays a role in the glycosylation of triterpenoid at the C-26 position under acidic conditions, but loses most of this activity under alkaline ones, suggesting that acidic solutions may enhance the catalytic activity of this and similar types of GTs toward triterpenoids.

Keywords: ganoderic acid A; glucosyltransferase; acidic; *Bacillus subtilis*; triterpenoid

1. Introduction

Ganoderma lucidum is a medicinal fungus that has been used to improve health and prevent certain diseases in Asia for thousands of years [1]. In modern ages, many bioactive compounds such as polysaccharides and triterpenoids [2,3] were identified and isolated from *G. lucidum*. These compounds were demonstrated to be effective for anti-cancer, anti-oxidant, anti-bacterial, anti-inflammation, and immune-regulation purposes [2,3].

Unlike triterpenoids from ginseng plants—which usually exist in the glycosidic form, called ginseng saponins—very few *Ganoderma* triterpenoid glycosides have been identified, despite the existence of many triterpenoids in *G. lucidum* [4]. The glycosidic form of triterpenoids might improve the bioactivity of the triterpenoids. For example, several ginseng saponins were found to exhibit more bioactivities involved in the central nervous system, cardiovascular system, and immune functions than ginseng triterpenoid aglycones were [5]. The glycosylation of flavonoids can also increase both water solubility and flavonoid stability [6–8]. It is, therefore, worthwhile to investigate the glycosylation of *Ganoderma* triterpenoids for potential medical and clinical purposes.

In nature, glycosylation is usually catalyzed by glycosyltransferase (GT, EC 2.4.x.y), a type of enzyme that uses a nucleotide-activated sugar donor, such as uridine diphosphate (UDP)-glucose, to transfer the sugar moiety to a sugar acceptor molecule [9]. Several GTs that catalyze the glycosylation of triterpenoids have already been discovered from plants, due to the accumulating knowledge on the metabolic pathways of triterpenoid glycosides [10]. However, plant GTs are not good candidates for the biotransformation of xenobiotics (such as *Ganoderma* triterpenoids) because plant GTs are usually involved in triterpenoid biosynthesis pathways and thus have very high substrate specificity. In contrast, GTs from bacterial sources usually have lower substrate specificity and have been demonstrated to be involved in the glycosylation of ginseng triterpenoids [11].

Among the *Ganoderma lucidum* bioactive compounds, ganoderic acid A (GAA) is one of the major triterpenoids and has been shown to prevent the proliferation of cancer cells and reduce inflammation activities [12–16]. Our previous study showed that the *Bacillus subtilis* ATCC (American type culture collection) 6633 strain can biotransform GAA into one major compound (**1**) and one minor compound (**2**) (Figure 1) [17]. In addition, two GTs—BsUGT398 and BsUGT489—were identified to catalyze the biotransformation of GAA into compound (**1**), which was later identified as GAA-15-*O*-β-glucoside [17]. However, the chemical structure of the compound (**2**) and its corresponding catalyzing enzyme remain elusive. In the present study, a GT enzyme that catalyzes the biotransformation of GAA to compound (**2**) was successfully identified, along with the optimal condition for producing compound (**2**) by the GT enzyme. The chemical structure of the previously-unknown compound (**2**) was also elucidated with the scaled-up production of the GT enzyme under an acidic condition.

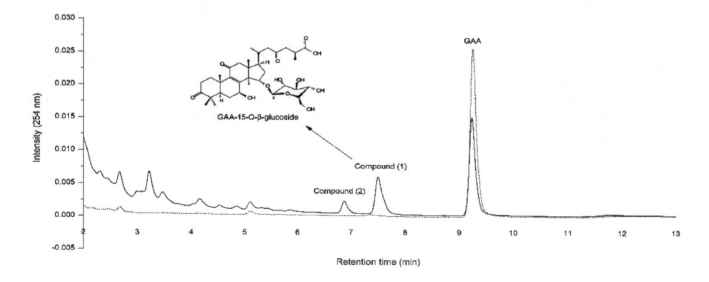

Figure 1. Biotransformation of ganoderic acid A (GAA) by *Bacillus subtilis* ATCC 6633 after 24 h of incubation (solid line). The figure was modified from Figure 1 of our previous study [17].

2. Results

2.1. Biotransformation of GAA by Recombinant BsGT110 from B. subtilis ATCC 6633

Our previous study showed that *B. subtilis* ATCC 6633 can biotransform GAA primarily into one major compound (**1**), GAA-15-O-β-glucoside, and one unknown minor compound (**2**) (Figure 1) [17]. To obtain enough unknown compound (**2**) through in vitro enzymatic biotransformation and then identify that compound's chemical structure, we strived to identify corresponding GT enzymes from the *B. subtilis* ATCC 6633 strain. In our previous work, we selected five GT genes—BsGT110, BsGT292, BsGT296, BsUGT398, and BsUGT489—and successfully overexpressed and purified them in *Escherichia coli* [17]. However, none of them were found to catalyze the biotransformation of GAA into compound (**2**) under a general GT reaction condition: 10 mM Mg^{2+}, 40 °C, and pH 8 [17]. We assayed the five recombinant BsGTs under different pH values and determined that BsGT110 produces a reasonable amount of compound (**2**) from the biotransformation of GAA under an acidic condition (pH 6), as shown in the solid line in Figure 2a. BsUGT398 and BsUGT489 produced only small amounts of compound (**2**) under the acidic condition (pH 6) (solid lines in Figure 2b,c). As expected, compound (**2**) was no longer produced from the biotransformation of GAA by any of the three GTs at pH 8 (dashed lines in Figure 2a–c). BsUGT398 and BsUGT489 produced large amounts of GAA-15-O-β-glucoside at pH 8. However, no metabolite was detected from the reactions with BsGT292 and BsGT296 at pH 6 or pH 8 (data not shown). We thus selected BsGT110 to produce compound (**2**) at pH 6 for further analysis. The amount of GAA-15-O-β-glucoside and compound (**2**) that can be catalyzed from GAA by BsGT110 at different pH values were indicated in Table 1. It is noted that the maximum amount of compound (**2**) was produced under 1 mg/mL GAA, 10 mM UDP-glucose, 15 µg/mL BsGT110, 10 mM $MgCl_2$, and 50 mM acetate buffer at pH 6.

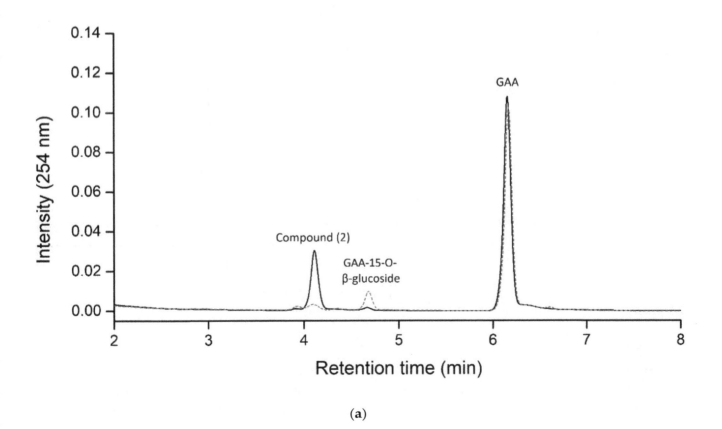

(a)

Figure 2. *Cont.*

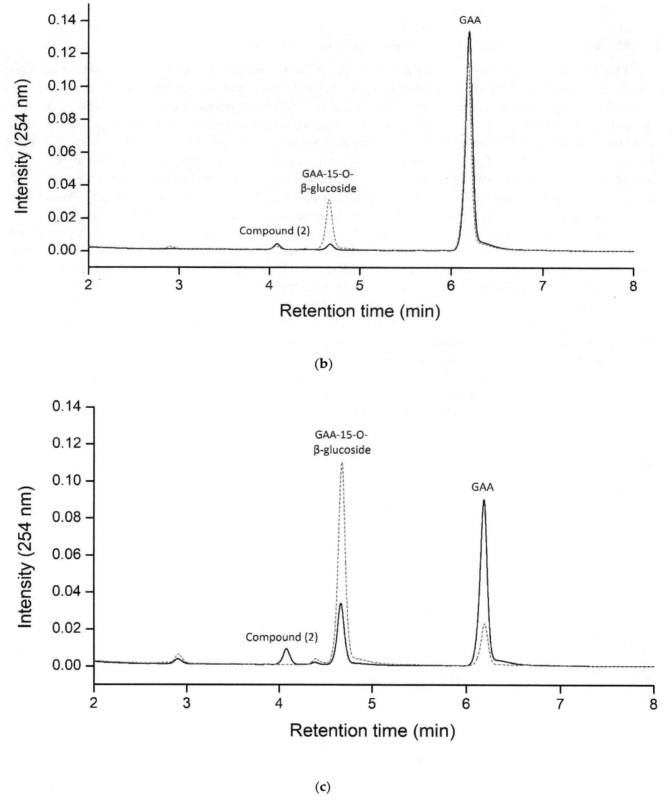

Figure 2. Ultra-performance liquid chromatography (UPLC) analysis of the biotransformation of GAA by BsGT110 (**a**), BsUGT398 (**b**), and BsUGT489 (**c**) at pH 6 (solid line) and pH 8 (dashed line). The biotransformation mixture contained 15 µg/mL purified enzyme, 1 mg/mL GAA, 10 mM uridine diphosphate (UDP)-glucose, 10 mM $MgCl_2$, and 50 mM acetate buffer at pH 6 or phosphate buffer (PB) at pH 8 and was incubated at 40 °C for 30 min. After incubation, the reaction was analyzed using UPLC. The UPLC operation procedure was described in the Materials and Methods section.

Table 1. Relative production [a] of GAA-15-O-β-glucoside and compound (**2**) catalyzed from GAA by BsGT110.

pH Value	Production of GAA-15-O-β-glucoside	Production of Compound (2)
5 [b]	1.13 ± 0.13	87.96 ± 4.67
6 [b]	4.31 ± 0.27	100.00 ± 7.85 [a]
6 [c]	4.85 ± 0.15	84.53 ± 9.49
7 [c]	15.26 ± 0.64	30.81 ± 0.71
8 [c]	34.02 ± 0.94	16.04 ± 1.04

[a] Relative production was normalized to the UPLC area of the peak of compound (**2**) in an acetate buffer of pH 6.
[b] 50 mM of acetate buffer. [c] 50 mM of PB.

To optimize the production of compound (**2**), a standard mixture was made of 1 mg/mL GAA, 10 mM UDP-glucose, 15 µg/mL BsGT110, and 50 mM acetate buffer at pH 6 under different temperature and metal ion conditions. After incubation, the amount of compound (**2**) produced was determined with UPLC (Figure 3). The results revealed that the optimal conditions for the production of compound (**2**) from GAA by the recombinant BsGT110 is pH 6, 40 °C, and 10 mM $MgCl_2$. The relative production of GAA-15-O-β-glucoside was less than 5% for all testing conditions.

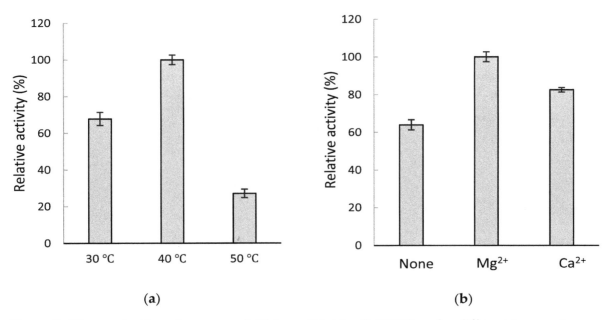

(a)   (b)

Figure 3. The production of compound (**2**) from GAA by BsGT110 under different temperature or metal ion conditions. The standard condition was set as 15 µg/mL purified enzyme, 1 mg/mL GAA, 10 mM $MgCl_2$, and 10 mM UDP-glucose in 50 mM acetate buffer at pH 6.0 and 40 °C. The tests were carried out by changing only the temperature (**a**) or metal ions (**b**) and maintaining all other settings. Relative activities were obtained by dividing the area summation of the UPLC reaction peak of the test condition by that of the standard condition. The data are expressed as mean ± SD, $n = 3$.

2.2. Identification of the Biotransformation Product

To resolve the chemical structure of compound (**2**), the biotransformation was scaled up to 25 mL, with 1 mg/mL GAA, 15 µg/mL BsGT110, 10 mM $MgCl_2$, and 10 mM UDP-glucose in 50 mM acetate buffer of pH 6 and 40 °C for a 30-min incubation. A total of 5.4 mg of compound (**2**) in the 25-mL reaction was purified with preparative high-performance liquid chromatography (HPLC). The chemical structure of the purified compound was then resolved using mass and nucleic magnetic resonance (NMR) spectral analyses. The molecular formula of compound (**2**) was established as $C_{36}H_{53}O_{12}$ by the electrospray ionization mass (ESI-MS) at m/z 679.67 $[M + H]^+$, indicating the presence of a glucose residue. The NMR spectra exhibit characteristic glucosyl signals: the anomeric carbon signal at δ_C 95.9, one CH_2 signal at δ_C 61.8, and four CH signals at δ_C 70.6, 73.9, 78.2, 79.1. The large coupling constant (8.1 Hz) of the anomeric proton H-1' (6.33 ppm) indicated the β-configuration. The cross peak of H-1' with C-26 (6.33/174.6 ppm) in the HMBC spectrum demonstrated the structure of compound (**2**) to be GAA-26-*O*-β-glucoside. The NMR spectra data are shown in Table S1 and Figures S1–S4. Figure 4 illustrated the biotransformation process of GAA by BsGT110.

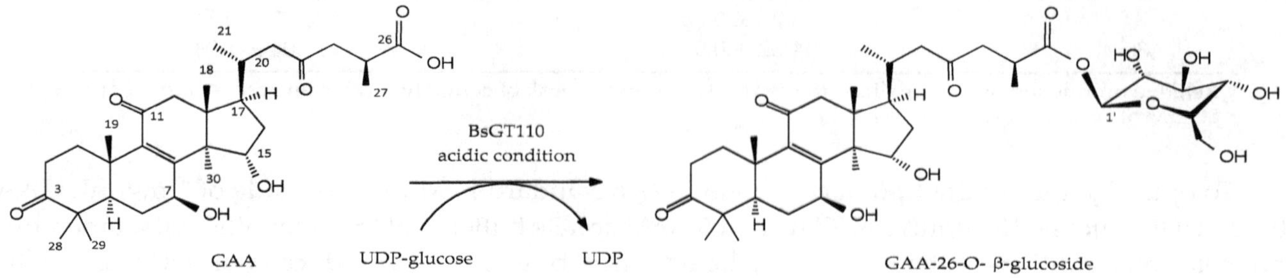

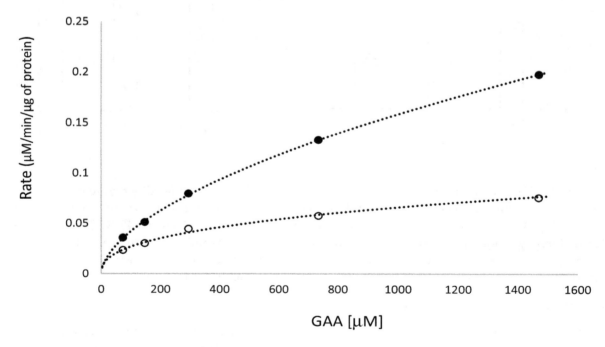

Figure 4. Biotransformation of GAA by BsGT110 in the acidic condition.

2.3. Kinetic Study of the Biotransformation of GAA by BsGT110

To study how pH affects the biotransformation activity of GAA by BsGT110, a kinetic study of the biotransformation was conducted at different concentrations of GAA, with 50 mM acetate buffer at pH 6 or PB at pH 7, 10 mM $MgCl_2$, and 10 mM UDP-glucose, at 40 °C. The results of the kinetic study were shown in Figure 5, and the calculated kinetic parameters were listed in Table 2. The results showed that the turnover number (k_{cat}) of BsGT110 toward GAA at pH 6 was 3-fold higher than that at pH 7.

Figure 5. Kinetic study of BsGT110 at pH 6 (closed symbols) and pH 7 (open symbols). Different concentrations of GAA were mixed with 15 µg purified BsGT110 protein, 10 mM UDP-glucose, 10 mM $MgCl_2$, and 50 mM PB (pH 7.0) or acetate buffer (pH 6) in 1 mL reaction mixture and incubated at 40 °C for 20 min. During the incubation, samples from each reaction were removed and analyzed by UPLC every 2 min. The reaction rate for each concentration of GAA was obtained from the slope of the plot of the amount of product over time. The UPLC operation procedure was described in the Materials and Methods section.

Table 2. Kinetic parameters of BsGT110 toward GAA at pH 6 and pH 7.

Reaction Condition	K_m (μM)	k_{cat} (min^{-1})	k_{cat}/K_m (min$^{-1}\mu$M^{-1})
pH 6	570.6 ± 29.4	11.2 ± 0.9	0.0196 ± 0.0007
pH 7	299.4 ± 84.4	3.8 ± 0.9	0.0149 ± 0.0074

3. Discussion

According to the Carbohydrate-Active Enzymes (CAZy) database, GTs can be classified into 107 families, in which GTs that catalyze the glycosylation of small molecules, such as flavonoids and triterpenoids, are classified as GT1 [18]. Although over 500 thousands of GT have been discovered, there are only six bacterial GTs reported to catalyze glycosylation of triterpenoids, including BsYjiC from *B. subtilis* 168 [11,19–22], UGT109A1 from *B. subtilis* CTCG 63501 [23,24], BsGT1 from *B. subtilis* KCTC 1022 [25], BsUGT398 and BsUGT489 from *B. subtilis* ATCC 6633 [17], and BsGT110 from *B. subtilis* ATCC 6633 [present study]. Among them, the BsYjiC group (BsYjiC, BsUGT489, UGT109A1, and BsGT1) were highly similar, sharing more than 90% identity in their amino acid sequences [17], BsGT110 and BsUGT398, however, were not grouped with the BsYjiC group, and only had 31% and 33% identity, respectively, with the BsYjiC group (Figure 6). On the other hand, some bacterial GT1 catalyzed glycosylation of flavonoids. Thus, BsGT110 was compared with the flavonoid-catalyzing GTs. The results showed that the amino acid identity between BsGT110 and other flavonoid-catalyzing bacterial GTs was also lower than 40% (Figure S5 and Table S2). Furthermore, the evolutionary tree is shown in Figure 6 also demonstrated the dissimilarity between BsGT110 and BsUGT398, indicating that BsGT110 is a unique bacterial GT with glycosylation activity toward triterpenoids.

There are two reaction mechanisms for GT, inverting and retaining reactions, depending on the outcome of the reaction [26]. There are two stereochemical outcomes for reactions that result in the formation of a new glycosidic bond: the anomeric configuration of the product is either retained or inverted with respect to the donor substrate. The mechanistic strategy for inverting GTs involves a side chain of a residue on the active-site of GT that serves as a base catalyst that deprotonates the incoming nucleophile of the acceptor, facilitating direct displacement of the activated (substituted) phosphate leaving the group of the sugar donor, UDP-glucose [26]. Up to now, all GT1s are inverting GTs and were not reported to show optimal activities in acidic conditions [26]. For example, the well-known triterpenoid-catalyzing BsGT1 [25] had optimal activity at pH 7, BsYjiC [20], BsUGT398, and BsUGT489 [17] had optimal activity at pH 8, and UGT109A1 [23] had optimal activity at pH 9–10. These triterpenoid-catalyzing GTs had a broad neutral-alkaline range in their triterpenoid glycosylation activity [17,20–25]. According to the reaction mechanism of the inverting GTs, the side-chain of a key residue in the catalytic site of the enzyme should be deprotonated to serve as a base during the reaction. Thus, it is reasonable that GT1 enzymes have optimal activities at neutral-alkaline conditions, which would favor the deprotonation of the side chain. Accordingly, we identified the glycosylation activity of the selected five BsGTs toward GAA under a standard GT reaction condition: 10 mM Mg^{2+}, 40 °C, and pH 8, and found that only BsUGT398 and BsUGT489 can catalyze C-15 glycosylation of GAA [17], but other candidates, including BsGT110, were unable to catalyze glycosylation of GAA under the standard GT reaction condition. Thus, the previous study did not observe the novel acidic glycosylation activity (C-26 glycosylation of GAA) of BsGT110 toward GAA. Hence, the BsGT110 that we identified in this work was much more capable of catalyzing GAA into pure triterpenoid glucoside (GAA-26-*O*-β-glucoside) under acidic conditions (pH 5–6) (Figures 2 and 4, and Table 1). In addition, the results of the kinetics study showed that the turnover number of BsGT110 toward GAA at pH 6 was 3-fold higher than that at pH 7 (Figure 5 and Table 2). Furthermore, the catalytic efficiency (k_{cat}/K_m) of BsGT110 toward GAA at pH 6 was 1.35-fold higher than that at pH 7. Taken together, our results are unique in that they indicate that BsGT110—unlike other GT1s, which are most active at regular neutral-alkaline pH—is most active at a narrow, more acidic range of pH values (pH 5–6), specifically toward the C-26 position of GAA.

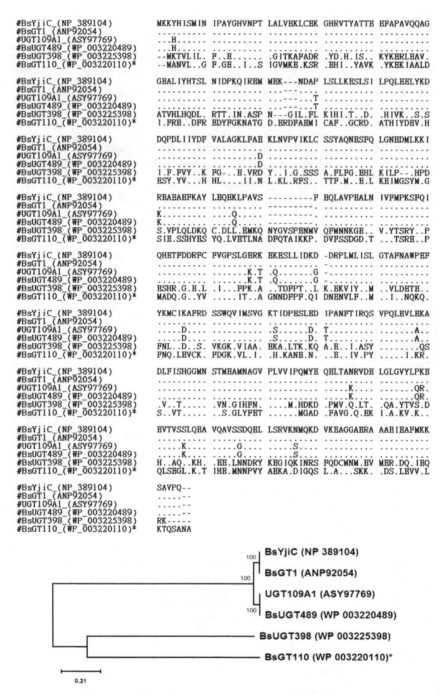

Figure 6. Aligned amino acid sequences and phylogenetic analysis using the Maximum Likelihood method. In total, 407 amino acids were aligned by Clustal W in MEGA X [27]. '.' denoted as identical amino acid, '-' denoted as indel(s). The phylogenetic tree was inferred using the Maximum Likelihood method and General Reversible Mitochondrial model [28]. The tree with the highest log likelihood (−3197.28) was shown. The percentage of trees in which the associated taxa clustered together was shown next to the branches. Initial tree(s) for the heuristic search were obtained automatically by applying Neighbor-Joining and BioNJ algorithms to a matrix of pairwise distances estimated using the JTT model, the topology with highest log-likelihood value was then selected. The tree was drawn to scale, with branch lengths measured based on the number of substitutions per site. This analysis involved six amino acid sequences. All positions with less than 95% site coverage were eliminated—i.e., less than 5% alignment gaps, missing data, and ambiguous bases were allowed at any position (partial deletion option). There were a total of 382 positions in the final dataset. Evolutionary analyses were conducted in MEGA X [27].

A few reports demonstrated that triterpenoid glycosides may improve the bioactivity of the triterpenoid aglycone [5]. Liang et al. produced an unusual ginsenoside, 3β, 12β-di-O-Glc-protopanaxadiol (PPD), from the glucosylation of PPD by UGT109A1, and showed that the ginsenoside had anti-cancer capabilities in the Lewis lung cancer xenograft mouse model [23]. Wang et al. used BsGT1 to produce 3β-O-Glc-ginsenoside F1, which inhibited melanin and tyrosinase activities [25]. Dai et al. reported the enzymatic synthesis of glycyrrhetinic acid (GA) glucosides—GA-30-O-β-glucoside and GA-3-O-β-glucoside—by BsYjiC and found that the two triterpenoid glucosides had higher water solubility and higher cytotoxicity against human liver cancer cells HepG2 and breast cancer cells MCF-7 than GA aglycone [20]. Therefore, the new GAA glucoside obtained in the present study, GAA-26-O-β-glucoside, warrants future investigation to determine whether it also has a higher bioactivity than GAA aglycone.

In summary, even though over 300 triterpenoids have been found in *G. lucidum*, very few triterpenoid glycosides have been identified [4]. Our study was the first to reveal that a single bacterium, the *Bacillus subtilis* ATCC 6633 strain, can biotransform GAA into both GAA-26-O-β-glucoside by BsGT110 in specific acidic conditions and GAA-15-O-β-glucoside by BsGT398 and BsGT489 in neutral-alkaline conditions.

4. Materials and Methods

4.1. Chemicals and Recombinant Enzymes

GAA was purchased from Baoji Herbest Bio-Tech (Xi-An, Shaanxi, China). UDP-glucose was obtained from Cayman Chemical (Ann Arbor, MI, USA). Recombinant BsGT enzymes (BsGT110, BsUGT398, BsUGT489, BsGT292, and BsGT296) were obtained from our previous studies [6,17]. The other reagents and solvents used were commercially available.

4.2. Glycosylation of GAA by Recombinant Enzymes

Glycosylation was performed in 0.1 mL reaction mixture containing 1 mg/mL GAA, 15 μg/mL the recombinant enzymes, 10 mM $MgCl_2$, and 10 mM UDP-glucose at pH 5-6 (50 mM acetate buffer) or pH 6–8 (50 mM PB). The reaction was performed at 40 °C for 30 min. Afterward, the reaction mixture was analyzed with UPLC. For optimization experiments, the above reaction mixture was incubated with 50 mM acetate buffer (pH 6) at different temperatures or with different metal ions.

For the kinetic experiments, different concentrations of GAA were mixture with 15 μg purified BsGT110 protein, 10 mM UDP-glucose, 10 mM $MgCl_2$, and 50 mM PB (pH 7.0) or acetate buffer (pH 6) in 1 mL reaction mixture and incubated at 40 °C for 20 min. During the incubation, samples from each reaction were removed every 2 min and analyzed by UPLC. The amount of GAA-26-O-β-glucoside production from the reaction was calculated from the peak area of UPLC analysis normalized with a standard curve. The rate of the reaction at each concentration of GAA was obtained from the slope of the plot of the amount of product over time. Kinetic parameters were obtained from the double-reciprocal plot of substrate GAA concentration versus the rate of reaction.

4.3. Ultra-Performance Liquid Chromatography (UPLC)

UPLC was performed with an Acquity® UPLC system (Waters, Milford, MA, USA). The stationary phase was a C18 column (Acquity UPLC BEH C18, 1.7 μm, 2.1 i.d. × 100 mm, Waters, MA, USA), and the mobile phase was 1% acetic acid in water (A) and methanol (B). The linear gradient elution condition was 0 min with 36% B to 7 min with 81% B at a flow rate of 0.2 mL/min. The detection condition was set at 254 nm.

4.4. Purification and Identification of the Glycosylated Product

Twenty-five mL of the reaction mixture (1 mg/mL GAA, 15 μg/mL BsGT110, 10 mM UDP-glucose, 10 mM MgCl$_2$, 50 mM acetate buffer at pH 6) was carried out at 40 °C for 30 min. Afterward, an equal volume of methanol was added into the reaction mixture to stop the reaction. Fifty mL of the reaction mixture with 50% methanol was applied to a preparative YL9100 HPLC system (YoungLin, Gyeonggi-do, Korea). The stationary phase was the Inertsil ODS 3 column (10 mm, 20 i.d. × 250 mm, GL Sciences, Eindhoven, The Netherlands), and the mobile phase was the same as that in the UPLC system, but with a flow rate of 15 mL/min. The detection condition was 254 nm, and the sample volume was 10 mL for each injection. The product of each run was collected, concentrated under a vacuum, and lyophilized with a freeze dryer. From the 25 mL of reaction, 5.4 mg of the product was purified. The chemical structure of the product compound was determined with mass and NMR spectral analyses. The mass spectral analysis was performed on a Finnigan LCQ Duo mass spectrometer (ThermoQuest Corp., San Jose, CA, USA) with ESI. ^{1}H- and ^{13}C-NMR, HSQC, and HMBC spectra were recorded on a Bruker AV-600 NMR spectrometer (Bruker Corp., Billerica, MA, USA) at ambient temperature. Standard pulse sequences and parameters were used for the NMR experiments, and all chemical shifts were reported in parts per million (ppm, δ).

5. Conclusions

A new GAA-26-*O*-β-glucoside was produced from the *O*-glucosylation of GAA with recombinant BsGT110 isolated from *B. subtilis* ATCC 6633 under acidic conditions (pH 6). BsGT110 was the first GT identified as catalyzing the glycosylation of triterpenoid at the C-26 position. Moreover, the optimal reaction condition of BsGT110 was at pH 6, and it lost most of this activity at pH 8, implying that such GTs might only catalyze other triterpenoid substrates under acidic conditions.

Supplementary Materials: Table S1. NMR spectroscopic data for compound (**2**) in pyridine-d$_5$ (600 MHz), Table S2. BsGT110 sequence comparison with candidate triterpenoid-catalyzing GTs and flavonoid-catalyzing GTs, Figure S1. The ^{1}H-NMR (600 MHz) spectrum of compound (**2**) in pyridine-d$_5$, Figure S2. The ^{13}C-NMR (150 MHz) spectrum of compound (**2**) in pyridine-d$_5$, Figure S3. The HSQC (600 MHz) spectrum of compound (**2**) in pyridine-d$_5$, Figure S4. The HMBC (600 MHz) spectrum of compound (**2**) i n pyridine-d$_5$, Figure S5. Phylogenetic analysis using the Maximum Likelihood method.

Author Contributions: Conceptualization: T.-S.C. and T.-Y.W.; data curation: T.-S.C. and Y.-H.K.; methodology: Y.-H.K. and C.-M.C.; project administration: T.-S.C.; writing—original draft: T.-S.C., T.-Y.W., and C.-M.C.; writing—review and editing: T.-S.C., T.-Y.W., J.-Y.W., Y.-W.W., and C.-M.C.

References

1. Ahmad, M.F. *Ganoderma lucidum*: Persuasive biologically active constituents and their health endorsement. *Biomed. Pharmacother.* **2018**, *107*, 507–519. [CrossRef] [PubMed]

2. Sohretoglu, D.; Huang, S. *Ganoderma lucidum* polysaccharides as an anti-cancer agent. *Anticancer Agents Med. Chem.* **2018**, *18*, 667–674. [CrossRef] [PubMed]

3. Wu, J.W.; Zhao, W.; Zhong, J.J. Biotechnological production and application of ganoderic acids. *Appl. Microbiol. Biotechnol.* **2010**, *87*, 457–466.

4. Xia, Q.; Zhang, H.; Sun, X.; Zhao, H.; Wu, L.; Zhu, D.; Yang, G.; Shao, Y.; Zhang, X.; Mao, X.; et al. A comprehensive review of the structure elucidation and biological activity of triterpenoids from *Ganoderma* spp. *Molecules* **2014**, *19*, 17478–17535. [CrossRef] [PubMed]

5. Shi, Z.Y.; Zeng, J.Z.; Wong, A.S.T. Chemical structures and pharmacological profiles of giseng saponins. *Molecules* **2019**, *24*, 2443. [CrossRef] [PubMed]

6. Chiang, C.M.; Wang, T.Y.; Yang, S.Y.; Wu, J.Y.; Chang, T.S. Production of new isoflavone glucosides from glycosylation of 8-hydroxydaidzein by glycosyltransferase from *Bacillus subtilis* ATCC 6633. *Catalysts* **2018**, *8*, 387. [CrossRef]

7. Shimoda, K.; Hamada, H.; Hamada, H. Synthesis of xylooligosaccharides of daidzein and their anti-oxidant and anti-allergic activities. *Int. J. Mol. Sci.* **2011**, *12*, 5616–5625. [CrossRef] [PubMed]

8. Cho, H.K.; Kim, H.H.; Seo, D.H.; Jung, J.H.; Park, J.H.; Baek, N.I.; Kim, M.J.; Yoo, S.H.; Cha, J.; Kim, Y.R.; et al. Biosynthesis of catechin glycosides using recombinant amylosucrase from *Deinococcus geothermalis* DSM 11300. *Enz. Microbial Tech.* **2011**, *49*, 246–253. [CrossRef] [PubMed]

9. Hofer, B. Recent developments in the enzymatic *O*-glycosylation of flavonoids. *Appl. Microbiol. Biotechnol.* **2016**, *100*, 4269–4281. [CrossRef]

10. Tiwari, P.; Sangwan, R.S.; Sangwan, N.S. Plant secondary metabolism linked glycosyltransferases: An update on expanding knowledge and scopes. *Biotechnol. Adv.* **2016**, *34*, 716–739. [CrossRef]

11. Dai, L.; Li, J.; Yang, J.; Zhu, Y.; Men, Y.; Zeng, Y.; Cai, Y.; Dong, C.; Dai, Z.; Zhang, X.; et al. Use of a promiscuous glycosyltransferase from *Bacillus subtilis* 168 for the enzymatic synthesis of novel protopanaxtriol-type ginsenosides. *J. Agric. Food Chem.* **2017**, *66*, 943–949. [CrossRef] [PubMed]

12. Liang, C.; Tian, D.; Liu, Y.; Li, H.; Zhu, J.; Li, M.; Xin, M.; Xia, J. Review of the molecular mechanisms of *Ganoderma lucidum* triterpenoids: Ganoderic acids A, C2, D, F, DM, X and Y. *Eur. J. Med. Chem.* **2019**, *174*, 130–141. [CrossRef] [PubMed]

13. Jiang, J.; Grieb, B.; Thyagarajan, A.; Sliva, D. Ganoderic acids suppress growth and invasive behavior of breast cancer cells by modulating AP-1 and NF-kB signaling. *Int. J. Mol. Med.* **2008**, *21*, 577–584. [PubMed]

14. Yao, X.; Li, G.; Xu, H.; Lu, C. Inhibition of the JAK-STAT3 signaling pathway by ganoderic acid A enhances chemosensitivity of HepG2 cells to cisplatin. *Planta Med.* **2012**, *78*, 1740–1748. [CrossRef] [PubMed]

15. Wang, X.; Sun, D.; Tai, J.; Wang, L. Ganoderic acid A inhibits proliferation and invasion, and promotes apoptosis in human hepatocellular carcinoma cells. *Mol. Med. Rep.* **2017**, *16*, 3894–3900. [CrossRef] [PubMed]

16. Akihisa, T.; Nakamura, Y.; Tagata, M.; Tokuba, H.; Yasukawa, K.; Uchiyama, E.; Suzukli, T.; Kimura, Y. Anti-inflammatory and anti-tumor-promoting effects of triterpene acids and sterols from the fungus *Ganoderma lucidum*. *Chem. Biod.* **2007**, *4*, 224–231. [CrossRef] [PubMed]

17. Chang, T.S.; Wu, J.J.; Wang, T.Y.; Wu, K.Y.; Chiang, C.M. Uridine diphosphate-dependent glycosyltransferases from *Bacillus subtilis* ATCC 6633 catalyze the 15-*O*-glycosylation of ganoderic acid A. *Int. J. Mol. Sci.* **2018**, *19*, 3469. [CrossRef] [PubMed]

18. Cantarel, B.; Coutinho, P.M.; Rancurel, C.; Bernard, T.; Lombard, V.; Henrissat, B. The Carbohydrate-Active EnZymes database (CAZy): An expert resource for Glycogenomics. *Nucleic Acids Res.* **2009**, *37* (Suppl. 1), D233–D238. [CrossRef] [PubMed]

19. Dai, L.; Li, J.; Yang, J.; Men, Y.; Zeng, Y.; Cai, Y.; Sun, Y. Enzymatic synthesis of novel glycyrrhizic acid glucosides using a promiscuous *Bacillus* glycosyltransferase. *Catalysts* **2018**, *8*, 615. [CrossRef]

20. Dai, L.; Li, J.; Yao, P.; Zhu, Y.; Men, Y.; Zeng, Y.; Yang, J.; Sun, Y. Exploiting the aglycon promiscuity of glycosyltransferase Bs-YjiC from *Bacillus subtilis* and its application in synthesis of glycosides. *J. Biotechnol.* **2017**, *248*, 69–76. [CrossRef]

21. Li, K.; Feng, J.; Kuang, Y.; Song, W.; Zhang, M.; Ji, S.; Qiao, X.; Ye, M. Enzymatic synthesis of bufadienolide *O*-glycosides as potent antitumor agents using a microbial glycosyltransferase. *Adv. Synth. Catal.* **2017**, *359*, 3765–3772. [CrossRef]

22. Chen, K.; He, J.; Hu, Z.; Song, W.; Yu, L.; Li, K.; Qiao, X.; Ye, M. Enzymatic glycosylation of oleanane-type triterpenoids. *J. Asia Nat. Prod. Res.* **2018**, *20*, 615–623. [CrossRef] [PubMed]

23. Liang, H.; Hu, Z.; Zhang, T.; Gong, T.; Chen, J.; Zhu, P.; Li, Y.; Yang, J. Production of a bioactive unnatural ginsenoside by metabolically engineered yeasts based on a new UDP-glycosyltransferase from *Bacillus subtilis*. *Metab. Eng.* **2017**, *44*, 60–69. [CrossRef] [PubMed]

24. Zhang, T.T.; Gong, T.; Hu, Z.F.; Gu, A.D.; Yang, J.L.; Zhu, P. Enzymatic synthesis of unnatural ginsenosides using a promiscuous UDP-glucosyltransferase from *Bacillus subtilis*. *Molecules* **2018**, *23*, 2797. [CrossRef] [PubMed]

25. Wang, D.D.; Jin, Y.; Wang, C.; Kim, Y.J.; Perez, J.E.J.; Baek, N.I.; Mathiyalagan, R.; Markus, J.; Yang, D.C. Rare ginsenoside Ia synthesized from F1 by cloning and overexpression of the UDP-glycosyltransferase gene from *Bacillus subtilis*: Synthesis, characterization, and in vitro melanogenesis inhibition activity in BL6B16 cells. *J. Gingeng. Res.* **2018**, *42*, 42–49. [CrossRef] [PubMed]

26.   Lairson, L.L.; Henrissat, B.; Davies, G.J.; Withers, S.G. Glycosyltransferases: Structures, functions, and mechanisms. *Annu. Rev. Biochem.* **2008**, *77*, 521–555. [CrossRef] [PubMed]

27.   Kumar, S.; Stecher, G.; Li, M.; Knyaz, C.; Tamura, K. MEGA X: Molecular Evolutionary Genetics Analysis across Computing Platforms. *Mol. Biol. Evol.* **2018**, *35*, 1547–1549. [CrossRef] [PubMed]

28.   Adachi, J.; Hasegawa, M. Model of amino acid substitution in proteins encoded by mitochondrial DNA. *J. Mol. Evol.* **1996**, *42*, 459–468. [CrossRef]

14

A Convenient, Rapid, Sensitive and Reliable Spectrophotometric Assay for Adenylate Kinase Activity

Kai Song [1], Yejing Wang [2,*], Yu Li [1], Chaoxiang Ding [1], Rui Cai [2], Gang Tao [1], Ping Zhao [1], Qingyou Xia [1] and Huawei He [1,3,*]

[1] Biological Science Research Center, Southwest University, Beibei, Chongqing 400715, China; Kaisong@email.swu.edu.cn (K.S.); liyu315@swu.edu.cn (Y.L.); ding7197767@email.swu.edu.cn (C.D.); taogang@email.swu.edu.cn (G.T.); zhaop@swu.edu.cn (P.Z.); xiaqy@swu.edu.cn (Q.X.)

[2] State Key Laboratory of Silkworm Genome Biology, College of Biotechnology, Southwest University, Beibei, Chongqing 400715, China; cairui0330@email.swu.edu.cn

[3] Chongqing Key Laboratory of Sericultural Science, Chongqing Engineering and Technology Research Center for Novel Silk Materials, Southwest University, Beibei, Chongqing 400715, China

* Correspondence: yjwang@swu.edu.cn (Y.W.); hehuawei@swu.edu.cn (H.H.);

Academic Editor: Stefano Serra

Abstract: Enzymatic activity assays are essential and critical for the study of enzyme kinetics. Adenylate kinase (Adk) plays a fundamental role in cellular energy and nucleotide homeostasis. To date, assays based on different principles have been used for the determination of Adk activity. Here, we show a spectrophotometric analysis technique to determine Adk activity with bromothymol blue as a pH indicator. We analyzed the effects of substrates and the pH indicator on the assay using orthogonal design and then established the most optimal assay for Adk activity. Subsequently, we evaluated the thermostability of Adk and the inhibitory effect of KCl on Adk activity with this assay. Our results show that this assay is simple, rapid, and precise. It shows great potential as an alternative to the conventional Adk activity assay. Our results also suggest that orthogonal design is an effective approach, which is very suitable for the optimization of complex enzyme reaction conditions.

Keywords: enzymatic activity assay; adenylate kinase; spectrophotometry; orthogonal experiment; bromothymol blue

1. Introduction

Adenylate kinase (Adk; ATP:AMP phosphotransferase, EC 2.7.4.3), also known as myokinase, is a conserved phosphoryl transferase, which catalyzes the translocation of a phosphoryl group between nucleotides in the reversible reaction (AMP + $Mg^{2+}\bullet$ATP $\leftrightarrow$ $Mg^{2+}\bullet$ADP + ADP) [1]. Adk is ubiquitous in different tissues of all living systems and plays a fundamental role in cellular energy and nucleotide homeostasis. The hydrogen bond between the adenine moiety and the backbone of Adk is critical for ATP selectivity and can help Adk recognize the correct substrates in the complex cellular environment [2]. Adk is involved in the regulation of cell differentiation, maturation, apoptosis, and oncogenesis. Adk mutations in humans cause a severe disease called reticular dysgenesis [3]. Adk is regarded as a potential target for medical diagnosis and treatment due to its close correlation with other diseases, such as aleukocytosis, hemolytic anemia, and primary ciliary dyskinesia [4]. To date, three methods have been proposed to determine Adk activity based on the detectable changes accompanied with this reaction, such as light absorption, acidity, or the coupled reaction products [5]. The manometric assay, established by Colowick and Kalckar [6], is used for the detection of Adk by measuring CO_2 liberation from a bicarbonate buffer. The reaction catalyzed by Adk is coupled to hexokinase, which specifically catalyzes the transformation of the terminal phosphate from ADP to glucose. The overall reaction is as follows:

$$ATP + AMP \xleftrightarrow{\text{Adk}} ADP + ADP \tag{1}$$

$$ADP + glucose \xrightarrow{\text{hexokinase}} glucose - 6 - P + AMP + H^+ \tag{2}$$

In the presence of excess hexokinase, the reaction rate is proportional to the Adk concentration. The forward direction reaction is defined as the formation of AMP and ATP from two ADPs, and the reverse direction is defined as the formation of two ADPs from AMP and ATP. Adk activity is conventionally measured in vitro by a spectrophotometric assay. For the forward direction reaction, Chiu et al. [7] have developed a modified assay of Oliver [8] to determine Adk activity by coupling the reaction to hexokinase and glucose-6-phosphate dehydrogenase in which the final product, NADPH, is measured spectrophotometrically at 340 nm. The overall reaction is as follows: H^+

$$ADP + ADP \xleftrightarrow{\text{Adk}} ATP + AMP \tag{3}$$

$$ATP + glucose \xrightarrow{\text{hexokinase}} glucose - 6 - P + ADP \tag{4}$$

$$glucose - 6 - P + NADP \xrightarrow[\text{dehydrogenase}]{\text{glucose} - 6 - \text{phosphate}} 6 - phosphogluconic\ acid + NADPH \tag{5}$$

Adk activity is also measured in the reverse direction by coupling the reaction to pyruvate kinase and lactate dehydrogenase and measuring the oxidation of NADH at 340 nm [5]. The principle of the assay is as follows:

$$ATP + AMP \xleftrightarrow{\text{Adk}} ADP + ADP \tag{6}$$

$$Phosphoenolpyruvate + ADP \xrightarrow{\text{Pyruvate kinase}} ATP + Pyruvate \tag{7}$$

$$Pyruvate + NADH + H + \xrightarrow{\text{Lactic dehydrogenase}} Lactate + NAD^+ \tag{8}$$

These assays have been used to determine Adk activity for the past decades. However, some disadvantages are also obvious for these assays. Firstly, these assays are time-consuming, multistep processes that require the assistance of other enzymes and are easily subject to errors at each step. Secondly, it is difficult to study the effects of activators and inhibitors on Adk activity with the aid of other enzymes. Finally, the real initial rate of Adk reaction cannot be determined accurately [9]. Therefore, it is necessary to develop a more convenient and accurate assay for Adk activity in vitro.

Acid–base indicators are usually applied in enzymatic assay for their extraordinary sensitivity to pH change. In 2002, Yu et al. [10] established an arginine kinase activity assay based on the light absorption of a complex acid–base indicator consisting of thymol blue and cresol red. In the reaction catalyzed by arginine kinase, the produced protons resulted in a decrease in pH of the reaction mixture, thus reducing the absorbance of the mixed indicator in the solution at 575 nm. The arginine kinase activity could be determined according to the change of the absorbance at 575 nm. In the same way, Dhale et al. developed a rapid and sensitive assay to measure L-asparaginase activity with methyl red as an indicator [11]. Bromothymol blue is an excellent indicator as it forms a highly conjugated structure while deprotonated in alkaline solution, resulting in an obvious color change from yellow to blue and the corresponding absorbance change [12].

Enzyme activity is typically influenced by many factors. The traditional method is to do a multifactor analysis that tests all possible combinations of the different factors. However, this takes up a lot of time and resources as the number of full factorial experiments is very large. As an alternative, the orthogonal design method has been proposed and established. The orthogonal experimental design [13] is a multifactor experiment design assay. It selects representative samples from a full factorial assay in a way that the samples are distributed uniformly within the test range, thus representing the overall situation. Therefore, it is highly efficient for the arrangement of multifactor experiments with optimal combination levels. The orthogonal design has three advantages: (1) The number of tests required to complete the experiment is relatively small. (2) The data points are evenly distributed. (3) The test results can be analyzed by mathematical calculation (e.g., range analysis and variance analysis), which is particularly useful to quantify the results.

In this study, we developed a one-step assay for Adk activity. It is based on proton generation after the addition of ATP and AMP as the substrates, which can be measured spectrophotometrically at 614 nm using bromothymol blue as a pH indicator. We investigated four factors affecting Adk activity—ATP, AMP, bromothymol blue, and glycine–NaOH buffer—at three levels and determined the best combination for Adk activity assay by an orthogonal experimental design. Finally, we evaluated the thermostability of Adk and the inhibitory effect of KCl on Adk activity with this assay. Our results suggest that this assay is simple, precise, less expensive, and a potential alternative to the conventional enzymes-coupled assay extensively used in clinical and research laboratories.

2. Results

In this study, Adk activity was determined by a direct and continuous spectrophotometric technique without coupled enzymes. In the enzymatic reaction catalyzed by Adk, the formation of two ADPs from ATP and AMP is accompanied by the generation of hydrogen ions. Bromothymol blue is an excellent acid–base indicator as it forms a highly conjugated structure while protonated in acid solution, resulting in an obvious color change from blue to yellow. The absorbance of bromothymol blue at 614 nm is associated with the hydrogen ion concentration in solution. Thus, Adk activity can be monitored in real time by the absorbance of bromothymol blue at 614 nm in solution, which can be detected by a sensitive spectrophotometer. The principle of this assay is illustrated in Figure 1.

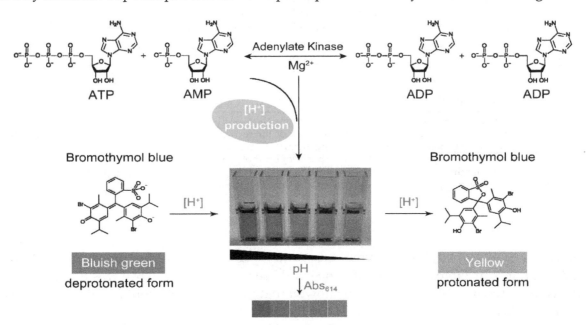

Figure 1. The principle of the spectrophotometric assay for adenylate kinase (Adk) activity.

The effects of substrates and the pH indicator on the assay were analyzed using an orthogonal design, which is key to establishing the most optimal assay for Adk activity. The factors and levels affecting Adk activity assay are shown in Table 1.

Table 1. Factors and levels affecting Adk activity assay.

Level	A ATP (mM)	B AMP (mM)	C Bromothymol Blue (mM)	D Glycine–NaOH (mM)
1	2.0	1.0	0.0930	0.1
2	2.5	1.5	0.1084	0.3
3	3.0	2.0	0.1238	0.5

2.1. The Maximum Absorption Wavelength of Reaction Mixture

A set of nine tests designed by orthogonal experiment is shown in Table 2. The absorption spectrum of each set was scanned from 450 to 800 nm in the presence of 5 mM $MgAC_2$. All spectra showed the same maximum absorption located at 614 nm without shift, as shown in Figure 2a.

Table 2. Orthogonal array $(9, 3^4)$ for the analysis of the effects of ATP, AMP, bromothymol blue, and glycine–NaOH buffer on the Adk activity assay.

No.	Combination	Factor A	B	C	D	ΔAbs [a] (0–30 s)
1	$A_1B_1C_1D_1$	2.0	1.0	0.0930	0.1	0.0914 ± 0.0020
2	$A_1B_2C_2D_2$	2.0	1.5	0.1084	0.3	0.0432 ± 0.0045
3	$A_1B_3C_3D_3$	2.0	2.0	0.1238	0.5	0.0112 ± 0.0017
4	$A_2B_1C_2D_3$	2.5	1.0	0.1084	0.5	0.0404 ± 0.0017
5	$A_2B_2C_3D_1$	2.5	1.5	0.1238	0.1	0.0221 ± 0.0020
6	$A_2B_3C_1D_2$	2.5	2.0	0.0930	0.3	0.0606 ± 0.0022
7	$A_3B_1C_3D_2$	3.0	1.0	0.1238	0.3	0.0341 ± 0.0009
8	$A_3B_2C_1D_3$	3.0	1.5	0.0930	0.5	0.0705 ± 0.0049
9	$A_3B_3C_2D_1$	3.0	2.0	0.1084	0.1	0.0382 ± 0.0017
T_1		0.1458	0.1659	0.2225	0.1517	
T_2		0.1231	0.1358	0.1218	0.1379	
T_3		0.1428	0.1100	0.0674	0.1221	
t_1		0.0486	0.0553	0.0742	0.0506	
t_2		0.0410	0.0453	0.0406	0.0460	
t_3		0.0476	0.0367	0.0225	0.0407	
	Range (R)	0.0076	0.0186	0.0517	0.0099	
	Order		C > B > D > A			
	Optimal level	A_1	B_1	C_1	D_1	
	Optimal combination		$A_1B_1C_1D_1$			

[a] Arithmetic mean of the absorbance changes of bromothymol blue at 614 nm (0–30 s) of three independent tests at each level under the same factor. Ti (T1, T2, T3) is the sum of the recorded absorbance changes (ΔAbs) at the same level and under the same factor, and ti (t1, t2, t3) is the arithmetic mean of Ti. The mean of ti represents the influence of different levels under the same factor on the absorbance of bromothymol blue. Range (R) is the difference between the maximum and the minimum of ti, indicating the effect of each factor on the absorbance of bromothymol blue. The greater the R value, the greater is the influence of this factor on the absorbance of bromothymol blue at 614 nm.

2.2. Optimization of Adk Activity Assay

The effects of ATP, AMP, bromothymol blue, and glycine–NaOH buffer on the Adk activity were analyzed, as shown in Table 2. The significance of these factors on the assay was determined by the range (R) value listed in Table 2. The results showed that the order of R value was RC > RB > RD > RA. Hence, the significance order of these factors on the assay was C > B > D > A, namely, bromothymol blue > AMP > glycine–NaOH buffer > ATP. Similarly, the significance of different levels on the assay was determined by the ti value listed in Table 2. The results showed that the orders were t1 > t3 > t2 for factor A, t1 > t2 > t3 for factor B, t1 > t2 > t3 for factor C, and t1 > t2 > t3 for factor D. Thus, the most optimal combination for Adk activity assay was A1B1C1D1, which was composed of 2 mM ATP, 1 mM AMP, 0.093 mM bromothymol blue, and 0.1 mM glycine–NaOH buffer.

Next, we conducted a full factorial design for two primary factors of bromothymol blue and AMP at three levels to further optimize the assay. The design matrix and parameters are listed in Table 3. The results showed that AB3C1D had the most significant absorbance change at 614 nm among all the tested conditions (Table 4), indicating AB3C1D (2 mM ATP, 0.6 mM AMP, 0.093 mM bromothymol blue, and 0.1 mM glycine-NaOH buffer) was the most optimal reaction condition for the Adk activity assay.

Table 3. Factors and levels affecting Adk activity assay with the constants of A and D.

Level	A ATP (mM)	B AMP (mM)	C Bomothymol Blue (mM)	D Glycine–NaOH (mM)
1	2.0	1.0	0.0930	0.1
2	2.0	0.8	0.0775	0.1
3	2.0	0.6	0.0620	0.1

Table 4. A full factorial design for the analysis of the effects of AMP, bromothymol blue on the Adk activity assay with the constants of A and D.

Run	Combination	Factor A	B	C	D	ΔAbs [b] (0–30 s)
1	AB_1C_1D	2.0	1.0	0.0930	0.1	0.0914 ± 0.0020
2	AB_1C_2D	2.0	1.0	0.0775	0.1	0.0926 ± 0.0030
3	AB_1C_3D	2.0	1.0	0.0620	0.1	0.0764 ± 0.0030
4	AB_2C_1D	2.0	0.8	0.0930	0.1	0.0914 ± 0.0020
5	AB_2C_2D	2.0	0.8	0.0775	0.1	0.0930 ± 0.0018
6	AB_2C_3D	2.0	0.8	0.0620	0.1	0.0680 ± 0.0041
7	AB_3C_1D	2.0	0.6	0.0930	0.1	0.0983 ± 0.0028
8	AB_3C_2D	2.0	0.6	0.0775	0.1	0.0850 ± 0.0054
9	AB_3C_3D	2.0	0.6	0.0620	0.1	0.0604 ± 0.0001

[b] Arithmetic mean of the absorbance changes of bromothymol blue at 614 nm (0–30 s) of three independent tests at each level under the same factor.

2.3. Effect of H^+ on the Absorbance of Bromothymol Blue

To determine the sensitivity of bromothymol blue on the Adk activity assay, we measured the response of bromothymol blue to hydrogen ion under the most optimal condition, AB3C1D. The absorbance spectrum of bromothymol blue was scanned from 450 to 800 nm in the presence of various concentrations of HCl. The results showed that the absorbance of bromothymol blue at 614 nm gradually declined with the increase in hydrogen ion concentration (Figure 2b). The absorbance change can be linearly fitted as the function of hydrogen ion concentration with the following equation (Figure 2c):

$$y = 0.138 * x - 0.066$$

The results suggested that the absorbance of bromothymol blue at 614 nm had a good response to pH change in the assay, and the absorbance change was positively correlated with the hydrogen ion concentration.

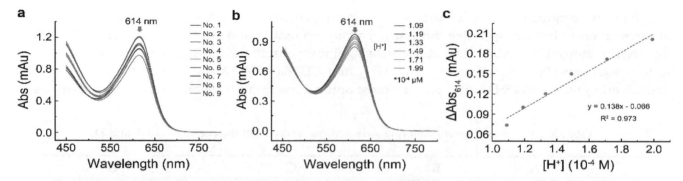

Figure 2. The maximum absorption wavelength of the reaction mixture and its relationship with the hydrogen ion concentration. (**a**) The absorption spectra of nine different combinations in the presence of 5 mM $MgAC_2$; (**b**) effect of hydrogen ion concentration on the absorption of the reaction system; (**c**) the correlation of the absorbance change of bromothymol blue at 614 nm with hydrogen ion concentration.

2.4. Effect of Adk Contents on the Reaction Velocity

The effect of Adk contents on the reaction velocity was determined as described in the Materials and Methods section. The results showed that, with the increase in Adk contents, the reaction velocity increased, and the reaction time required to reach equilibrium shortened (Figure 3a). In the first 5 s, the absorbance of bromothymol blue at 614 nm (Abs_{614}) declined linearly with time; thus, the slope of the reaction in the first 5 s was defined as the initial reaction velocity. The plot in Figure 3b shows that the absorbance change of bromothymol blue at 614 nm could be linearly fitted as the function of Adk contents.

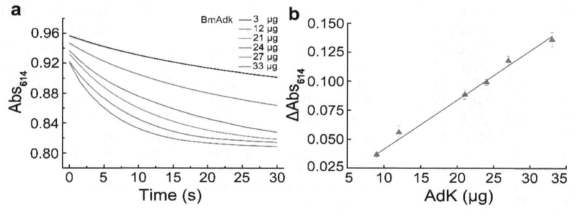

Figure 3. The effect of Adk contents on the assay. (**a**) Effect of different Adk contents on the assay; (**b**) the correlation of the absorbance change of bromothymol blue at 614 nm with different Adk contents.

2.5. Effect of Temperature and KCl on Adk Activity

The effect of temperature on Adk activity was investigated to characterize the thermostability of Adk. The results showed that Adk activity was almost unaffected under 45 °C. However, when the temperature was increased to 60 °C, Adk quickly lost its activity (Figure 4a). Adk from the muscle has a half-life of 30 min in 0.1 N hydrochloric acid at 100 °C [14]. Our results indicate that the thermostability of Adk from *Bombyx mori* (BmAdk) is much lower than that of Adk from the muscle.

Allan Hough et al. proved that KCl can almost completely inhibit myokinase activity [15]. Here, the effect of KCl on Adk activity was assessed with the assay. The results showed that low concentration of KCl (<5 mM) had a slight inhibitory effect on Adk activity. With the increase in KCl concentration, the inhibitory effect of KCl on Adk activity became more and more obvious. About 70 mM KCl resulted in 50% loss of Adk activity (Figure 4b). Compared with the "three-minute" method [15], the inhibitory effect of KCl on Adk activity could be assessed more easily with our developed assay.

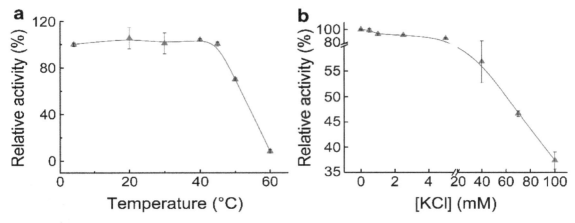

Figure 4. The thermostability of Adk and the inhibition of KCl on Adk activity. (**a**) Effect of temperature on Adk activity; (**b**) effect of KCl concentration on Adk activity.

3. Materials and Methods

3.1. Chemicals and Materials

ATP and AMP were purchased from Aladdin (Shanghai, China) in the form of sodium salt. Magnesium acetate was from Sigma (St. Louis, MO, USA). Bromothymol blue sodium salt, glycine, and other reagents all came from Sangon Biotech Corp. (Shanghai, China). Plastic cuvettes were purchased from Centome Corp. (Chengdu, China).

3.2. Adk Preparation and Concentration Determination

The DNA fragment encoding Adk was obtained from the cDNA library by PCR from the midgut of *Bombryx mori* strain Dazao using primer sets (5′-ATGGCACCGGCCGCTGC-3′ and 5′-TTACAAA GCAGACCGTGCTCTGCTG-3′). The amplification product was gel-purified, recovered, and inserted into plasmid vectors pSKB2. The bacterial transformants containing error-free inserts were identified. Adk was expressed in *Escherichia coli* BL21(DE3) and purified by Ni-NTA affinity chromatography (GE Healthcare, Chicago, IL, USA). The fused polyhistidine tag was cleaved by Prescission protease (GE Healthcare, USA) and removed as described by Liu et al. [16]. Protein concentration was determined using the extinction coefficient of 12,950 $M^{-1} \cdot L \cdot cm^{-1}$ at 280 nm on a NanoDrop 2000C spectrophotometer (Thermo Fisher, Waltham, MA, USA).

3.3. Adk Activity Assay

ATP and AMP were used as the substrates in the assay. The reaction mixture (1 mL) was composed of 2 mM ATP, 0.6 mM AMP, 0.1 mM glycine–NaOH (pH 9.0), 0.093 mM bromothymol blue, and 5 mM $MgAC_2$. The initial absorbance of the freshly prepared reaction mixture at 614 nm was adjusted to 1.05 with approximately 0.5 M NaOH so that the absorbance of the mixture would not be changed after the addition of Adk. The purified Adk was exchanged into buffer A (20 mM Tris-HCl, pH 7.6, 150 mM NaCl, 5% glycerol, and 0.1 mM dithiothreitol (DTT)) via gel filtration. The reaction was triggered by adding 15 μg Adk into the mixture. The reaction velocity was defined as the slope of the absorbance change of bromothymol blue at 614 nm in the initial 30 s, which was recorded on a DU 800 nucleic acid/protein analyzer (BeckmanCoulter, Brea, CA, USA) using a 1-cm light path plastic cuvette. For the control reaction, Adk was replaced with buffer A. All measurements were carried out at 25 °C. Each test was replicated at least three times.

3.4. Orthogonal Design

The concentration of ATP, AMP, bromothymol blue, and glycine–NaOH buffer is vital for Adk activity assay. Consequently, these factors were considered in the orthogonal design to screen the most optimal conditions for Adk activity. Assuming there were three levels for each factor, and the interaction among the factors were not taken into account, the orthogonal experiment of four factors at three levels was designed.

4. Discussion

Adk plays a crucial role in maintaining a balance of cellular energy and nucleic acid metabolism. Human Adk isoenzymes specifically expressed in organs are regarded as important indicators of organ dysfunction [17] or differentiation stages [18]. The Adk activity assays that have been developed so far are largely dependent on the coupled secondary enzymes hexokinase and glucose-6-phosphate dehydrogenase or adenosine monophosphate deaminase [19]. In addition, the different conditions between the coupled enzymes and Adk affect the application of these assays [14]. Therefore, developing a convenient, rapid, sensitive, reliable, and economic assay for Adk activity is of great significance.

Here, we developed a spectrophotometric assay for Adk activity using bromothymol blue as a pH indicator without coupled enzymes. The effective range of bromothymol blue is pH 6.0–7.6. Adk is optimally active at pH 7.6 [14]. Hence, Adk was dissolved in pH 7.6 buffer to keep it active. The correspondence of pH and the absorbance of the system are listed in Table 5. Here, we set the initial absorbance of 1.05 as the beginning of the reaction, which represented pH 6.8 of the reaction system. When 5 μL buffer was added into the system (995 μL), there was almost no change in the pH of the system. At the same time, the assay system could ensure bromothymol blue had a sensitive and stable response to pH change caused by Adk-catalyzed reaction.

Table 5. Correspondence between pH and the absorbance of bromothymol blue at 614 nm.

pH	6.47	6.54	6.73	6.79	6.85	7.07	7.21	7.28
Abs	0.8145	0.9197	0.9438	1.0375	1.0847	1.2385	1.3850	1.4188

The effects of ATP, AMP, bromothymol blue, and glycine–NaOH buffer on the Adk activity assay were determined by a rational orthogonal design. For a full factorial assay with four factors at three levels, the number of tests is up to 81 (3^4). However, our rational orthogonal design greatly reduced the unnecessary experiments and effectively achieved significant results (Table 2). The orthogonal experimental design is a rational design method for multifactor experiment, which selects representative points from a full factorial assay to represent the overall situation. Therefore, it is highly efficient for the design of multifactor experiments with optimal combination levels. By means of the orthogonal design, we greatly reduced the number of required experiments and achieved significant results with the least number of experiments.

The optimal temperature for the growth of *Bombyx mori* is about 25 °C. Adk from *Bombyx mori* was relatively stable below 45 °C, implying its significance on the growth and development of *Bombyx mori*. The Adk structure suggests that the hydrophobic core packing is important for the stability and activity of Adk [20]. AMP has a strong inhibition effect on the reaction with ADP as the substrate, thus resulting in a rapid decrease in the reaction rate with time [6,19]. Here, we found that high concentration of AMP could inhibit Adk activity, which is in line with a previous report. Slater et al. [21] found that the inhibitory effect of AMP was easily observed even when excessive hexokinase and glucose were present. The inhibition was ascribed to the affinity of ADP to Adk, which is lower than that of AMP with Adk. However, in this study, the substrates were ATP and AMP. Therefore, the inhibition of AMP on Adk activity may be attributed to the noncompetitive inhibition of AMP with respect to ATP [22].

To summarize, we developed a simple and rapid assay to determine Adk activity with bromothymol blue as an indicator instead of coupled enzymes. The assays that have been developed so far require the assistance of other enzymes to convert the reaction product to other detectable signals; thus, they consist of multiple reaction steps and are discontinuous. The assays are not accurate as they are easily subject to errors at each step. However, this new assay only relies on the protons produced in the reaction and the corresponding absorbance changes of bromothymol blue in the reaction solution. Small changes in pH can lead to significant changes in the absorbance of bromothymol blue at 614 nm. It does not need the assistance of any other enzyme. Bromothymol blue is less expensive than the coupled enzymes, and the assay can be done in just one step, thus reducing the chance of errors and improving reliability. Compared with the existing assays, it is more simple, sensitive, and precise. In addition, it can be applied to research the activation or inhibition of Adk as it is continuous. However, although this assay is simple and precise, the reaction substrate must be prepared freshly as carbon dioxide in the air may interfere with the assay. ATP spontaneously and slowly hydrolyzes in solution, which can also cause an interference on the assay. Nevertheless, it can be a good alternative to the conventional enzymes-coupled Adk activity assay extensively used in clinical and research laboratories.

Author Contributions: K.S., Y.W., Y.L., C.D., and H.H. conceived and designed the experiments; K.S. and Y.L. performed the experiments; K.S., Y.L., and Y.W. analyzed the data; R.C., T.G., and P.Z. contributed reagents/materials/analysis tools; K.S. and Y.W. wrote the draft; Y.W. and H.H. supervised the research; Q.X. and H.H. revised the manuscript.

Abbreviations

Adk	Adenylate kinase
ATP	adenosine 5′-triphosphates
ADP	adenosine 5′-diphosphates
AMP	adenosine 5′-monophosphate
NADPH	reduced nicotinamide adenine dinucleotide phosphate
Abs	absorbance

References

1. Krishnamurthy, H.; Lou, H.F.; Kimple, A.; Vieille, C.; Cukier, R.I. Associative mechanism for phosphoryl transfer: A molecular dynamics simulation of Escherichia coli adenylate kinase complexed with its substrates. *Proteins* **2005**, *58*, 88–100. [CrossRef] [PubMed]
2. Rogne, P.; Rosselin, M.; Grundstrom, C.; Hedberg, C.; Sauer, U.H.; Wolf-Watz, M. Molecular mechanism of ATP versus GTP selectivity of adenylate kinase. *Proc. Natl. Acad. Sci. USA* **2018**, *115*, 3012–3017. [CrossRef]
3. Oshima, K.; Saiki, N.; Tanaka, M.; Imamura, H.; Niwa, A.; Tanimura, A.; Nagahashi, A.; Hirayama, A.; Okita, K.; Hotta, A.; et al. Human AK2 links intracellular bioenergetic redistribution to the fate of hematopoietic progenitors. *Biochem. Biophys. Res. Commun.* **2018**, *497*, 719–725. [CrossRef]
4. Wujak, M.; Czarnecka, J.; Gorczycka, M.; Hetmann, A. Human adenylate kinases-classification, structure, physiological and pathological importance. *Postep. Hig. Med. Dosw.* **2015**, *69*, 933–945. [CrossRef] [PubMed]
5. Bucher, T.; Pfleiderer, G. Pyruvate Kinase from Muscle. *Method Enzymol.* **1955**, *1*, 435–440.
6. Colowick, S.P.; Kalckar, H.M. The role of myokinase in transphosphorylations I. The enzymatic phosphorylation of hexoses by adenyl pyrophosphate. *J. Biol. Chem.* **1943**, *148*, 117–126.
7. Chiu, C.S.; Su, S.; Russell, P.J. Adenylate kinase from baker's yeast. I. Purification and intracellular location. *Biochimica Biophys. Acta* **1967**, *132*, 361–369. [CrossRef]
8. Oliver, I.T. Spectrophotometric Method for the Determination of Creatine Phosphokinase and Myokinase. *Biochem. J.* **1955**, *61*, 116–122. [CrossRef] [PubMed]
9. Haslam, R.J.; Mills, D.C.B. Adenylate Kinase of Human Plasma Erythrocytes and Platelets in Relation to Degradation of Adenosine Diphosphate in Plasma. *Biochem. J.* **1967**, *103*, 773. [CrossRef] [PubMed]
10. Yu, Z.; Pan, J.; Zhou, H.M. A direct continuous PH-spectrophotometric assay for arginine kinase activity. *Protein Peptide Lett.* **2002**, *9*, 545–552.
11. Dhale, M.A.; Mohan-Kumari, H.P. A comparative rapid and sensitive method to screen L-asparaginase producing fungi. *J. Microbiol. Meth.* **2014**, *102*, 66–68. [CrossRef] [PubMed]

12. De Meyer, T.; Hemelsoet, K.; Van Speybroeck, V.; De Clerck, K. Substituent effects on absorption spectra of pH indicators: An experimental and computational study of sulfonphthaleine dyes. *Dyes Pigments* **2014**, *102*, 241–250. [CrossRef]

13. Zhu, J.J.; Chew, D.A.S.; Lv, S.N.; Wu, W.W. Optimization method for building envelope design to minimize carbon emissions of building operational energy consumption using orthogonal experimental design (OED). *Habitat. Int.* **2013**, *37*, 148–154. [CrossRef]

14. Colowick, S.P. Adenylate Kinase (Myokinase, Adp Phosphomutase). *Method Enzymol.* **1955**, *2*, 598–604.

15. Bowen, W.J.; Kerwin, T.D. The Kinetics of Myokinase. 1. Studies of the Effects of Salts and Ph and of the State of Equilibrium. *Arch. Biochem. Biophys.* **1954**, *49*, 149–159. [CrossRef]

16. Liu, L.; Li, Y.; Wang, Y.; Zhao, P.; Wei, S.; Li, Z.; Chang, H.; He, H. Biochemical characterization and functional analysis of the POU transcription factor POU-M2 of Bombyx mori. *Int. J. Biol. Macromol.* **2016**, *86*, 701–708. [CrossRef] [PubMed]

17. Bernstei, L.H.; Horenstein, J.M.; Sybers, H.D.; Russell, P.J. Adenylate Kinase in Human Tissue. 2. Serum Adenylate Kinase and Myocardial-Infarction. *J. Mol. Cell Cardiol.* **1973**, *5*, 71–85. [CrossRef]

18. Russell, P.J.; Horenstein, J.M.; Goins, L.; Jones, D.; Laver, M. Adenylate Kinase in Human Tissues. 1. Organ Specificity of Adenylate Kinase Isoenzymes. *J. Biol. Chem.* **1974**, *249*, 1874–1879.

19. Kalckar, H.M. The role of myokinase in transphosphorylations II. The enzymatic action of myokinase on adenine nucleotides. *J. Biol. Chem.* **1943**, *148*, 127–137.

20. Moon, S.; Kim, J.; Bae, E. Structural analyses of adenylate kinases from Antarctic and tropical fishes for understanding cold adaptation of enzymes. *Sci. Rep.* **2017**, *7*, 16027. [CrossRef]

21. Slater, E.C. A Method of Measuring the Yield of Oxidative Phosphorylation. *Biochem. J.* **1953**, *53*, 521–530. [CrossRef]

22. Palella, T.D.; Andres, C.M.; Fox, I.H. Human Placental Adenosine Kinase-Kinetic Mechanism and Inhibition. *J. Biol. Chem.* **1980**, *255*, 5264–5269. [PubMed]

15

Engineering the Enantioselectivity of Yeast Old Yellow Enzyme OYE2y in Asymmetric Reduction of (E/Z)-Citral to (R)-Citronellal

Xiangxian Ying [1,*], Shihua Yu [1], Meijuan Huang [1], Ran Wei [1], Shumin Meng [1], Feng Cheng [1], Meilan Yu [2], Meirong Ying [3], Man Zhao [1] and Zhao Wang [1]

[1] Key Laboratory of Bioorganic Synthesis of Zhejiang Province, College of Biotechnology and Bioengineering, Zhejiang University of Technology, Hangzhou 310014, China; yushihuafx@163.com (S.Y.); meyroline.huang@gmail.com (M.H.); weiranzjut@163.com (R.W.); mengshuminzjut@163.com (S.M.); fengcheng@zjut.edu.cn (F.C.); mzhao@zjut.edu.cn (M.Z.); hzwangzhao@163.com (Z.W.)

[2] College of Life Sciences, Zhejiang Sci-Tech Univeristy, Hangzhou 310018, China; meilanyu@zstu.edu.cn

[3] Grain and Oil Products Quality Inspection Center of Zhejiang Province, Hangzhou 310012, China; hz85672100@163.com

* Correspondence: yingxx@zjut.edu.cn.

Academic Editor: Stefano Serra

Abstract: The members of the Old Yellow Enzyme (OYE) family are capable of catalyzing the asymmetric reduction of (E/Z)-citral to (R)-citronellal—a key intermediate in the synthesis of L-menthol. The applications of OYE-mediated biotransformation are usually hampered by its insufficient enantioselectivity and low activity. Here, the (R)-enantioselectivity of Old Yellow Enzyme from *Saccharomyces cerevisiae* CICC1060 (OYE2y) was enhanced through protein engineering. The single mutations of OYE2y revealed that the sites R330 and P76 could act as the enantioselectivity switch of OYE2y. Site-saturation mutagenesis was conducted to generate all possible replacements for the sites R330 and P76, yielding 17 and five variants with improved (R)-enantioselectivity in the (E/Z)-citral reduction, respectively. Among them, the variants R330H and P76C partly reversed the neral derived enantioselectivity from 32.66% *e.e.* (S) to 71.92% *e.e.* (R) and 37.50% *e.e.* (R), respectively. The docking analysis of OYE2y and its variants revealed that the substitutions R330H and P76C enabled neral to bind with a flipped orientation in the active site and thus reverse the enantioselectivity. Remarkably, the double substitutions of R330H/P76M, P76G/R330H, or P76S/R330H further improved (R)-enantioselectivity to >99% *e.e.* in the reduction of (E)-citral or (E/Z)-citral. The results demonstrated that it was feasible to alter the enantioselectivity of OYEs through engineering key residue distant from active sites, e.g., R330 in OYE2y.

Keywords: asymmetric reduction; citral; citronellal; enantioselectivity; Old Yellow Enzyme; site-saturation mutagenesis; substrate binding mode

1. Introduction

(R)-citronellal is a valuable intermediate for the synthesis of L-menthol through an acidic ene-cyclization and subsequent hydrogenation [1–3]. The potential of (R)-citronellal was also explored for the synthesis of natural vitamin E—a kind of fat-soluble vitamin with relatively high antioxidant ability [4,5]. The commercial Takasago process of (R)-citronellal began with myrcene to form an allylic amine, which underwent asymmetric isomerization in the presence of a 2,2′-bis(diphenylphosphino)-1,1′-binaphthyl(BINAP)-Rh complex and subsequent hydrolysis with acid to give enantiomerically pure (R)-citronellal [6]. In contrast to the three-step asymmetric synthesis from myrcene, the one-step enantioselective reduction of natural citral (the crude mixture of 60% geranial and 40% neral) was a simplified process for the synthesis of (R)-citronellal [7]. The enantioselective hydrogenation of (E/Z)-citral to afford an identical enantiomer remained challenging since the reduction of the geometric isomers geranial and neral by the same catalyst usually yielded the enantiocomplementary products. In organocatalysis, the enantioselective hydrogenation of (E/Z)-citral to yield (R)-citronellal required the use of a dual catalyst system comprising of Pd/BaSO4 and chiral 2-diarylmethylpyrrolidine [8]. However, the obtained (R)-citronellal with 89% *e.e.* was insufficient for broad industrial applications.

To develop a greener and cost-effective alternative to organocatalysis, Old Yellow Enzymes (OYEs; EC 1.6.99.1) as biocatalysts have been widely investigated, which are capable of catalyzing the C=C bond reduction of α,β-unsaturated compounds such as (E/Z)-citral [9–13]. Past efforts have been made on the discovery of new, improved biocatalysts for suitable enantioselectivity and activity. Bacterial OYEs commonly produced (S)-citronellal from (E/Z)-citral reduction, while the counterparts from yeasts mainly afforded to (R)-enantiomer [10]. Representative yeast OYEs have been well characterized, including OYE2.6 from *Pichia stipites* [9], OYE1 from *Saccharomyces pastorianus*, and OYE2 and OYE3 from *Saccharomyces cerevisiae* [14,15]. So far, the application of OYE-mediated citral reduction still suffers from insufficient enantioselectivity and activity. Protein engineering has emerged as an attractive and powerful strategy for improving enzyme activity and selectivity [16–19]. The circular permutation of OYE1 from *S. pastorianus* yielded the variants exhibiting over an order of magnitude improved catalytic activity [20]. The activity improvement in the protein engineering of yeast OYEs commonly varied by substrate. The variant P295A of OYE1 from *S. pastorianus* showed three- and seven-fold activity for (R)- and (S)-carvone higher than those of wild type enzyme, respectively; however, it had no significant improvement for geranial and neral [21].

With regard to the alteration of OYE enantioselectivity, one of representative examples was the variant W116F of OYE1 from *S. pastorianus* partly reversed the enantioselectivity in the neral reduction from 19% *e.e.* (S) to 65% *e.e.* (R) as compared to the wild type [22]. In contrast to the substrate binding mode of the wild type enzyme, the W116F mutation enabled the substrate to bind with a flipped orientation in the active site, and thus reverse the enantioselectivity, while maintaining the same mechanism of trans-hydrogenation of C=C bond [23,24]. W116 is not the sole determinant of enantioselectivity in OYEs, and the enantioselectivity switches seemed to vary by enzyme: Y78, I113, and F247 in OYE2.6 [25]; C26, I69, and H167 in ene reductases YqjM [26]; and W66 and W100 in OYE from *Gluconobacter oxydans* (Gox0502) [27]. The study of enantioselectivity alteration in OYEs rarely use citral as substrate. The latest example was the NADH-dependent cyclohexenone ene reductase from *Zymomonas mobilis* (NCR), in which W66 was critical for controlling the orientation of (E/Z)-citral binding and, thus, the variant W66A/I231A of NCR reversed the geranial derived enantioselectivity from 99% *e.e.* (S) to 63% *e.e.* (R) [28].

The study aims to develop the asymmetric reduction of (E/Z)-citral to (R)-citronellal using engineered OYE coupled with formate dehydrogenase for NADH regeneration (Scheme 1). The Old Yellow Enzyme from *S. cerevisiae* CICC1060 (OYE2y) was cloned, overexpressed, and purified, which reduced geranial and neral to citronellal with 82.87% *e.e.* (R) and 32.66% *e.e.* (S), respectively. OYE2y was chosen for enantioselectivity alteration since the wild type enzyme showed higher enantioselectivity than OYE1 from *S. pastorianus* and OYE2 and OYE3 from *S. cerevisiae* in the (E/Z)-citral reduction [29]. The key residues for the enantioselectivity of OYE2y were identified through the combination of sequence alignment and single-point mutations. Relying on subsequent site-saturation mutagenesis, the OYE2y variants with double substitutions exhibited full (R)-enantioselectivity in the reduction of (E)-citral or (E/Z)-citral. In addition, the role of key residues and the substrate binding modes were examined via homology modeling and molecular docking.

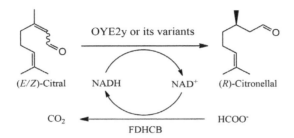

Scheme 1. OYE-mediated asymmetric reduction of (E/Z)-citral to (R)-citronellal coupled with formate dehydrogenase from *Candida boidinii* (FDHCB)-catalyzed NADH regeneration. The reactions were conducted at 37 °C and 200 rpm for 11 h.

2. Results

2.1. OYE2y-Mediated Reduction of (E/Z)-Citral

The yeast Old Yellow Enzyme OYE2y was heterologously expressed in *Echerichia coli* BL21(DE3), and the resulting recombinant OYE2y with *N*-terminal His tag was purified using affinity chromatography. The enzyme OYE2y with 400 amino acid residues shared the sequence identities of 91.50%, 98.75%, and 81.25% to OYE1 from *S. pastorianus* and OYE2 and OYE3 from *S. cerevisiae* [14], respectively. To investigate the enantioselectivity of OYE2y in citral reduction, the purified OYE2y rather than the whole-cell biocatalyst was used to avoid side reactions. OYE2y accepted NADH or NADPH as coenzyme, and formate/formate dehydrogenase system was used for NADH regeneration in this study. The composition of (E/Z)-citral was determined to contain 58.45% geranial and 41.55% neral. During the first 3 h, the concentration of geranial decreased significantly faster than that of neral. Meanwhile, the concentration of (R)-citronellal increased rapidly, and the *e.e.* value was kept at a higher level of >60% (R) (Figure 1A). Then, the conversion rate of neral turned faster from 3 h to 5 h, resulting in the decreasing of *e.e.* value from 65.02% (R) to 40.26% (R). After 6 h, the conversion rate of neral was nearly parallel to that of geranial, meanwhile the *e.e.* values were kept almost constant. The 11 h reaction was completed with 89.51% yield, giving (R)-citronellal with an *e.e.* value of 38.13% (R). The time course of (E/Z)-citral reduction clearly indicated that the ratio of geranial and neral significantly affected the *e.e.* value of (R)-citronellal, which was consistent with previous observations [7,14]. In addition, the isomerization of geranial and neral occurred under some conditions [9,30]. Thus, the use of freshly prepared geranial and neral with high purity was necessary to study the enantioselectivity of OYEs.

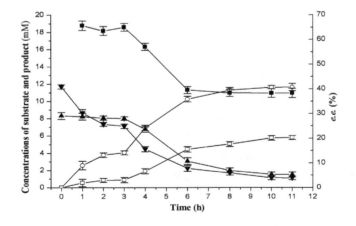

(A)

Figure 1. *Cont.*

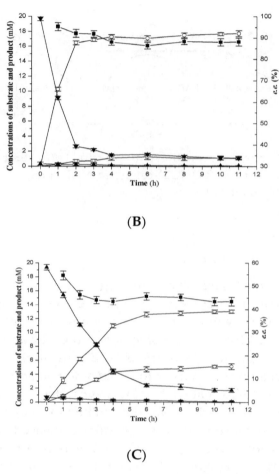

(B)

(C)

Figure 1. Asymmetric reduction of (*E*/*Z*)-citral (**A**), (*E*)-citral (**B**), and (*Z*)-citral (**C**) using the purified OYE2y. □, (*R*)-citronellal; ○, (*S*)-citronellal; ▲, geranial; ▼, neral; ■, the *e.e.* value of hydrogenated product. Data present mean values ± SD from three independent experiments.

Considering the high cost of commercial products, geranial and neral with high purity were prepared according to the previous procedure with improvements [31]. Based on the optimized conditions, the yields of (*E*)-citral and (*Z*)-citral were increased up to 99.39% and 99.35%, respectively. The obtained (*E*)-citral sample contained 98.38% geranial and 1.62% neral, and the obtained (*Z*)-citral sample contained 96.84% neral and 3.16% geranial. When the enzyme OYE2y was tested with newly prepared (*E*)-citral or (*Z*)-citral for 4 h, the enantioselectivity of OYE2y stayed at a relatively constant level (Figure 1B,C). The (*E*)-citral and (*Z*)-citral-derived *e.e.* values after 11 h reduction were 82.87% (*R*) and 32.66% (*S*), respectively.

2.2. Identification of Key Residues for the Enantioselectivity of OYE2y

It was previously reported that the variant W116F of OYE1 from *S. pastorianus* partly reversed the enantioselectivity in the neral reduction [22]. However, the same substitution at site W117 corresponding to W116 in OYE1 even lowered the *e.e.* value from 38.13% (*R*) to 24.01% (*R*) when (*E*/*Z*)-citral was tested as substrate, indicating that the enantioselectivity switch for OYE1 and OYE2y seemed different. Furthermore, the NAD(P)H-dependent enoate reductase (OYE2p) from *S. cerevisiae* YJM1341 was newly discovered for asymmetric reduction of (*E*/*Z*)-citral to (*R*)-citronellal with the *e.e.* value of 88.8% (*R*), with four amino acid residues—G13, A59, I289, and H330—in OYE2p different from S13, S59, V289, and R330 in OYE2y (Figure 2). Considering the difference of enantioselectivity between OYE2p and OYE2y, it was expected that S13, S59, V289, and/or R330 might be critical for the enantioselectivity. Then, the single substitutions were conducted to evaluate this expectation. The catalytic performance of the variants S13G, S59A, and V289I was similar to that of OYE2y (Table 1).

The substitution R330 to H significantly increased the (*R*)-enantioselectivity from 38.13% to 86.88% when (*E*/*Z*)-citral was used as substrate. Remarkably, the enantioselectivity was reversed from 32.66% (*S*) to 71.92% (*R*) when (*Z*)-citral was tested. Except for sequence alignment, the identification of key residues was conducted through the single mutation on the randomly-selected residues. Through multiple mutation attempts, the variant P76M was discovered to benefit the (*R*)-enantioselectivity of OYE2y in the citral reduction (Table 1). The substitution P76 to M increased the (*E*/*Z*) citral-derived *e.e.* value from 38.13% (*R*) to 57.60% (*R*), while the enantioselectivity in the reduction of (*Z*)-citral was lowered from 32.66% (*S*) to 4.76% (*S*). Thus, both R330 and P76 were chosen as the targets for subsequent site saturation mutagenesis.

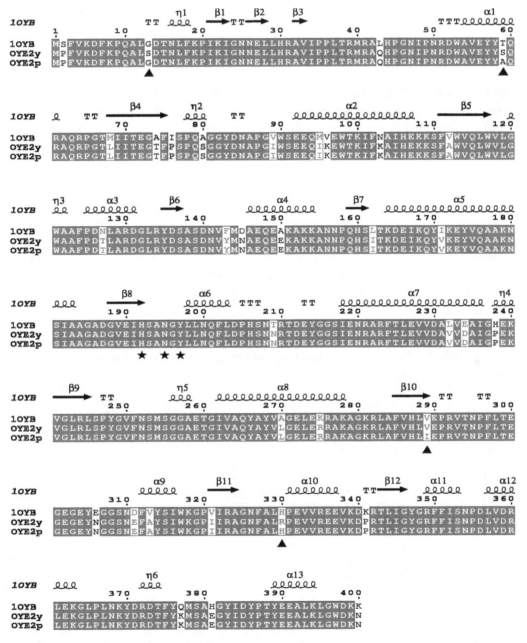

Figure 2. Structure-related sequence alignment between OYE2y and its homologous OYEs. 1OYB: PDB code of OYE1 from *S. pastorianus*; OYE2p: NAD(P)H-dependent enoate reductase from *S. cerevisiae* YJM1341. The secondary structural elements of 1OYB (α-helices, β-strands, T-turns, and η-helices) were indicated above the aligned sequences. The numbering shown was from 1OYB. A red background highlights conserved residues. ▲, the positions where the amino acid residues differed between OYE2y and OYE2p; ★, key residues for catalytic activity. The figure was produced using ESPript 3.0 [32].

Table 1. The catalytic performance of OYE2y and its variants S13G, S59A, P76M, V289I, and R330H [a].

Enzyme	(E)-Citral		(Z)-Citral		(E/Z)-Citral	
	e.e. (%)	Yield (%)	e.e. (%)	Yield (%)	e.e. (%)	Yield (%)
S13G	83.19 ± 1.57 (R)	95.12 ± 2.10	45.29 ± 1.46 (S)	94.60 ± 1.22	37.35 ± 1.63 (R)	96.34 ± 1.87
S59A	80.69 ± 2.21 (R)	90.32 ± 1.54	30.96 ± 1.25 (S)	90.88 ± 2.23	41.66 ± 1.59 (R)	93.48 ± 1.07
V289I	75.18 ± 1.70 (R)	94.88 ± 1.81	37.45 ± 1.55 (S)	91.71 ± 2.10	35.52 ± 2.42 (R)	90.78 ± 0.93
R330H	88.08 ± 1.39 (R)	71.23 ± 0.85	71.92 ± 1.34 (R)	64.12 ± 1.33	86.88 ± 1.36 (R)	52.83 ± 0.75
P76M	86.22 ± 0.75 (R)	80.39 ± 0.64	4.76 ± 1.50 (S)	55.19 ± 0.95	57.60 ± 0.92 (R)	61.53 ± 0.89
OYE2y	82.87 ± 0.98 (R)	92.20 ± 1.07	32.66 ± 1.77 (S)	88.65 ± 1.49	38.13 ± 1.55 (R)	89.51 ± 1.68

[a] Data present mean values ± SD from three independent experiments. (E)-citral contained 98.38% geranial and 1.62% neral, (Z)-citral contained 96.84% neral and 3.16% geranial, and (E/Z)-citral contained 58.45% geranial and 41.55% neral.

2.3. Site-Saturation Mutagenesis of R330 in OYE2y

All R330X variants of OYE2y (X = one of the other 19 amino acids) were successfully expressed in *E. coli*. After the cells were harvested by centrifugation and then disrupted by ultrasonication, each variant with N-terminal His tag was purified using affinity chromatography. As shown in Figure S1, all 19 variant proteins remained in the soluble fraction, revealing that these substitutions did not decrease the solubility. In comparison with the wild type OYE2y, the R330 variants of OYE2y fell into three categories: R330P without catalytic activity, R330Y with similar (R)-stereoselectivity to OYE2y, and the other 17 variants with improved (R)-stereoselectivity (Table 2). R330P did not retain the yellow color, suggesting that its substitution might deactivate the coenzyme binding. When (E/Z)-citral was tested as substrate, the variants except R330Y and R330P increased the (R)-stereoselectivity but decreased the product yield to some extent. In contrast to the reduction of (E)-citral, those 17 variants showed more significant improvement of (R)-enantioselectivity in the reduction of (Z)-citral. Among them, the variants R330H, R330D, and R330W had superior catalytic performance in terms of activity and enantioselectivity.

Table 2. The catalytic performance of OYE2y and its R330X variants [a].

Enzyme	(E)-Citral		(Z)-Citral		(E/Z)-Citral	
	e.e. (%)	Yield (%)	e.e. (%)	Yield (%)	e.e. (%)	Yield (%)
R330H	88.08 ± 1.39 (R)	71.23 ± 0.85	71.92 ± 1.34 (R)	64.12 ± 1.33	86.88 ± 1.36 (R)	52.83 ± 0.75
R330D	92.42 ± 2.30 (R)	63.12 ± 1.93	73.56 ± 0.51 (R)	36.41 ± 1.94	80.30 ± 1.85 (R)	62.70 ± 0.85
R330W	95.05 ± 1.85 (R)	47.32 ± 1.11	72.19 ± 1.09 (R)	53.38 ± 1.12	79.71 ± 1.03 (R)	70.98 ± 1.15
R330I	87.38 ± 1.11 (R)	31.45 ± 0.40	20.93 ± 1.21 (R)	13.19 ± 0.23	74.52 ± 0.69 (R)	28.17 ± 0.68
R330L	89.64 ± 0.79 (R)	75.14 ± 1.25	45.59 ± 0.75 (R)	55.72 ± 0.59	72.58 ± 1.32 (R)	67.45 ± 1.21
R330F	89.40 ± 1.05 (R)	89.86 ± 1.74	57.68 ± 0.86 (R)	78.39 ± 1.34	72.50 ± 1.46 (R)	80.49 ± 1.69
R330E	89.22 ± 0.61 (R)	80.92 ± 0.80	47.80 ± 1.14 (R)	47.82 ± 1.11	69.42 ± 2.34 (R)	72.46 ± 1.02
R330A	88.10 ± 1.37 (R)	83.10 ± 2.11	51.07 ± 1.23 (R)	49.12 ± 1.63	69.42 ± 1.11 (R)	69.61 ± 0.72
R330T	85.74 ± 2.33 (R)	83.33 ± 2.06	27.97 ± 0.48 (R)	43.83 ± 0.87	69.12 ± 2.32 (R)	46.61 ± 0.29
R330N	92.15 ± 2.49 (R)	90.55 ± 1.85	41.87 ± 0.73 (R)	72.44 ± 1.61	68.19 ± 1.56 (R)	78.89 ± 1.98
R330V	89.66 ± 1.50 (R)	91.10 ± 1.01	24.39 ± 1.55 (R)	65.15 ± 1.69	67.68 ± 1.42 (R)	75.77 ± 1.35
R330S	89.60 ± 1.48 (R)	85.62 ± 1.38	33.95 ± 0.67 (R)	69.67 ± 1.92	64.39 ± 2.05 (R)	71.80 ± 1.57
R330C	87.52 ± 1.56 (R)	80.50 ± 1.47	33.80 ± 1.41 (R)	63.79 ± 1.27	64.35 ± 2.33 (R)	70.16 ± 0.86
R330K	86.18 ± 0.84 (R)	84.82 ± 2.19	5.45 ± 0.59 (R)	63.69 ± 1.51	61.53 ± 1.76 (R)	80.64 ± 0.43
R330Q	84.81 ± 1.41 (R)	90.74 ± 2.55	10.90 ± 0.84 (R)	76.88 ± 0.66	61.17 ± 0.71 (R)	85.23 ± 1.44
R330G	92.39 ± 1.66 (R)	10.96 ± 0.29	5.07 ± 1.22 (R)	8.49 ± 0.15	60.97 ± 1.56 (R)	14.82 ± 0.33
R330M	87.88 ± 1.23 (R)	92.40 ± 1.77	0.37 ± 0.65 (R)	85.18 ± 2.08	58.40 ± 1.08 (R)	86.71 ± 1.37
R330Y	87.03 ± 1.12 (R)	96.61 ± 2.16	40.92 ± 1.53 (S)	88.24 ± 1.89	38.65 ± 1.11 (R)	91.56 ± 1.72
R330P [b]	/	/	/	/	/	/
OYE2y	82.87 ± 0.98 (R)	92.20 ± 1.07	32.66 ± 1.77 (S)	88.65 ± 1.49	38.13 ± 1.55 (R)	89.51 ± 1.68

[a] X represents any of 20 amino acids. Data present mean values ± SD from three independent experiments. (E)-citral contained 98.38% geranial and 1.62% neral, (Z)-citral contained 96.84% neral and 3.16% geranial, and (E/Z)-citral contained 58.45% geranial and 41.55% neral. [b] "/" represents no catalytic activity.

2.4. Site-Saturation Mutagenesis of P76 in OYE2y

Similar to R330X variants, all P76X variants of OYE2y were successfully expressed in *E. coli* and purified (Figure S2). However, the number of the P76 variants in the category without catalytic activity (P76Y, P76Q, P76D, P76E, P76R, P76H, P76F, P76W, and P76K) was obviously greater than that of R330 variants, suggesting that P76 could be also critical for the activity. The category with significantly improved enantioselectivity included P76C, P76S, P76M, P76G, and P76N, whereas the other five variants (P76A, P76V, P76T, P76L, and P76I) showed similar catalytic performance to that of OYE2y (Table 3). Similar to the trend in the R330X variants, higher (*R*)-stereoselectivity of OYE2y variants was accompanied by lower product yields. When (*E/Z*)-citral was used as substrate, the substitution of P76 to C increased the *e.e.* value from 44.13% (*R*) to 69.92% (*R*), but the yield was lowered from 89.51% to 49.65%. Particularly, the *e.e.* value in the reduction of (*Z*)-citral was partly reversed from 32.66% (*S*) to 37.50% (*R*).

Table 3. The catalytic performance of OYE2y and its P76X variants [a].

Enzyme	(E)-Citral e.e. (%)	(E)-Citral Yield (%)	(Z)-Citral e.e. (%)	(Z)-Citral Yield (%)	(E/Z)-Citral e.e. (%)	(E/Z)-Citral Yield (%)
P76C	85.39 ± 2.70 (R)	65.32 ± 0.76	37.50 ± 1.55 (R)	38.54 ± 1.61	69.92 ± 2.46 (R)	49.65 ± 1.40
P76S	81.23 ± 1.79 (R)	81.84 ± 0.62	10.19 ± 0.73 (S)	48.15 ± 0.88	62.80 ± 0.97 (R)	64.48 ± 0.90
P76M	86.22 ± 0.75 (R)	80.39 ± 0.64	4.76 ± 1.50 (S)	55.19 ± 0.95	57.60 ± 1.92 (R)	61.53 ± 0.89
P76G	85.52 ± 2.55 (R)	83.94 ± 0.78	3.63 ± 1.47 (S)	57.54 ± 1.84	53.19 ± 0.81 (R)	78.04 ± 1.52
P76N	86.45 ± 1.33 (R)	85.36 ± 0.49	13.05 ± 0.67 (S)	69.39 ± 1.43	49.59 ± 1.55 (R)	72.59 ± 0.77
P76A	86.55 ± 0.91 (R)	90.57 ± 1.98	21.21 ± 1.95 (S)	88.96 ± 0.87	41.88 ± 0.95 (R)	90.43 ± 2.22
P76V	87.54 ± 1.55 (R)	92.50 ± 2.51	30.15 ± 1.67 (S)	82.30 ± 2.07	40.49 ± 1.01 (R)	87.37 ± 1.28
P76T	86.12 ± 1.35 (R)	93.79 ± 2.26	28.10 ± 1.69 (S)	91.24 ± 2.12	37.05 ± 1.13 (R)	91.50 ± 0.94
P76L	84.28 ± 1.62 (R)	90.95 ± 1.83	38.01 ± 2.27 (S)	88.36 ± 1.46	35.05 ± 1.64 (R)	88.46 ± 1.32
P76I	86.10 ± 1.03 (R)	94.76 ± 1.74	40.23 ± 1.81 (S)	89.95 ± 1.33	34.51 ± 1.17 (R)	91.88 ± 2.45
P76Y [b]	/	/	/	/	/	/
P76Q [b]	/	/	/	/	/	/
P76D [b]	/	/	/	/	/	/
P76E [b]	/	/	/	/	/	/
P76R [b]	/	/	/	/	/	/
P76H [b]	/	/	/	/	/	/
P76F [b]	/	/	/	/	/	/
P76W [b]	/	/	/	/	/	/
P76K [b]	/	/	/	/	/	/
OYE2y	82.87 ± 0.98 (R)	92.20 ± 1.07	32.66 ± 1.77 (S)	88.65 ± 1.49	38.13 ± 1.55 (R)	89.51 ± 1.68

[a] X represents one of the other 19 amino acids. Data present mean values ± SD from three independent experiments. (*E*)-citral contained 98.38% geranial and 1.62% neral, (*Z*)-citral contained 96.84% neral and 3.16% geranial, and (*E/Z*)-citral contained 58.45% geranial and 41.55% neral. [b] "/" represents no catalytic activity.

2.5. Evaluation of Double Substitution at Sites P76 and R330 of OYE2y

To investigate the effect of the double substitutions on P76 and R330, we firstly conducted site-directed mutagenesis of P76 to C and R330 to H, D, W, and C, resulting in the four variants (Table 4). The *e.e.* values and yields of the resulting variants were higher than those of the variant P76C, but similar to those of the corresponding R330H, R330D, R330W, and R330C. On the other hand, the other set of double substitutions was created by site-directed mutagenesis of R330 to H and P76 to M, G, and S (Table 4). Among them, the (*E*)-citral-derived *e.e.* values were significantly increased up to >99% (*R*) with relatively higher yields (64.09%~73.88%). In the (*E/Z*)-citral reduction, the variants P76M/R330H, P76M/R330H, and P76M/R330H also exhibited full (*R*)-enantioselectivity despite lower product yields (9.12–15.83%).

Table 4. The catalytic performance of double substitution variants of OYE2y at sites P76 and R330 [a].

Enzyme	(E)-Citral		(Z)-Citral		(E/Z)-Citral	
	e.e. (%)	Yield (%)	e.e. (%)	Yield (%)	e.e. (%)	Yield (%)
P76C	85.39 ± 2.70 (R)	65.32 ± 0.76	37.50 ± 1.55 (R)	38.54 ± 1.61	69.92 ± 2.46 (R)	49.65 ± 1.40
P76C/R330H	88.00 ± 2.04 (R)	78.09 ± 1.16	76.16 ± 0.88 (R)	54.53 ± 0.45	82.32 ± 1.22 (R)	57.74 ± 0.84
P76C/R330D	88.77 ± 1.81 (R)	80.00 ± 1.52	76.45 ± 1.03 (R)	52.23 ± 1.25	81.70 ± 2.18 (R)	63.28 ± 1.36
P76C/R330W	89.53 ± 2.23 (R)	70.30 ± 1.87	70.32 ± 1.16 (R)	56.49 ± 0.68	81.44 ± 1.31 (R)	65.31 ± 1.72
P76C/R330C	87.43 ± 1.37 (R)	90.10 ± 1.96	49.17 ± 2.44 (R)	51.14 ± 0.79	77.37 ± 1.04 (R)	66.19 ± 1.42
R330H	88.08 ± 1.39 (R)	71.23 ± 0.85	71.92 ± 1.34 (R)	64.12 ± 1.33	86.88 ± 1.36 (R)	52.83 ± 0.75
P76M/R330H	>99 (R)	73.88 ± 2.66	75.01 ± 1.59 (R)	13.99 ± 0.23	>99 (R)	15.83 ± 0.45
P76G/R330H	>99 (R)	64.09 ± 1.32	77.60 ± 2.06 (R)	12.64 ± 0.31	>99 (R)	11.61 ± 0.38
P76S/R330H	>99 (R)	64.74 ± 1.57	74.74 ± 1.84 (R)	8.90 ± 0.16	>99 (R)	9.12 ± 0.22
OYE2y	82.87 ± 0.98 (R)	92.20 ± 1.07	32.66 ± 1.77 (S)	88.65 ± 1.49	38.13 ± 1.55 (R)	89.51 ± 1.68

[a] Data present mean values ± SD from three independent experiments. (E)-citral contained 98.38% geranial and 1.62% neral, (Z)-citral contained 96.84% neral and 3.16% geranial, and (E/Z)-citral contained 58.45% geranial and 41.55% neral.

3. Discussion

For the synthesis of (R)-citronellal from OYE-mediated citral reduction, low-cost (E/Z)-citral is the industrial desire in contrast to geranial and neral [9]. However, (E/Z)-citral reduction remains challenging due to limited chemoselectivity and enantioselectivity. On the one hand, the presence of multiple C=C and C=O double bonds of citral makes the whole-cell biocatalyst impossible to avoid side reactions [33]. Thus, the purified enzyme is commonly required as biocatalyst. On the other hand, the hydrogenated products from the geometric isomers geranial and neral are usually enantiocomplementary when the wild type OYE was used as biocatalyst, reducing the reaction's enantioselectivity. The features of (E/Z)-citral reduction make it difficult to implement a high-throughput screening (HTS) method for determining the enantioselectivity of the large variant libraries [34–36]. Typically, the samples must be examined individually by chiral GC, which requires at least thirty minutes per sample. To keep the variant library as minimum as possible, the best strategy for enantioselectivity alteration turns to be site-saturation mutagenesis of individual key residue(s) rather than the HTS-based directed evolution [18,37,38]. Based on the strategy of site-saturated mutagenesis, several groups have successfully increased (R)-enantioselectivity in the (E/Z)-citral reduction, and our study further demonstrated that it was feasible to achieve the full (R)-enantioselectivity in the OYE-mediated (E/Z)-citral reduction through protein engineering.

The R330X and P76X variant libraries yielded 17 and five variants with improved (R)-enantioselectivity, respectively, indicating that the subtle change of structure would significantly affect the enantioselectivity. In the variants R330H and P76C, the amino acid pair R and H possessed electrically charged side group, while the side groups of the amino acid pair P and C was polar and uncharged. It was suggested that the substitution could give priority to amino acid residue(s) with similar side group in terms size, polarity and charge, if no clear structure–function relationship was available. Furthermore, the (Z)-citral-derived e.e. value of R330H (71.92%, R) was much higher than that of OYE2p (26.5%, R) [7], suggesting that the role of S13, S59, and V289 could not be ignored for the (R)-enantioselectivity. For the single substitutions, the (R)-enantioselectivity improvement in the (E/Z)-citral reduction was mainly attributed to the enantioselectivity inversion in the (Z)-citral reduction, meanwhile, the same variant reduced (E)-citral to (R)-citronellal with similar e.e. values. Different from the single substitutions, the double substitutions led to further improved (R)-stereoselectivity in the (E)-citral reduction rather than (Z)-reduction, which was rarely observed in the enantioselectivity alternation of OYEs.

To get detailed insights into the molecular mechanism, the models of OYE2y were created from the OYE1 structure (PDB code: 1OYB) using SWISS-MODEL and molecular docking was conducted using the program AutoDock Vina. The substrate and FMN were docked in silico into the models of OYE2y and its variants. In the model of wild type OYE2y, a conserved H192/N195 pair formed hydrogen bonds with the carbonyl oxygen of α,β-unsaturated carbonyl compounds; a hydride was enantioselectively transferred to the substrate C_β atom from $FMNH_2$; and the Y197 residue provided a proton to the substrate C_α atom as an electron acceptor [17]. From the model of wild type OYE2y, the distances from C_β of neral to the side groups of P76 and R330 were calculated to be 14.92 and 20.84 Å, respectively (Figure 3). Similarly, the distances from C_β of geranial to the side groups of P76 and R330 were calculated to be 14.49 and 21.16 Å, respectively. It was assumed that the residues affecting the enantioselectivity of OYEs might directly interact with the substrate [25]. Our results indicated that the residue distant from active sites, e.g., R330 in OYE2y, could also be pivotal for determining the enantioselectivity of OYEs. Furthermore, substrate modeling into the wild type enzyme revealed two different binding modes for the two isomers geranial and neral, leading to products with different enantioselectivity. The docking analyses of variants P76C, R330H, and P76M/R330 suggested that the reversed enantioselectivity in the neral reduction was due to the flipped binding orientation that placed the opposite face of the alkene above the *si* face of the FMN cofactor [24]. Meanwhile, the same variant reduced (*E*)-citral with preserved (*R*)-enantioselectivity derived from the same binding orientation as wild type OYE2y (Figure 4). Although it was acceptable that enantioselectivity was controlled by tuning the orientation of substrate in the binding sites, how subtle changes control the orientation of substrate binding remains to be unraveled, and X-ray crystallography of the variants in the future study would benefit to clarify the subtle structural differences at the substrate-binding site.

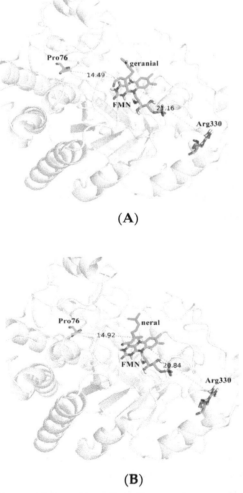

(A)

(B)

Figure 3. The residues targeted for site-saturation mutagenesis in the homology model of OYE2y. The model structure of OYE2y was constructed with the crystal structure of OYE1 (PDB code: 1OYB) as template. Distances between the C_β atom of geranial (**A**) and neral (**B**) and side chains of P76 and R330 were determined. Green, carbon atom; blue, nitrogen atom; tangerine, oxygen atom; white, hydrogen atom; orange, phosphorus atom.

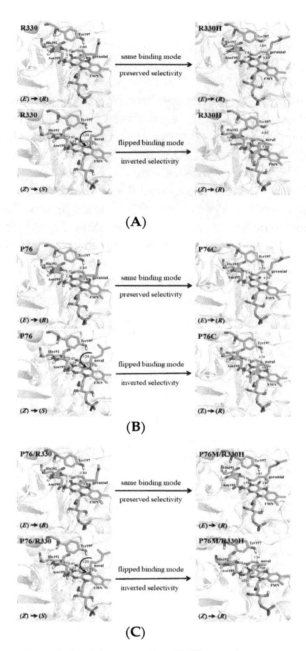

Figure 4. The binding modes of citral isomers in OYE2y and its variants R330H (**A**), P76C (**B**), and P76M/R330H (**C**) leading to either (*R*)- or (*S*)-citronellal. The catalytic residues H192, N195, Y197, and the prosthetic group FMN were depicted. Green, carbon atom; blue, nitrogen atom; tangerine, oxygen atom; white, hydrogen atom; orange, phosphorus atom.

4. Materials and Methods

4.1. Organisms and Chemicals

The organism *S. cerevisiae* CICC1060 was purchased from the China Center of Industrial Culture Collection (CICC, Beijing, China). *S. cerevisiae* CICC1060 was cultured with YPD medium (tryptone 20 g/L, yeast extract 10 g/L, and glucose 20 g/L) at 30 °C for 24 h. The pEASY-E1 expression vector from TransGen Biotech Co., Ltd (Beijing, China) was used for overexpression of the enzyme OYE2y, and the *E. coli* strain BL21(DE3) was used as the host. *E. coli* cultures were grown routinely in Luria Bertani (LB) medium at 37 °C for 12 h.

The standards (*S*)-citronellal, (*R*)-citronellal, and (*S*/*R*)-citronellal were obtained from Sigma-Aldrich (Shanghai) Trading Co., Ltd. (Shanghai, China). Other chemicals of analytical grade were purchased from Sangon Biotech Co. Ltd (Shanghai, China) or Shanghai Jingchun Reagent Co., Ltd (Shanghai, China). The site-directed mutagenesis kit and the restriction enzyme *Dpn* I were obtained from Vazyme Biotech Co., Ltd. (Nanjing, China). KOD DNA polymerase was purchased from TransGen Biotech Co., Ltd (Beijing, China). The Ni-NTA-HP resin column for protein purification was obtained from GE Healthcare Life Sciences (Shanghai, China).

4.2. Preparation of (Z)-Citral and (E)-Citral

(Z)-citral and (E)-citral were prepared by a modification of the procedure described previously [38]. Activated MnO_2 (1.15 g) was added into a 50-mL three-necked bottom round flask which atmosphere was replaced with N_2. One-hundred milligrams of geraniol or nerol was dissolved in 16 mL dry hexane and was charged in the flask to initiate the alcohol oxidation. The reaction was maintained at 0 °C and 450 rpm for 6 h. Then, the reaction solution was filtered through a filter paper and hexane in the filtrate was removed by vacuum evaporation at 45 °C. Finally, an aliquot of the collected product (E)-citral or (Z)-citral was dissolved in ethyl acetate and subjected to the analyses of gas chromatography (GC) and gas chromatography–mass spectrometry (GC–MS).

4.3. Cloning, Expression, and Purification of OYE2y

The gene encoding OYE2y was PCR-amplified from the genomic DNA of *S. cerevisiae* CICC1060 using a set of primers: Forward, 5′-ATGCCATTTGTTAAGGACTTTAAGCCAC-3′; Reverse, 5′-TTAATTTTTGTCCCAACCGAGTTTTAGAGC-3′. The conditions for PCR amplification of the *oye2y* gene were 94 °C for 2 min for initial denaturalization, 30 cycles of 94 °C for 30 s, 57 °C for 30 s, 72 °C for 80 s, and 72 °C for 10 min for the final extension.

Following the procedure of expression and purification of ReBDH [39], the PCR products were purified and then ligated with the expression vector pEASY-E1 through the AT ligation strategy. The recombinant plasmid harboring the *oye2y* gene, designated as pEASY-E1-*oye2y*, was verified by DNA sequencing (Sangon Biotech, Shanghai, China) and then transformed into E. coli BL21 (DE3) competent cells, resulting in the recombinant strain E. coli BL21(DE3)/pEASY-E1-*oye2y*. The recombinant cells containing pEASY-E1-*oye2y* were grown in the LB medium with 100 μg/mL ampicillin at 37 °C and 200 rpm until the OD_{600} was 0.6–0.8, and then 0.2 mM isopropyl β-D-1-thiogalactopyranoside (IPTG) was supplemented to initiate the induction at 23 °C and 160 rpm. After 12 h of growth, recombinant E. coli cells were harvested by centrifugation and further washed using 50 mM Tris-HCl buffer (pH 8.0). The cells were disrupted through ultrasonication for 10 min, and the cell debris and cell lysate were removed by centrifugation to result in a clear cell extract. The crude cell extracts containing OYE2y was applied to a Ni-NTA chelating affinity column equilibrated with the binding buffer (5 mM imidazole and 300 mM NaCl dissolved in 50 mM Tris-HCl, pH 8.0). Unbound proteins were washed off by the application of the binding buffer. The recombinant OYE2y was eluted with 100 mM imidazole in 50 mM Tris-HCl (pH 8.0), desalted with 50 mM Tris-HCl buffer (pH 8.0) by ultrafiltration and then stored at −20 °C for further study.

4.4. Construction of OYE2y Variants by Site-Directed Mutagenesis

The variants with single or double substitutions were constructed by site-directed mutagenesis according to the QuikChange Mutagenesis Kit. PCR amplification to introduce substitution was performed in 50 μL of standard PCR mixture with 50 ng of template plasmid DNA and 15 pmol each of the appropriate set of primers using the following temperature cycle; 30 s at 95 °C, followed by 30 cycles of 95 °C for 15 s, appropriate annealing temperature (55~61 °C) for 15 s, and 72 °C for 80 s, and the final extension of 5 min at 72 °C. The plasmid pEASY-E1-*oye2y* was used as template DNA in the single substitution, while the creation of double mutants was based on the generated OYE2y P76C or R330H gene as the template. The primer sets for single or double substitutions were shown in Tables S1 and S2. The amplified PCR fragments were digested with the restriction enzyme *Dpn* I at 37 °C for 1 h, and then the digested DNA was directly introduced into E. coli strain BL21(DE3).

Each constructed plasmid was confirmed by sequencing. Expression and purification of the resulting OYE2y variants were conducted using the same procedure as OYE2y.

4.5. Homology Modeling and Molecular Docking

Using the crystal structure of OYE1 from *S. pastorianus* (PDB number: 1OYB) as a template [40], the structural model of OYE2y was obtained by a homology modeling strategy [41,42]. Molecular docking simulation was performed by AutoDock Vina [43] when the binding package was set at a distance of 15 Å from FMN N_5 atom. (*E*)-citral or (*Z*)-citral acted as a ligand to molecular docking with OYE2y, and the calculation of geometric parameters and ligand structure was performed by ChemBioDraw 12.0 (CambridgeSoft, Cambridge, MA, USA). To make the results more accurate, 100 consecutive runs were performed and the highest ranked score from each run was used to calculate the average score of each flexible ligand configuration. The optimal configuration and the resulting substrate–enzyme complexes were further processed using the PYMOL software [44]. The candidate complexes were acceptable when they met both criteria: (1) the substrate carbonyl oxygen should be capable of forming hydrogen bonds with the side chains of both H192 and N195 and (2) the distance between the FMN N_5 atom and the β-unsaturated carbon of the substrate molecule was be in an appropriate range from 3.5 Å to 4.1 Å [22,45,46].

4.6. Asymmetric Reduction of Citral Mediated by OYE2y or Its Variants

The reaction mixture (1 mL) contained 50 mM PIPES buffer (pH 7.0), 20 mM substrate, 1 mg OYE2y or its variant, 0.6 U formate dehydrogease from *Candida boidinii* (FDHCB), and 100 mM sodium formate, 0.96 mM NAD^+. The substrate included (*Z*)-citral, (*E*)-citral, or (*E/Z*)-citral, which stock solution was 200 mM substrate in isopropanol. The overexpression and purification of FDHCB were conducted according to the procedure described previously [39]. The reaction was conducted at 30 °C and 200 rpm for 11 h, unless otherwise specified. The reaction mixture was centrifuged to remove the cells, and the resulting supernatant was extracted with equal volume of ethyl acetate at 30 °C and 200 rpm for 2 h. Finally, the solvent phase was collected, dried over anhydrous sodium sulfate and subjected to the analyses of GC and GC–MS.

4.7. Analyses of GC and GC–MS

(*Z*)-citral, (*E*)-citral, (*S*)-citronellal, and (*R*)-citronellal were determined by GC (Agilent 6890N) equipped with an FID detector and chiral capillary BGB-174 column (BGB Analytik, Böckten, Switzerland, 30 m × 250 μm × 0.25 μm). The flow rate and split ratio of N_2 as the carrier gas were set as 1.38 mL/min and 1:100, respectively. Both injector and detector were kept at 250 °C. The column temperature program was listed as follows; initial temperature of 90 °C for 25 min, 20 °C/min ramp to 150 °C for 3 min, and 30 °C/min ramp to 180 °C for 3 min. The injection volume was 1 μL. The retention times of (*S*)-citronellal, (*R*)-citronellal, (*Z*)-citral, and (*E*)-Citral were 22.459 min, 23.067 min, 29.164 min and 30.398 min, respectively (Figure S3).

(*S*)-citronellal, (*R*)-citronellal, (*Z*)-citral, and (*E*)-Citral were validated through GC–MS analysis (Figure S4). The GC–MS analysis (Agilent7890A/5975C, Agilent Technologies Inc., Santa Clara, CA, USA) comprised the following parameters; auxiliary heating zone temperature, 250 °C; MS quadrupole temperature, 150 °C; ion source temperature, 230 °C; scan quality range, 30–500 amu; emission current, 200 μA; and electron energy, 70 eV.

4.8. Nucleotide Sequence Accession Number

The gene encoding OYE2y has been deposited in the GenBank database under the accession numbers of MK372229.

5. Conclusions

In summary, significant increase of (*R*)-enantioselectivity in the (*E*/*Z*)-citral reduction was achieved by saturation mutagenesis of P76 and R330 in OYE2y. Remarkably, the variants P76M/R330H, P76G/R330H, and P76S/R330H exhibited full (*R*)-enantioselectivity in the reduction of (*E*)-citral or (*E*/*Z*)-citral. The variants with improved (*R*)-enantioselectivity usually came along with lower catalytic activities, indicating that the sites P76 and R330 were important for enantioselectivity as well as activity. In contrast to P76, R330 was relatively distant from active sites and its substitutions brought more beneficial impacts on enantioselectivity. Our results proved that it was reasonable to alter the enantioselectivity of OYE2y by saturation mutagenesis of key residue distant from active sites.

Supplementary Materials: Table S1: The primer information of site saturation mutation of P76 in OYE2y; Table S2: The primer information of site saturation mutation of R330 in Oye2y; Figure S1: SDS-PAGE (12%) analysis of the purified OYE2y R330X variants; Figure S2: SDS-PAGE (12%) analysis of the purified OYE2y P76X variants; Figure S3: Gas chromatography analysis for standards (*S*)-citronellal (22.459 min), (*R*)-citronellal (23.067 min), (*Z*)-citral (29.164 min), and (*E*)-citral (30.398 min); Figure S4: Gas chromatography–mass spectrometry analysis for (*S*)-citronellal (A), (*R*)-citronellal (B),(*Z*)-citral (C), and (*E*)-citral (D) in the asymmetric reduction of (*E*/*Z*)-citral.

Author Contributions: Conceptualization, X.Y.; Data Curation, X.Y., M.Y. (Meilan Yu), and M.Y. (Meirong Ying); Formal Analysis, X.Y., M.Y. (Meilan Yu), and M.Y. (Meirong Ying); Funding Acquisition, X.Y.; Investigation, S.Y., M.H., R.W., S.M., F.C., and M.Z.; Supervision, Z.W.; Writing—Original Draft, X.Y.

References

1. Itoh, H.; Maeda, H.; Yamada, S.; Hori, Y.; Mino, T.; Sakamoto, M. Kinetic resolution of citronellal by chiral aluminum catalysts: L-menthol synthesis from citral. *Org. Chem. Front.* **2014**, *1*, 1107–1115. [CrossRef]

2. Lenardão, E.J.; Botteselle, G.V.; de Azambuja, F.; Perin, G.; Jacob, R.G. Citronellal as key compound in organic synthesis. *Tetrahedron* **2007**, *63*, 6671–6712. [CrossRef]

3. Nie, Y.; Chuah, G.-K.; Jaenicke, S. Domino-cyclisation and hydrogenation of citronellal to menthol over bifunctional Ni/Zr-Beta and Zr-beta/Ni-MCM-41 catalysts. *Chem. Commun.* **2006**, 790–792. [CrossRef] [PubMed]

4. Coffen, D.L.; Cohen, N.; Pico, A.M.; Schmid, R.; Sebastian, M.J.; Wang, F. A microbial lipase based stereoselective synthesis of (d)-α-tocopherol from (*R*)-citronellal and (*S*)-(6-hydroxy-2,5,7,8-tetramethylchroman-2-yl)acetic acid. *Heterocycles* **1994**, *39*, 527–552. [CrossRef]

5. Eggersdorfer, M.; Laudert, D.; Létinois, U.; McClymont, T.; Medlock, J.; Netscher, T.; Bonrath, W. One hundred years of vitamins—a success story of the natural sciences. *Angew. Chem. Int. Ed.* **2012**, *51*, 12960–12990. [CrossRef] [PubMed]

6. Tani, K.; Yamagata, T.; Akutagawa, S.; Kumobayashi, H.; Taketomi, T.; Takaya, H.; Miyashita, A.; Noyori, R.; Otsuka, S. Highly enantioselective isomerization of prochiral allylamines catalyzed by chiral diphosphine Rhodium(I) complexes: Preparation of optically active enamines. *J. Am. Chem. Soc.* **1984**, *106*, 5208–5217. [CrossRef]

7. Zheng, L.; Lin, J.; Zhang, B.; Kuang, Y.; Wei, D. Identification of a yeast old yellow enzyme for highly enantioselective reduction of citral isomers to (*R*)-citronellal. *Bioresour. Bioprocess.* **2018**, *5*, 9. [CrossRef]

8. Maeda, H.; Yamada, S.; Itoh, H.; Hori, Y. A dual catalyst system provides the shortest pathway for L-menthol synthesis. *Chem. Commun.* **2012**, *48*, 1772–1774. [CrossRef] [PubMed]

9. Bougioukou, D.J.; Walton, A.Z.; Stewart, J.D. Towards preparative-scale, biocatalytic alkene reductions. *Chem. Comm.* **2010**, *46*, 8558–8560. [CrossRef] [PubMed]

10. Müller, A.; Hauer, B.; Rosche, B. Enzymatic reduction of the α,β-unsaturated carbon bond in citral. *J. Mol. Catal. B Enzym.* **2006**, *38*, 126–130. [CrossRef]

11. Toogood, H.S.; Gardiner, J.M.; Scrutton, N.S. Biocatalytic reductions and chemical versatility of the old yellow enzyme family of flavoprotein oxidoreductases. *ChemCatChem* **2010**, *2*, 892–914. [CrossRef]

12. Toogood, H.S.; Scrutton, N.S. New developments in 'ene'-reductase catalysed biological hydrogenations. *Curr. Opin. Chem. Biol.* **2014**, *19*, 107–115. [CrossRef]

13. Toogood, H.S.; Scrutton, N.S. Discovery, characterization, engineering, and applications of ene-reductases for industrial biocatalysis. *ACS Catal.* **2018**, *8*, 3532–3549. [CrossRef]

14. Müller, A.; Hauer, B.; Rosche, B. Asymmetric alkene reduction by yeast old yellow enzymes and by a novel *Zymomonas mobilis* reductase. *Biotechnol. Bioeng.* **2007**, *98*, 22–29. [CrossRef]

15. Brenna, E.; Gatti, F.G.; Monti, D.; Parmeggiani, F.; Serra, S. Stereochemical outcome of the biocatalysed reduction of activated tetrasubstituted olefins by old yellow enzymes 1–3. *Adv. Synth. Catal.* **2012**, *354*, 105–112. [CrossRef]

16. Amato, E.D.; Stewart, J.D. Applications of protein engineering to members of the old yellow enzyme family. *Biotechnol. Adv.* **2015**, *33*, 624–631. [CrossRef]

17. Kataoka, M.; Miyakawa, T.; Shimizu, S.; Tanokura, M. Enzymes useful for chiral compound synthesis: Structural biology, directed evolution, and protein engineering for industrial use. *Appl. Microbiol. Biotechnol.* **2016**, *100*, 5747–5757. [CrossRef]

18. Li, R.; Wijma, H.J.; Song, L.; Cui, Y.; Otzen, M.; Tian, Y.; Du, J.; Li, T.; Niu, D.; Chen, Y.; et al. Computational redesign of enzymes for regio- and enantioselective hydroamination. *Nature Chem. Biol.* **2018**, *14*, 664–670. [CrossRef]

19. Ying, X.; Zhang, J.; Wang, C.; Huang, M.; Ji, Y.; Cheng, F.; Yu, M.; Wang, Z.; Ying, M. Characterization of a carbonyl reductase from *Rhodococcus erythropolis* WZ010 and its variant Y54F for asymmetric synthesis of (S)-N-Boc-3-hydroxypiperidine. *Molecules* **2018**, *23*, 3117. [CrossRef]

20. Daugherty, A.B.; Govindarajan, S.; Lutz, S. Improved biocatalysts from a synthetic circular permutation library of the flavin-dependent oxidoreductase old yellow enzyme. *J. Am. Chem. Soc.* **2013**, *135*, 14425–14432. [CrossRef]

21. Quertinmont, L.T.; Lutz, S. Cell-free protein engineering of Old Yellow Enzyme 1 from *Saccharomyces pastorianus*. *Tetrahedron* **2016**, *72*, 7282–7287. [CrossRef]

22. Padhi, S.K.; Bougioukou, D.J.; Stewart, J.D. Site-saturation mutagenesis of tryptophan 116 of *Saccharomyces pastorianus* old yellow enzyme uncovers stereocomplementary variants. *J. Am. Chem. Soc.* **2009**, *131*, 3271–3280. [CrossRef]

23. Brenna, E.; Crotti, M.; Gatti, F.G.; Monti, D.; Parmeggiani, F.; Powell, R.W., III; Santangelo, S.; Stewart, J.D. Opposite enantioselectivity in the bioreduction of (Z)-β-aryl-bcyanoacrylates mediated by the tryptophan 116 mutants of old yellow enzyme 1: Synthetic approach to (R)- and (S)-β-aryl-glactams. *Adv. Synth. Catal.* **2015**, *357*, 1849–1860. [CrossRef]

24. Pompeu, Y.A.; Sullivan, B.; Stewart, J.D. X-ray crystallography reveals how subtle changes control the orientation of substrate binding in an alkene reductase. *ACS Catal.* **2013**, *3*, 2376–2390. [CrossRef]

25. Walton, A.Z.; Sullivan, B.; Patterson-Orazem, A.C.; Stewart, J.D. Residues controlling facial selectivity in an alkene reductase and semirational alterations to create stereocomplementary variants. *ACS Catal.* **2015**, *4*, 2307–2318. [CrossRef]

26. Rüthlein, E.; Classen, T.; Dobnikar, L.; Schölzel, M.; Pietruszka, J. Finding the selectivity switch—A rational approach towards stereocomplementary variants of the ene reductase YqjM. *Adv. Synth. Catal.* **2015**, *357*, 1775–1786. [CrossRef]

27. Yin, B.; Deng, J.; Lim, L.; Yuan, Y.A.; Wei, D. Structural insights into stereospecific reduction of α, β-unsaturated carbonyl substrates by old yellow enzyme from *Gluconobacter oxydans*. *Biosci. Biotechnol. Biochem.* **2015**, *79*, 410–421. [CrossRef]

28. Kress, N.; Rapp, J.; Hauer, B. Enantioselective reduction of citral isomers in NCR ene reductase: Analysis of an active site mutant library. *ChemBioChem* **2017**, *18*, 717–720. [CrossRef]

29. Hall, M.; Stueckler, C.; Hauer, B.; Stuermer, R.; Friedrich, T.; Breuer, M.; Kroutil, W.; Faber, K. Asymmetric bioreduction of activated C=C bonds using *Zymomonas mobilis* NCR enoate reductase and old yellow enzymes OYE 1-3 from yeasts. *Eur. J. Org. Chem.* **2008**, 1511–1516. [CrossRef]

30. Wolken, W.A.M.; ten Have, R.; van der Werf, M.J. Amino acid-catalyzed conversion of citral: *Cis-trans* isomerization and its conversion into 6-methyl-5-hepten-2-one and acetaldehyde. *J. Agric. Food Chem.* **2000**, *48*, 5401–5405. [CrossRef]

31. Tsuboi, S.; Ishii, N.; Sakai, T.; Tari, I.; Utaka, M. Oxidation of alcohols with electrolytic manganese dioxide. Its application fro the synthesis of insect pheromones. *Bull. Chem. Soc. Jpn.* **1990**, *63*, 1888–1893. [CrossRef]

32. Robert, X.; Gouet, P. Deciphering key features in protein structures with the new ENDscript server. *Nucleic Acids Res.* **2014**, *42*, W320–W324. [CrossRef]

33. Hall, M.; Hauer, B.; Stuermer, R.; Kroutil, W.; Faber, K. Asymmetric whole-cell bioreduction of an α,β-unsaturated aldehyde (citral): Competing prim-alcohol dehydrogenase and C-C lyase activities. *Tetrahedron: Asymmetry* **2006**, *17*, 3058–3062. [CrossRef]

34. Cheng, F.; Tang, X.; Kardashliev, T. Transcription factor-based biosensors in high-throughput screening: Advances and applications. *Biotechnol. J.* **2018**, *13*, 1700648. [CrossRef]

35. Cheng, F.; Zhu, L.; Schwaneberg, U. Directed evolution 2.0: Improving and deciphering enzyme properties. *Chem. Commun.* **2015**, *51*, 9760–9772. [CrossRef]

36. Deng, J.; Yao, Z.; Chen, K.; Yuan, Y.A.; Lin, J.; Wei, D. Towards the computational design and engineering of enzyme enantioselectivity: A case study by a carbonyl reductase from *Gluconobacter oxydans*. *J. Biotechnol.* **2016**, *217*, 31–40. [CrossRef]

37. Sullivan, B.; Walton, A.Z.; Stewart, J.D. Library construction and evaluation for site saturation mutagenesis. *Enzyme Microb. Technol.* **2013**, *53*, 70–77. [CrossRef]

38. Valetti, F.; Gilardi, G. Improvement of biocatalysts for industrial and environmental purposes by saturation mutagenesis. *Biomolecules* **2013**, *3*, 778–811. [CrossRef]

39. Yu, M.; Huang, M.; Song, Q.; Shao, J.; Ying, X. Characterization of a (2R,3R)-2,3-butanediol dehydrogenase from *Rhodococcus erythropolis* WZ010. *Molecules* **2015**, *20*, 7156–7173. [CrossRef]

40. Fox, K.M.; Karplus, P.A. Old yellow enzyme at 2 Å resolution: Overall structure, ligand binding, and comparison with related flavoproteins. *Structure* **1994**, *2*, 1089–1105. [CrossRef]

41. Arnold, K.; Bordoli, L.; Kopp, J.; Schwede, T. The SWISS-MODEL workspace: A web-based environment for protein structure homology modelling. *Bioinformatics* **2006**, *22*, 195–201. [CrossRef]

42. Schwede, T.; Kopp, J.; Guex, N.; Peitsch, M.C. SWISS-MODEL: An antomated protein homology-modeling server. *Nucleic Acids Res.* **2003**, *31*, 3381–3385. [CrossRef]

43. Trott, O.; Olson, A.J. AutoDock Vina: Improving the speed and accuracy of docking with a new scoring function, efficient optimization, and multithreading. *J. Comput. Chem.* **2010**, *31*, 455–461. [CrossRef]

44. Seeliger, D.; de Groot, B.L. Ligand docking and binding site analysis with PyMOL and Autodock/Vina. *J. Comput. Aided Mol. Des.* **2010**, *24*, 417–422. [CrossRef]

45. Breithaupt, C.; Strassner, J.; Breitinger, U.; Huber, R.; Macheroux, P.; Schaller, A.; Clausen, T. X-Ray structure of 12-oxophytodienoate reductase 1 provides structural insight into substrate binding and specificity within the family of OYE. *Structure* **2001**, *9*, 419–429. [CrossRef]

46. Fraaije, M.W.; Mattevi, A. Flavoenzymes: Diverse catalysts with recurrent features. *Trends Biochem. Sci.* **2000**, *25*, 126–132. [CrossRef]

PERMISSIONS

LIST OF CONTRIBUTORS

Mihir V. Shah, Andrew C. Warden, Carol J. Hartley, Hadi Nazem-Bokaee and Colin Scott
CSIRO Synthetic Biology Future Science Platform, Canberra 2601, Australia

James Antoney, Suk Woo Kang and Colin J. Jackson
CSIRO Synthetic Biology Future Science Platform, Canberra 2601, Australia
Research School of Chemistry, Australian National University, Canberra 2601, Australia

Raneem Ahmad, Jordan Shanahan, Sydnie Rizaldo, Daniel S. Kissel and Kari L. Stone
Department of Chemistry, Lewis University, Romeoville, IL 60446, USA

Noelia Losada-Garcia, Carla Garcia-Sanz, Alicia Andreu and Jose M. Palomo
Department of Biocatalysis, Institute of Catalysis (CSIC), Marie Curie 2, Cantoblanco, Campus UAM, 28049 Madrid, Spain

Zaida Cabrera and Paulina Urrutia
School of Biochemical Engineering, Pontificia Universidad Católica de Valparaíso, Avda. Brasil, 2085 Valparaíso, Chile

Nadya Dencheva, Joana Braz and Zlatan Denchev
IPC-Institute for Polymers and Composites, University of Minho, 4800-056 Guimarães, Portugal

Dieter Scheibel
Department of Chemistry, State University of New York-ESF, Syracuse, NY 13210, USA

Marc Malfois
ALBA Synchrotron Facility, Cerdanyola del Valés, 0890 Barcelona, Spain

Ivan Gitsov
Department of Chemistry, State University of New York-ESF, Syracuse, NY 13210, USA
The Michael M. Szwarc Polymer Research Institute, Syracuse, NY 13210, USA

Grigorios Dedes, Anthi Karnaouri and Evangelos Topakas
Industrial Biotechnology & Biocatalysis Group, School of Chemical Engineering, National Technical University of Athens, 9 Iroon Polytechniou Str., Zografou Campus, 15780 Athens, Greece

Xiangxian Ying, Jie Zhang, Can Wang, Meijuan Huang, Yuting Ji, Feng Cheng and Zhao Wang
Key Laboratory of Bioorganic Synthesis of Zhejiang Province, College of Biotechnology and Bioengineering, Zhejiang University of Technology, Hangzhou 310014, China

Paulo Castro, Leonora Mendoza, Paz Cornejo Pereira, Freddy Navarro, Karin Lizama and Milena Cotoras
Laboratorio de Micología, Facultad de Química y Biología, Universidad de Santiago de Chile, Avenida Libertador Bernardo O'Higgins 3363, Santiago 518000, Chile

Claudio Vásquez
Laboratorio de Microbiología Molecular, Departamento de Biología, Facultad de Química y Biología, Universidad de Santiago de Chile, Santiago 518000, Chile

Rocío Santander
Departamento de Ciencias del Ambiente, Facultad de Química y Biología, Universidad de Santiago de Chile, Casilla 40 Correo 33, Santiago 518000, Chile

Arūnas Krikštaponis and Rolandas Meškys
Department of Molecular Microbiology and Biotechnology, Institute of Biochemistry, Life Sciences Center, Vilnius University, Sauletekio al. 7, LT-10257 Vilnius, Lithuania

Xiaoyu Lei, Shuangshuang Gao, Zhicheng Huang, Wen Huang and Ying Liu
College of Food Science and Technology, Huazhong Agricultural University, Wuhan 430070, China

Xi Feng
Department of Nutrition, Food Science and Packaging, California State University, San Jose, CA 95192, USA

Yinbing Bian
Institute of Applied Mycology, Huazhong Agricultural University, Wuhan 430070, China

Stefano Serra and Davide De Simeis
C.N.R. Istituto di Chimica del Riconoscimento Molecolare, Via Mancinelli 7, 20131 Milano, Italy

Lijuan Zhu, Linhu Zhu, Ayesha Murtaza, Yan Liu, Junjie Li, Aamir Iqbal, Xiaoyun Xu, Siyi Pan and Wanfeng Hu
College of Food Science and Technology, Huazhong Agricultural University, No.1, Shizishan Street, Hongshan District, Wuhan 430070, China
Key Laboratory of Environment Correlative Dietology, Huazhong Agricultural University, Ministry of Education, Wuhan 430070, China

Siyu Liu
Key Laboratory of Structural Biology, School of Chemical Biology & Biotechnology, Peking University Shenzhen Graduate School, Shenzhen 518055, China

Xinhong Liang, Wanli Zhang, Junjian Ran, Junliang Sun, Lingxia Jiao, Longfei Feng and Benguo Liu
School of Food Science, Henan Institute of Science and Technology, Xinxiang 453003, China

Te-Sheng Chang and Yu-Han Kao
Department of Biological Sciences and Technology, National University of Tainan, Tainan 70005, Taiwan

Chien-Min Chiang
Department of Biotechnology, Chia Nan University of Pharmacy and Science, No. 60, Sec. 1, Erh-Jen Rd., Jen-Te District, Tainan 71710, Taiwan

Jiumn-Yih Wu
Department of Food Science, National Quemoy University, Kinmen County 892, Taiwan

Yu-Wei Wu
Graduate Institute of Biomedical Informatics, College of Medical Science and Technology, Taipei Medical University, Taipei 106, Taiwan

Clinical Big Data Research Center, Taipei Medical University Hospital, Taipei 110, Taiwan

Tzi-Yuan Wang
Biodiversity Research Center, Academia Sinica, Taipei 115, Taiwan

Kai Song, Yu Li, Chaoxiang Ding, Gang Tao, Ping Zhao and Qingyou Xia
Biological Science Research Center, Southwest University, Beibei, Chongqing 400715, China

Yejing Wang and Rui Cai
State Key Laboratory of Silkworm Genome Biology, College of Biotechnology, Southwest University, Beibei, Chongqing 400715, China

Huawei He
Biological Science Research Center, Southwest University, Beibei, Chongqing 400715, China
Chongqing Key Laboratory of Sericultural Science, Chongqing Engineering and Technology Research Center for Novel Silk Materials, Southwest University, Beibei, Chongqing 400715, China

Shihua Yu, Ran Wei, Shumin Meng and Man Zhao
Key Laboratory of Bioorganic Synthesis of Zhejiang Province, College of Biotechnology and Bioengineering, Zhejiang University of Technology, Hangzhou 310014, China

Meilan Yu
College of Life Sciences, Zhejiang Sci-Tech Univeristy, Hangzhou 310018, China

Meirong Ying
Grain and Oil Products Quality Inspection Center of Zhejiang Province, Hangzhou 310012, China

Index

Printed in the USA
CPSIA information can be obtained
at www.ICGtesting.com
JSHW051623061123
51533JS00005B/84

9 781647 403898